Bird and Reptile Species in Environmental Risk Assessment Strategies

Issues in Toxicology

Editor-in-chief:
Diana Anderson, *University of Bradford, UK*

Series editors:
Michael D. Waters, *Michael Waters Consulting, USA*
Timothy C. Marrs, *Edentox Associates, UK*
Alok Dhawan, *CSIR-Indian Institute of Toxicology Research, India*

Titles in the Series:
1: Hair in Toxicology: An Important Bio-Monitor
2: Male-mediated Developmental Toxicity
3: Cytochrome P450: Role in the Metabolism and Toxicity of Drugs and other Xenobiotics
4: Bile Acids: Toxicology and Bioactivity
5: The Comet Assay in Toxicology
6: Silver in Healthcare
7: *In Silico* Toxicology: Principles and Applications
8: Environmental Cardiology
9: Biomarkers and Human Biomonitoring, Volume 1: Ongoing Programs and Exposures
10: Biomarkers and Human Biomonitoring, Volume 2: Selected Biomarkers of Current Interest
11: Hormone-Disruptive Chemical Contaminants in Food
12: Mammalian Toxicology of Insecticides
13: The Cellular Response to the Genotoxic Insult: The Question of Threshold for Genotoxic Carcinogens
14: Toxicological Effects of Veterinary Medicinal Products in Humans: Volume 1
15: Toxicological Effects of Veterinary Medicinal Products in Humans: Volume 2
16: Aging and Vulnerability to Environmental Chemicals: Age-related Disorders and their Origins in Environmental Exposures
17: Chemical Toxicity Prediction: Category Formation and Read-Across
18: The Carcinogenicity of Metals: Human Risk Through Occupational and Environmental Exposure
19: Reducing, Refining and Replacing the Use of Animals in Toxicity Testing
20: Advances in Dermatological Sciences
21: Metabolic Profiling: Disease and Xenobiotics
22: Manganese in Health and Disease
23: Toxicology, Survival and Health Hazards of Combustion Products
24: Masked Mycotoxins in Food: Formation, Occurrence and Toxicological Relevance
25: Aerobiology: The Toxicology of Airborne Pathogens and Toxins

26: Chemical Warfare Toxicology, Volume 1: Fundamental Aspects
27: Chemical Warfare Toxicology, Volume 2: Management of Poisoning
28: Toxicogenomics in Predictive Carcinogenicity
29: Human Stem Cell Toxicology
30: The Comet Assay in Toxicology, 2nd edition
31: Computational Systems Pharmacology and Toxicology
32: Ecotoxicology and Genotoxicology: Non-traditional Terrestrial Models
33: Ecotoxicology and Genotoxicology: Non-traditional Aquatic Models
34: Oxidative Stress and Redox Signalling in Parkinson's Disease
35: Nanotoxicology: Experimental and Computational Perspectives
36: Regulatory Toxicology in the European Union
37: Formaldehyde: Exposure, Toxicity and Health Effects
38: Chemical Health Threats: Assessing and Alerting
39: The Micronucleus Assay in Toxicology
40: Nanoparticle–Protein Corona: Biophysics to Biology
41: Big Data in Predictive Toxicology
42: Challenges in Endocrine Disruptor Toxicology and Risk Assessment
43: Bisphenol A: A Multi-modal Endocrine Disruptor
44: Marsupial and Placental Mammal Species in Environmental Risk Assessment Strategies
45: Bird and Reptile Species in Environmental Risk Assessment Strategies

How to obtain future titles on publication:
A standing order plan is available for this series. A standing order will bring delivery of each new volume immediately on publication.

For further information please contact:
Book Sales Department, Royal Society of Chemistry, Thomas Graham House, Science Park, Milton Road, Cambridge, CB4 0WF, UK
Telephone: +44 (0)1223 420066, Fax: +44 (0)1223 420247
Email: booksales@rsc.org
Visit our website at www.rsc.org/books

Bird and Reptile Species in Environmental Risk Assessment Strategies

Edited by

Guillermo Eli Liwszyc
University of Helsinki, Finland
Email: guillermo.liwszyc@kolumbus.fi

and

Marcelo L. Larramendy
National University of La Plata, Argentina
Email: marcelo.larramendy@gmail.com

ROYAL SOCIETY
OF **CHEMISTRY**

Issues in Toxicology No. 45

Print ISBN: 978-1-83916-710-2
PDF ISBN: 978-1-83767-076-5
EPUB ISBN: 978-1-83767-077-2
Print ISSN: 1757-7179
Electronic ISSN: 1757-7187

A catalogue record for this book is available from the British Library

The Royal Society of Chemistry is a charity, registered in England and Wales, Number 207890, and a company incorporated in England by Royal Charter (Registered No. RC000524), registered office: Burlington House, Piccadilly, London W1J 0BA, UK, Telephone: +44 (0) 20 7437 8656.

For further information see our web site at www.rsc.org

Printed in the United Kingdom by CPI Group (UK) Ltd, Croydon, CR0 4YY, UK

Preface

This volume is the second one of a series of four books to be published by the Royal Society of Chemistry; the first in the series is that on Mammals (Marsupials and Placentals), and the second one is to be followed by those on Fish and Amphibians.

The chapters included in this book are just a mere enumeration of some practical examples. There are many more species that can be used as non-conventional experimental models and the list should be expanded by different research groups and academics all over the world. We hope that many more scientists realize that it is more important to tackle subjects related to the status of the environment using autochthonous, or non-target species, which are truly exposed to locally used xenobiotic agents. On the other hand, this research plan would also increase the chances of independent scientists and research institutions in getting access to grants and attracting the attention of local authorities, who would be more interested in financing projects that really are within their sphere of political interest and pride, not to mention their country or province/district.

Without running the risk of being repetitive, we would like to recap on some important concepts as previously mentioned in the two volumes published by the Royal Society of Chemistry, *i.e.*, *Ecotoxicology and Genotoxicology: Non-traditional Terrestrial Models* and *Ecotoxicology and Genotoxicology: Non-traditional Aquatic Models*, by the same editor group. We strongly recommend the perusal of both volumes, which are not overlapping subjects, in order to gain the full benefit of this series and have a more holistic and stimulating approach to the matter.

We share our planet with thousands of living species. Understanding how environmental stressors affect them will also help us in our quest for solutions of problems humanity is facing at present and which will be more relevant in the near future. Thinking out of silos and crosslinking medical

Issues in Toxicology No. 45
Bird and Reptile Species in Environmental Risk Assessment Strategies
Edited by Guillermo Eli Liwszyc and Marcelo L. Larramendy

Published by the Royal Society of Chemistry, www.rsc.org

with applied science will lead to a faster pace of understanding many of the factors related to the emergence of a disease and thus help us in our search for finding a cure.

New emerging studies in several reputable publications (*e.g.*, Worldwide Trends 1975–2015, *The Lancet*, 2016, **384**(10064), 37–55) of common worldwide human diseases such as arterial hypertension, among others, seem to suggest that not only classical and lifestyle factors should be addressed in the search for adequate treatment. From a medical perspective, it is not only a disease of affluence. One in eight deaths worldwide is due to this condition and its corresponding factors (heart disease, kidney disease and stroke). These studies seem to suggest that a closer look at other, so far unrelated factors is required. Early-life nutrition and exposure to air pollution, heavy metals and even noise are factors implicated, which may push blood pressure up later in life.

Many researchers have contributed to the publication of this book. We hope that it serves as a herald in order to bolster enthusiasm for the use of native, or easily available local species in order to widen our knowledge on the subject. We would like to especially thank the authors of the chapters for their positive response, their time, contribution and feedback, making possible the compilation of this book. The task had been further complicated by the global pandemic of COVID-19. Many, if not most, of our institutions had to close their doors or cease all academic activity for long periods of time. Access to data banks, field trips, *etc.*, was curtailed; nevertheless, against all odds, we were able to compile this piece of work which we hope would be of interest, of use and above all would continuously nurture the appetite for future lines of research. Last, but not least, our most sincere gratitude for the trust deposited upon us by the Editor-in-Chief Diana Anderson, Series Editors Michael D. Waters, Timothy C. Marrs and Alok Dhawan and the Royal Society of Chemistry in this project. We hope to have fulfilled the reader's expectations.

Guillermo Eli Liwszyc
University of Helsinki, Helsinki, Finland

Marcelo L. Larramendy
Universidad Nacional de La Plata, Argentina
Consejo Nacional de Investigaciones Científicas y Técnicas (CONICET), Buenos Aires, Argentina

Contents

Chapter 1 General Aspects – Current and Further Perspectives 1
Marcelo L. Larramendy and Guillermo Eli Liwszyc

Chapter 2 Development of Aquatic Bird Indicators of Sub-lethal Mercury Exposure and Risk in Wild Populations of Water Birds in the Everglades (Florida, United States of America) 6
J. Zabala and P. Frederick

2.1 Background and Study Area Description 6
2.1.1 Mercury: Local Emissions, Worldwide Contamination 6
2.1.2 The Everglades 7
2.1.3 Mercury Contamination in the Everglades 11
2.1.4 Development of a Field Sampling Protocol 12
2.1.5 Development of Mercury Exposure Indicators – Tissues 13
2.2 Results 15
2.2.1 Sub-lethal Effects: Results from Experimental Studies 15
2.2.2 Evidence of Mercury Effects in Field Conditions 16
2.3 General Discussion, Lessons Learnt and Pros and Cons of Our and Alternative Approaches 18
2.3.1 Evidence and Estimation of Hg Effects in Natural Populations 18
2.3.2 Indicator Tissues: Comparative Advantages, Limitations and Uncertainty 19

Issues in Toxicology No. 45
Bird and Reptile Species in Environmental Risk Assessment Strategies
Edited by Guillermo Eli Liwszyc and Marcelo L. Larramendy

Published by the Royal Society of Chemistry, www.rsc.org

2.4 Conclusion and Advice for Similar Cases or Final Remarks 24
Acknowledgements 24
References 24

Chapter 3 The Importance of Ecological Traits in Assessing Seabird Vulnerability to Environmental Risks 33
Can Zhou, Joan A. Browder and Yan Jiao

3.1 Introduction 33
3.2 Failings of the Standard Approach 34
3.3 Vulnerability to Anthropogenic and Natural Risks 34
3.4 Challenges for an Observational Study 35
3.5 A Trait-based Approach 37
3.5.1 An Ecological Dimension Reduction Technique 37
3.5.2 Useful Traits for Ecological Risk Evaluation 39
3.5.3 Trait-based Prediction 41
3.6 Other Challenges 43
3.6.1 Correlation *vs.* Causation 43
3.6.2 The Default of No Risk 43
3.6.3 Ecological Regulation 44
3.6.4 Climate Change and Variability 44
3.6.5 Other Approaches 47
Disclaimer 47
Acknowledgements 47
References 47

Chapter 4 A Review of the Levels and Distribution Patterns of Organochlorine Pesticides in the Eggs of Wild Birds in India 54
Dhananjayan Venugopal, Jayakumar Samidurai, Jayanthi Palaniyappan, Jayakumar Rajamani and Muralidharan Subramanian

4.1 Introduction 54
4.1.1 Organochlorine Pesticides 54
4.1.2 Organochlorine Pesticides – Marketing and Consumption 56
4.2 Pesticides – Indian Scenario 56
4.3 Impact of Pesticides in Eggs of Wild Birds in India 57
4.3.1 OCP Residues in Birds' Eggs 57

4.3.2 Variation in Residue Levels Based on Species and Food Habits 64
4.3.3 Eggshell Thinning and Reproductive Impairment 66
4.4 Conclusions and Further Recommendations 66
References 67

Chapter 5 Impacts of Agricultural Intensification on Farmland Birds and Risk Assessment of Pesticide Seed Treatments 73
Julie Céline Brodeur and Maria Belén Poliserpi

5.1 The Intensification of Agriculture 73
5.2 Agricultural Intensification and Bird Declines 74
5.3 Impact of Pesticides on Birds: Direct *vs.* Indirect Effects 77
5.3.1 Direct Effects 77
5.3.2 Indirect Effects 78
5.4 Seed Treatment With Pesticides: Impacts on Birds 79
5.4.1 Agricultural Intensification Through Seed Treatment 79
5.4.2 Bird Exposure to Pesticide-treated Seeds 80
5.5 Assessment of Risks of Pesticide-treated Seeds to Birds 81
5.5.1 Regulatory Environmental Toxicity Testing 81
5.5.2 Tier I Risk Assessment 83
5.5.3 Refinements and Weight-of-evidence Risk Assessment 83
5.5.4 Future Directions 85
References 85

Chapter 6 Teratological Effects of Pesticides in Reptiles – A Review 97
A. Garcês and I. Pires

6.1 Introduction 97
6.2 Reptiles as Sentinel Species 99
6.3 Teratological Effects of Pesticides on Reptiles 99
6.3.1 Order Testudinata 99
6.3.2 Order Rhynchocephalia 102
6.3.3 Order Crocodilia 102
6.3.4 Order Squamata 103
6.4 Final Remarks 104
6.5 Conclusion 105
Acknowledgements 105
References 105

Chapter 7 Combined Impact of Pesticides and Other Environmental Stressors on Reptile Diversity in Irrigation Ponds Compared to Other Animal Taxa **110**
Hiroshi C. Ito and Noriko Takamura

7.1 Introduction 110
7.2 Biological Communities in Irrigation Ponds 112
7.2.1 Biological Communities 112
7.2.2 Reptiles 112
7.3 Combined Impact of Multiple Stressors on Reptile Diversity Compared to Other Animal Taxa 114
7.4 Impact of Each Stressor on Reptile Diversity Compared to Other Animal Taxa 116
7.4.1 Concrete Bank Protection, Water Depth Reduction and Macrophyte Decline 116
7.4.2 Eutrophication and Pesticide Pollution 119
7.4.3 Invasive Alien Species 120
7.5 Perspective: Usefulness of Turtles as Bioindicators 121
7.6 Conclusion 123
Acknowledgements 123
References 123

Chapter 8 Current Progress in Developing Standardized Methods for Reptilian Toxicity Testing to Inform Ecological Risk Assessment **130**
Scott M. Weir, Monica R. Youssif, Taylor Anderson and Christopher J. Salice

8.1 Background on Reptile Toxicity Testing 130
8.1.1 Lack of Reptile Toxicity Data in the Literature 130
8.1.2 Lack of Standardized Methods 131
8.2 Progress in Standardized Methods for Reptile Ecotoxicology/Ecological Risk Assessment 131
8.2.1 Early Efforts in Reptile Toxicology 131
8.2.2 Oral Dosing Methods Using Gelatin Capsules to Accommodate Small Reptiles 133
8.2.3 Standardization of Methods and Developing Breeding Assay Using *Anolis* Species in Reptile Ecotoxicology 137
8.2.4 Moving Beyond Oral Exposure Dosing: The Potential Importance of Dermal Exposure and Toxicity 144

8.3 Conclusions 144
Acknowledgements 145
References 146

Chapter 9 Morphological and Molecular Evidence of Active Principle Glyphosate Toxicity on the Liver of the Field Lizard *Podarcis siculus* 151
Mariailaria Verderame, Teresa Chianese and Rosaria Scudiero

9.1 Introduction 151
9.2 Lizards in Contaminated Environments 152
9.3 Effects of Pure GLY on the Liver of *P. siculus* Specimens 154
9.4 GLY-induced Changes in *P. siculus* Liver Histology 155
9.5 GLY-induced Changes in the Expression and Synthesis of Proteins in the *P. siculus* Liver 158
9.6 Conclusion 161
Acknowledgements 162
References 162

Chapter 10 What Is *Caiman latirostris* Teaching Us About Endocrine Disruptors? 169
M. Durando, G. H. Galoppo, Y. E. Tavalieri, M. V. Zanardi and M. Muñoz-de-Toro

10.1 Ecophysiological Characteristics of the Broad-snouted Caiman (*Caiman latirostris*) 169
10.1.1 Sex Determination 170
10.1.2 Reproduction 170
10.1.3 Hatchling Growth and Development 171
10.1.4 Feeding Habits, Social Behavior and Longevity 171
10.1.5 Sexual Dimorphism 171
10.2 Endocrine-disrupting Chemicals 172
10.2.1 Mechanism of Action of EDCs 173
10.2.2 Types of EDCs 173
10.3 *Caiman latirostris* as a Sentinel of Environmental Pollution 174
10.4 EDCs and Their Effects on Reproductive Features 175
10.4.1 Natural Exposure 175
10.4.2 Experimental Exposure 177
10.5 EDCs and Their Effects on the Thyroid Histofunctional Characteristics 182

10.6 Conclusions 183
Acknowledgements 183
References 184

Chapter 11 The Broad-snouted Caiman *(Caiman latirostris):* A Model Species for Environmental Pesticide Contamination Assessment Through Molecular Markers 196
L. M. Odetti, M. F. Simoniello, P. A. Siroski and G. L. Poletta

11.1 *Caiman latirostris*: Life History Characteristics and Population Situation in Argentina 197
11.1.1 Geographical Distribution and Sustainable Use Programs 197
11.1.2 Environmental Problems Associated With Agricultural Expansion and the Use of Pesticides 198
11.2 Why Use *C. latirostris* as a Sentinel of Pesticide Contamination? 200
11.2.1 Evidence of Pesticide Effects on the Broad-snouted Caiman 200
11.2.2 Identification and Development of Gene Expression Markers 202
11.3 Future Perspectives 209
Acknowledgements 209
References 209

Chapter 12 Epilogue and Final Remarks 217
Guillermo Eli Liwszyc and Marcelo L. Larramendy

Reference 226

Subject Index 227

CHAPTER 1

General Aspects – Current and Further Perspectives

MARCELO L. LARRAMENDY*[a,b] AND
GUILLERMO ELI LIWSZYC[c]

[a] Cátedra de Citología, Facultad de Ciencias Naturales y Museo, Universidad Nacional de La Plata, Calle 64 No. 3, B1904AMA La Plata, Argentina; [b] Consejo Nacional de Investigaciones Científicas y Técnicas (CONICET), Buenos Aires, Argentina; [c] Former Guest Scientist at the Department of Pathology, University of Helsinki, Helsinki, Finland
*Email: marcelo.larramendy@gmail.com

By definition, it is accepted worldwide that environmental risk assessment is a scientific practice that identifies and evaluates the likelihood of an activity or substance to threaten the environment in general, and/or living organisms, natural habitats and ecosystems, in particular. In other words, it is a term used to describe the overall process or method allowing (a) the identification of hazards and risk factors that have the potential to cause harm (hazard identification) and (b) the analysis and evaluation of the risk associated with that particular hazard (risk analysis and risk evaluation). It also suggests all the appropriate ways to eliminate or minimize the hazard, or control the risk when the hazard cannot be eliminated (risk control).

The Food and Agriculture Organization of the United Nations (FAO) predicted a few years ago that the global population will be over 9.1 billion by the middle of the 21st century. Accordingly, food production will have to increase by about 70% above current levels to maintain pace with demand. Disruptions in transport and production of basic commodities due to unforeseen

Issues in Toxicology No. 45
Bird and Reptile Species in Environmental Risk Assessment Strategies
Edited by Guillermo Eli Liwszyc and Marcelo L. Larramendy

Published by the Royal Society of Chemistry, www.rsc.org

circumstances, such as the present war in Ukraine, has revealed once more the caveats of this model. One plausible method for achieving this target would be to increase yields and the amount of arable land. However, the conversion of natural forests and/or other wild habitats engenders a number of well-known negative impacts on climate change and global biodiversity. Furthermore, it is accepted worldwide that such an expansion of agriculture could be responsible for approximately 12% of global warming. Several anthropogenic activities introduce extensive amounts of chemicals and pollutants into the environment on a daily basis. In this sense, the use of fossil fuels, and the mining, manufacturing and construction industries, among others, are the main contributors to water, air and soil pollution. Increases in improper disposal of waste, chemicals, heavy use of air-polluting equipment and machinery, asbestos handling, soil contamination and industrial noise need to be factored in too. Even our clothes, in many cases composed of artificial fibres that do not naturally degrade, when thrown away, or sent to some far-away country, as a sort of "help" from more economically privileged nations, find their way to landfills. In some cases these are located close to seashores and eventually end up polluting oceans. To this, we must add overfishing and overexploitation of marine resources, among others.

Most of the xenobiotics in the different environmental compartments exert their effects through cytotoxicity and genotoxicity, and by interfering with metabolic pathways. Accordingly, the current awareness of the real/potential hazards of pollutants in the environment has high interest in the use of terrestrial invertebrate and vertebrate species as indicators for monitoring pollutant-induced environmental deleterious effects. Notwithstanding, it is known that agrochemicals, among others, not only affect target organisms but also concomitantly exert negative effects on non-target species.

In environmental risk assessment studies, there is increasing interest in biomonitoring markers to provide practical indicators of biological exposure to pollutants. To achieve this goal, several endpoints for testing toxicity have been developed. Cyto- and genotoxicity tests have been employed on aquatic and terrestrial organisms to assess the impact of pollution on contaminated areas (*in situ* assays) and for screening xenobiotics after direct or indirect exposure (*in vivo* assays).

This book covers many non-traditional bird and reptile species models employed to analyse the jeopardizing effects of environmental pollutants, as well as examples of applications of models in integrated assessment solutions. We also hope it will provide answers or at least some clues to the inherent potential risk of the presence of these xenobiotics in our environment. To achieve this goal, considerations about exposure, hazard, risk, mitigation, and short- and long-term monitoring are required. This information will give all interested parties, *e.g.*, regulators and safety advisory bodies, the tools needed for a better understanding of the matter in a broader context.

Many important ecosystems around the world are being constantly challenged due to growing human and industrial pressure exerted upon them.

More than one million animals and plants are facing an unprecedented pattern of extinction. We are experiencing the largest biological disappearance wave since the extinction of dinosaurs and humans could be the main cause of this event.

Various biomarkers in local, easily available species can be of use to evaluate the response of the biota to such pollutants. Several biological parameters mirror the interactions between toxic agents and biotic matrices. These are powerful tools that can be applied to environmental monitoring tests and studies. Their responses may reveal general deleterious effects on organisms in general, pinpointing alterations at the cellular, biochemical and molecular levels, as well as at higher levels of organization.

The health of our planet and its ecosystems has deteriorated severely. There is no question about this statement. Mankind has reached for the first time in its history an unprecedented level of food security and availability of consumer goods. Unfortunately, this comes with a price tag. The different geographical areas, species and substances studied show that no region, even if far away from polluting centres, is spared from this unfortunate phenomenon. We hope that the contributions of this book will stimulate further work along these lines, while at the same time raising several flags. As scientists we have an obligation to present evidence of the damage that our planet, Earth, is being subject to. A transversal approach is needed, even in matters that do not seem to be related, but in reality are. An example of this is light pollution in populated areas which disrupts the normal behaviour of birds and insects, altering their feeding habits and the pollination role they play in the ecosystem. This information, in turn, will help launch the debate at a civil society level, to change, in turn, many of its consumer patterns, press producers and manufacturers, while at the same time demanding more stringent action on behalf of politicians and law makers to protect this planet, our home, before it is too late.

Our global society needs to list down actions and set rules to evaluate and considerably reduce the real and potentially hazardous factors in the environment that can, as previously stated, result in health risks for all forms of life (including *Homo sapiens sapiens*). Despite the major positive contributions in the field of health, due to the immense progress achieved in science, technology and industrialization, the interaction between environmental risks and health is an often intricate, not self-evident equation, which involves a variety of not only social, political and economic but also life style factors. This cannot be emphasized enough. Health depends on the good quality of environmental "basic ingredients", such as air, water, soil and food, among others. We believe that the ultimate challenge in this matter is to weigh in short-term positive gains, while, at the same time, taking into account long-term effects of substances used. Available information about the toxicity of heterogeneous xenobiotics, continuously released into human habitats, inadvertently, deliberately or by non-regulated industrial discharges on biological components of the environment, is inconclusive.

There is no clear-cut definition of the concept of environmental health. Various openings help us in the understanding of this concept. According to the World Health Organization (WHO), it is defined by "...all the physical, chemical and biological factors external to a person and all the related factors impacting upon behaviours. It encompasses the assessment and control of those environmental factors that can potentially affect health. It is targeted towards preventing disease and creating health-supportive environments...". For the National Environmental Health Association, this concept refers to "the protection against environmental factors that may adversely impact human health or the ecological balances essential to long term human health and environmental quality, whether in the natural or man-made environment". A third definition by the National Institute of Environmental Health Sciences also involves the criterion that "the social environment encompasses lifestyle factors like diet and exercise, socioeconomic status, and other societal influences that may affect health".

In general terms, our health and the health of many other species are negatively affected by five broad categories of environmental hazards, namely, electromagnetic fields (produced by high-power lines, electrical wiring, appliances, mobile phones, computers, TV sets, *etc.*), radiation (including nuclear fallout from weapon testing, fission materials from nuclear power plants and their respective accidents, leaking radioactive disposal sites, air travel and X-rays), toxic chemicals (some organochlorines, phthalates, polybrominated flame retardants, perfluorinated substances, bisphenol-A) and several toxic metals, among others), which have been shown to have endocrine-disrupting properties, and soil mineral depletion as a complex environmental hazard.

By definition, health risk assessment in its quantitative and/or qualitative determinations includes variants such as the type of risk involved and the severity of response, within or without a probabilistic context. In this regard, risk-based methods of analysis play a strategic role in identifying and ranking adverse responses or the structure of the effects of exposure *vis-à-vis* environmental factors.

Many compounds can be hazardous if not used appropriately and may present a real risk to the environment by contaminating soil, water and air. Most of the pollutants in the different environmental compartments exert their effects through cytotoxic, genotoxic and metabolically toxic mechanisms. In pollution studies, there is increasing interest in biomonitoring markers of biological exposure to pollutants. To achieve this goal, several endpoints for the three afore-mentioned factors have been used in aquatic and terrestrial invertebrate and vertebrate species in contaminated areas (*in situ* assays) and to screen for xenobiotics after direct or indirect exposure (*in vivo* assays).

The use of species as indicators for monitoring pollutant-induced deleterious environmental effects will raise the current awareness of real and potential hazards. It is also known that most environmental pollutants not only affect target organisms but also concomitantly exert negative effects on non-target species.

Invertebrate and vertebrate animal models have been used for decades in acute and chronic toxicity tests for hazard identification. They can be very efficient screening systems that have a major role to play in toxicity research, because certain aspects of their biological, physiological and genetic characteristics make them suitable models in ecotoxicological and genotoxicological studies.

This volume aims to shed some light on the matter, while offering relevant tools for evaluating risks and providing a framework of practical discussions. These will foster decisions and actions required to reduce the environmental health risk against environmental factors. This book presents some real-life examples, extending concepts (of hazardous factors) to living species that may stimulate new research ideas and trends in relevant fields.

Available information has been compiled from a diversity of sources, in an attempt to achieve a representative global and geographical balance, as far as possible, while at the same time aiming at high-quality studies. We believe that this book is unique in this sense.

The scientific community is starting to envisage that the time has come for a shift in emphasis away from basic towards applied research. While basic research is needed in order to shed more light on the fundamentals, applied research is required in order to find solutions to the many problems the world faces, *e.g.*, overpopulation, excessive use of the Earth's natural resources and pollution.

Many researchers from different parts of the world have contributed to the chapters of this book. We hope that this volume will meet the expectations and needs of all those interested in the environmental risk assessment field of study by the use of widely available species worldwide. Finally, we also hope that the examples included in the different chapters of this book will awaken the ability to search for new organisms in local and regional ecosystems to pursue further studies in ecotoxicology and genotoxicology.

CHAPTER 2

Development of Aquatic Bird Indicators of Sub-lethal Mercury Exposure and Risk in Wild Populations of Water Birds in the Everglades (Florida, United States of America)

J. ZABALA*[a,b] AND P. FREDERICK[a]

[a] Department of Wildlife Ecology and Conservation, University of Florida, Gainesville, Florida, USA; [b] Department of Zoology and Animal Cell Biology, University of the Basque Country, UPV/EHU, C/Paseo de la Universidad 7, 01006 Vitoria-Gasteiz, Araba, Basque Country, Spain
*Email: j.zabala@ehu.eus

2.1 Background and Study Area Description

2.1.1 Mercury: Local Emissions, Worldwide Contamination

Human activities have resulted in increased releases of metals and other pollutants to the atmosphere, with a marked increase following the Industrial Revolution. Atmospheric re-distribution of several pollutants has resulted in a global distribution, spreading risk to even remote pristine ecosystems. Historical deposition may create a legacy of pollutants that can

Issues in Toxicology No. 45
Bird and Reptile Species in Environmental Risk Assessment Strategies
Edited by Guillermo Eli Liwszyc and Marcelo L. Larramendy

Published by the Royal Society of Chemistry, www.rsc.org

affect organisms decades after emissions have been reduced. In addition, global change will affect biochemical cycles, potentially resulting in higher release rates of some pollutants, like mercury (Hg).[1–4]

Elevated levels of Hg can be found in ecosystems all around the globe, including remote areas otherwise untouched by industrial activities, like the Arctic and Antarctic areas.[4–7] Only a small fraction of Hg emitted to the atmosphere deposits locally (<10%) but this can increase the local deposition background rates by 10–100 times.[8] The remaining is dispersed regionally or globally. Hg may remain in atmospheric residence from months to years and is uniformly distributed in the troposphere. In the atmosphere volatile Hg can be oxidized, which can then precipitate with rainfall.[8] Once Hg enters a terrestrial or aquatic system, it can remain at high levels for long periods of time after emissions ceased.[9,10]

Methylmercury (MeHg), the organic form of Hg, is of particular concern because it can be readily assimilated from food and can bioaccumulate and reside for long periods within the organism. In natural ecosystems it gets into living organisms and bioaccumulates and biomagnifies along food webs, especially in aquatic ones. Therefore, top predators of aquatic ecosystems are often used as biomonitors of MeHg risk, and to study its effects. At high concentrations, MeHg can cause ataxia, muscle weakness and severe problems with the process of the nervous system, poor vision and hearing, strongly delayed responses to stimuli and ultimately death. At sub-lethal doses, such as those found in living organisms in the wild, MeHg causes a wide array of subtler effects. Sub-lethal effects reported include neurological dysfunction, endocrine disruption, reproductive impairment, developmental malformation and carcinogenesis. Of these, breeding impairment is the most commonly used measure to assess exposure effects and develop dose–response relationships.[11–15]

2.1.2 The Everglades

The Everglades of Florida is a large (*cf.* 20 000 km^2) extremely flat subtropical wetland system subject to seasonal inundation to depths of 0–150 mm (see Figure 2.1). Its soils are mostly peat and the hydroperiod varies from a few months to permanent depending on the location and year. In any case, the water table is usually close to the surface. The marsh is dominated by sawgrass (*Cladium mariscus jamaicense*), interspersed with tree islands of various sizes, and mainly composed of emergent willows (*Salix caroliniana*), bald cypress (*Taxodium distichum*) and pond apple (*Annona glabra*). The Everglades is strongly oligotrophic, with extremely low levels of phosphorus in soils and water, which drive both vegetation patterns and animal communities. In areas around colonies, where bird feces enrich the soil, or areas where agricultural runoff from northern sugar cane cultures enters the area, sawgrass is substituted by cattail (*Typha* spp.).

The Everglades is a unique and iconic ecosystem, well known for its large breeding populations of long-legged wading birds. There are breeding populations of ibises, spoonbills, storks and up to 10 breeding species of herons, and remarkable breeding colonies of white ibises (*Eudocimus albus*;

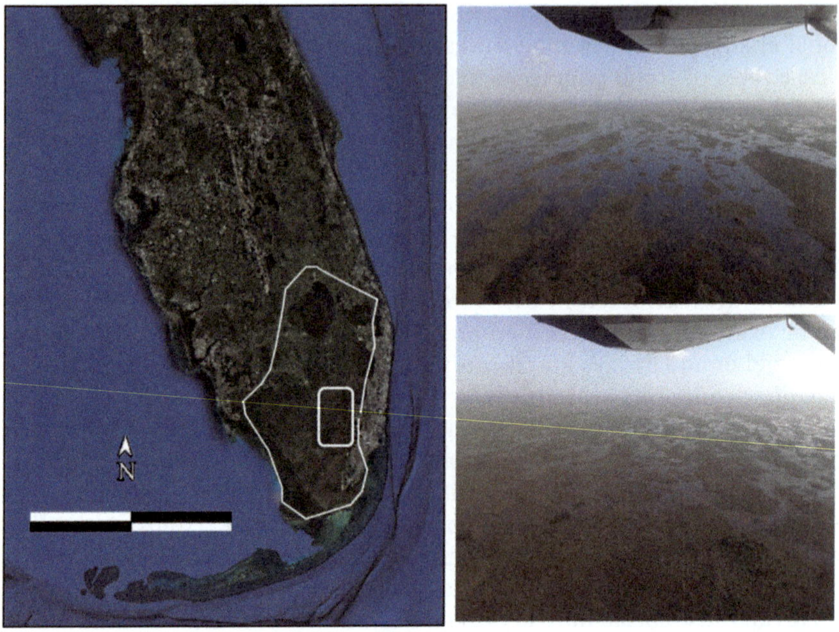

Figure 2.1 Study area in Florida (USA). The white polygon indicates the approximate extent of the Greater Everglades area and the inner rectangle our study area. The bar represents 200 km. Each black and white bar sections corresponds to 100 km. To the right are two aerial pictures of the study area. The top one shows high water conditions, and most of the vegetation observable is emergent sawgrass. The bottom one shows drier conditions. Note the reduced flooded surface and presence of isolated pools on the edges of flooded areas where fish are concentrated and become more vulnerable to predation. A tree island is visible on the bottom right of the picture. Note that both images are not from the same area. The Florida Peninsula image is adapted from Google Earth Pro 7.3.4.8248, Data SIO, NOAA, US Navy, NGA, GEBCO, Image Landsat/Copernicus.

see Figure 2.2), great egrets (*Ardea alba*; see Figure 2.3) and wood storks (*Mycteria americana*; see Figure 2.4).

The Everglades is also famous for harboring populations of flagship and endangered species including vulnerable West Indian manatees (*Trichechus manatus*), endangered Florida panthers (*Puma concolor cougar*), American alligators (*Alligator mississippiensis*), vulnerable American crocodiles (*Crocodylus acutus*), snail kites (*Rostrhamus sociabilis*), ospreys (*Pandion haliaetus*) and bald eagles (*Haliaeetus leucocephalus*), among others.

Wading bird populations are supported by some key ecological characteristics of this large regional wetland. The first is seasonal fluctuation of water levels, driven by a subtropical summer monsoonal rainfall pattern, and an extremely flat landscape. In June–July, the summer rainy season brings heavy rainfall that raises water levels by 0.5–1.5 m and high water levels may last until December–January, creating flooded habitats and

Figure 2.2 Nest of white ibises, one of our model species, with two nestlings of approximately 10 days of age. Florida Everglades.

Figure 2.3 Great egrets, our other model species, in breeding plumage sitting on nest. Wakodahatchee Wetlands, Florida.

Figure 2.4 Wood stork perching on a pine tree in the Everglades. Wood storks are an indicator species for the restoration of the Everglades and a top predator in the system. Yet, we did not select them as model species because they do not nest in large numbers every year and nestling sampling is harder than with other species.

conditions for growth and reproduction of fish and macroinvertebrate populations. With the start of the dry season, the water levels start receding. As the water recession proceeds during the dry season (November through May), inundated areas decrease and fish and other types of prey are concentrated into a succession of shallow pools and ponds. Wading birds can only forage effectively in shallow water and many of the foraging behaviors are efficient only when prey is concentrated. Foraging during the dry season is therefore highly focused on shallow areas where fish are vulnerable and breeding only occurs during this period of high food availability.[16–18] As the recession progresses, new shallow pools with concentrated fish are continuously created, resulting in constant supply of food for breeding birds. The process stops with the onset of the rainy season (June–November) when the rainfall raises water levels, allowing fish to disperse and take refuge in deeper waters.[19,20]

Large breeding bird colonies are located on small islands of trees located within the vast graminoid wetland, and the locations of active colonies shift with annual inundation patterns. These colonies are successful only if they remain inundated and become targets for mammalian nest predators if they are not. When inundated, alligators (*A. mississippiensis*) are attracted to take advantage of a seasonal food pulse in the form of dropped nestlings. The presence of alligators is also a strong deterrent to mammalian predators, creating a symbiosis between nesting birds and alligators.[21–23]

2.1.3 Mercury Contamination in the Everglades

The Everglades is relatively isolated from polluting industries and has few recognized environmental contaminants. However, mercury is a major contaminant, which is received primarily through wet deposition. Sources are both global and local, with evidence that some of the Hg, particularly in the 1970s–1990s, resulted from local waste incineration in the nearby Miami metropolitan area. The Everglades is also known to offer excellent conditions for methylation of Hg (as much as 100% in 24 h) as a result of communities of sulfur-reducing bacterial mats at the soil–water interface.[24,25] Methylation is probably strongly influenced by additions of agricultural sulfur upstream of the Everglades, creating intermediate to high sulfate conditions that may favor methylating bacteria.[25,26] In addition to its high methylation potential, the Everglades has high bioaccumulation potential ($1–3 \times 10^6$ times between water and predatory fish) as a result of long aquatic food webs. Further, frequent drying and wetting of the soil surface, which are characteristic of this shallow wetland, mean that any Hg deposited is rarely buried through soil building and made available through repeated oxidation at the soil surface.

By the early 1990s, scientists and public health agencies were aware that Hg exposure rates in the Everglades could pose a threat to wildlife and humans,[27–29] with the Hg values in bird feathers showing 5–7 fold increases by comparison with the 1960s.[30]

Thereafter, Hg has declined rapidly, probably as a result of both decreased local sources and reduced inputs of sulfate, creating less widespread conditions for methylation. The Hg values measured in great egret nestling feathers decreased >70% between 1994 and 2000.[31] The Hg concentrations in bird tissues decreased generally in several species, and the numbers of breeding birds showed a distinct and persistent uptick in the late 1990s; Hg by that point had been banned from inclusion in paints, batteries and other household items typically incinerated in waste facilities in nearby Miami/Dade County. The temporal match between declining Hg in avian tissues and the increase in nesting suggested sub-lethal Hg effects as a possible cause for population fluctuations in the Everglades.[31,32]

These emerging patterns have increasingly emphasized the need for unbiased animal indicators of Hg in the Everglades as well as the need to assess the effects of Hg on wildlife populations. No other toxic contaminants are known to occur in relevant quantities in the Everglades,[33] making this an exceptional area to study and understand the sub-lethal effects of exposure to Hg in natural settings.

In order to monitor the patterns and trends of Hg exposure in the Everglades and assess sub-lethal effects we favored bird species over other mammals or other groups because birds are more abundant, can be more easily captured and sampled, their total breeding numbers can be estimated using standardized techniques[34] and it is comparatively easy to monitor their reproductive success. Model avian species used in laboratory research and toxicological assessments include the Japanese quail (*Coturnix japonica*), northern bobwhite tail (*Colinus virginianus*), mallard duck (*Anas platyrhynchos*), zebra finch (*Taeniopygia guttata*), European starling (*Sturnus vulgaris*) and chicken (*Gallus gallus domesticus*).[35] Of these, none are present in significant numbers in the Everglades. Further, we needed a monitoring tool that would represent the aquatic food web, which is the dominant web in this ecosystem.

Therefore, we had to establish local model species and develop tissue indicators and reference values for them. Of the available species we discarded great blue herons and anhingas because they nest in small numbers, and wood storks because it is difficult to access their nests to gather biological samples and they do not nest every year in large numbers. While egrets and herons of the genus *Egretta* (Little Blue Herons, Tricolored Herons, Snowy Egrets) can be abundant and widespread nesters, it is difficult to estimate their nesting numbers either because of dark coloration and poor visibility in aerial surveys, or because it is difficult to distinguish them from other white-colored species.

Great egrets and white ibises, in turn, present several characteristics that make them suitable candidates for model species in our system. The species have similar sizes, lifespans and ages of first breeding, and mostly prey on fish or macroinvertebrates.[16,36–41] The breeding cycle from egg-laying to fledgling lasts 44–52 days in great egrets and 32–47 days in white ibises, allowing biologists to monitor breeding success within the conditions imposed by the Everglades (discussed later). Further, great egrets nest (almost) every year in several colonies of different sizes across the Everglades, thus allowing for annual samples from several colonies and to assess geographical variation in exposure. Both species have also been recognized as indicators for the availability of food within the Everglades system and therefore of the ecological dynamics of a healthy functioning system.[42,43]

2.1.4 Development of a Field Sampling Protocol

Beginning in 1994, we monitored the breeding success of great egrets and white ibises in the Everglades of Dade, Broward and Monroe counties, Florida, USA (see Figure 2.1). Great egrets form breeding colonies on tree islands set widely apart (*cf.* 2–15 km) within the extensive graminoid Everglades wetland (9200 km^2). We used monthly systematic, 100% coverage aerial surveys with ground verification checks to detect breeding colonies and flew in circles around colonies to estimate the breeding numbers and took high resolution pictures to later count nests on them.[34,44] From 1998 onwards we selected 5–10 colonies each year to monitor breeding success and Hg exposure.

We selected colonies based on location to ensure a geographically representative sample of variation in breeding success and Hg exposure within the study area.[31,45] Once we detected, or estimated, nesting initiation from aircraft surveys, we visited selected islands to monitor the breeding success in individual nests once a week. Visiting breeding colonies requires 2–3 h of traveling by road and boat from field houses to colony islands. To prevent heat exposure related to human visits to colonies, we limited visits to one hour per week, usually in the early morning hours. In consequence, one team could sample a single colony, or two if they are close by, in a day. In the first visit we established trails from the edge to center of the islands across nesting areas and searched for nest platforms within 4 m on either side of the trail. We labeled nests along the trail with unique IDs and monitored the number of eggs and chicks on each visit. White ibises nest on dense concentrations. We aimed to include in nest monitoring at least two islands with white ibis breeding colonies and label all white ibis nests in a sector of the breeding colony in addition to all great egrets along the transect. To monitor exposure to Hg, we sampled nestling feathers from ≥10 randomly selected nests per colony. In some colonies low breeding numbers or low breeding success resulted in smaller sample sizes. In selected nests, when nestlings were 20–28 days of age we collected 10 fully grown scapular furthers (see the rationale and discussion for this given later). We stored collected feathers in sealed paper envelopes, one envelope for each individual sampled, in a dry place until analyses. Feather samples were collected over the entire study period, but we also collected other tissues for comparison during short-term studies (discussed later).

2.1.5 Development of Mercury Exposure Indicators – Tissues

To compare the efficacy of tissues to use for sampling, we combined information from controlled dose-concentration studies, measures of concentrations in tissues of wild birds and captive dose–response experiments.[31,46,47] In 1996, 16 first-hatched great egret fledglings were captured in the Everglades and raised in captivity in control, low (0.5 mg kg^{-1} ww) and high (5 mg kg^{-1} ww) doses of dietary methylmercury chloride. Birds were dosed from about 8 days to 14 weeks of age, when the experiment was finished and the remaining birds were sacrificed to analyze Hg in internal tissues. Growing scapular feathers correlated most closely with cumulative Hg consumed per weight, followed by mature scapular feathers.[46] Blood [Hg] also correlated well with cumulative Hg intake, yet its values were more dynamic and were below the average while feathers were developing and they increased faster after the feather growth stopped.[46]

Field samples collected in 1994 and 1995 showed a good correlation between blood [Hg] and nestling feather [Hg] ($R^2 = 0.72$, $P = 0.0001$; $N = 32$[48]). Based on these results we focused the long-term monitoring on nestling scapular feathers, because feather samples could be more easily and quickly collected from nestlings than blood samples, and the sampling process was less invasive and did not require extensive training to perform (see further discussion on its pros and cons given later). Analyses of feather samples from different colonies and years showed geographical and inter-annual variation in exposure to Hg within the Everglades.[31] Further, the comparison with feathers from museum specimens showed a marked increase in [Hg] in feathers in the 1980s and 1990s, peaking in the 1990s at six times the concentrations from earlier decades. Thereafter, a decline in [Hg] in feathers was observable during the late 1990s.[48]

In 2015–2017 we re-assessed the association between Hg in different tissues of great egrets. In a random selection of nests and colonies we non-destructively sampled albumen from the second laid egg;[49] later we sampled blood from the second largest nestling in the nest when nestlings were around 14 days of age and identified the nestling with plastic alpha-numeric rings, and finally, when nestlings were around 21 days of age, we collected fully grown scapular feathers from them and removed the rings.[50] We found significant but low correlations in Hg concentrations in different samples from the same individuals or individuals from the same nest, and blood samples from the same individual taken 7–14 days apart showed high variation in time and were not significantly correlated.[50] That was consistent with the reported dynamic variation in fledgling blood [Hg] with growth physiology.[49] The comparison of values within tissues and between colonies yielded higher correlations, supporting (a) the geographical structure of variation in exposure to Hg within the Everglades and (b) the use of average [Hg] in nestling feathers in a colony to approximate the [Hg] in other tissues and the exposure to Hg in earlier stages of the nestling cycle. To further investigate the influence of variation among individuals from the same colony and the reliability of nest and colony exposure estimates based on scapular feathers, in 2018 we focused on three colonies and sampled multiple nests. In selected colonies, we sampled fully grown scapular feathers of 111 great egret nestlings from 14, 21 and 22 nests. We compared variation in [Hg] among replicated analyses of feathers of the same individual, among siblings, within colonies and between colonies.[51] Variation in [Hg] among samples of the same individual was very small (the deviation of each sample was within 2% of the average [Hg] of all samples of the individual), and variation among siblings was also small (<6%) and it did not significantly differ from variation among replicates of the same individual. Thus, the [Hg] levels in grown scapular feathers of any nestling were representative of feather [Hg] of the whole brood.[51] We found considerable variation among nests in the same colony, and significant variation among colonies, supporting the geographical structure of the exposure to Hg in the Everglades, and variation in exposure among breeders in the same colony. We applied resampling techniques to the data to analyze the relationship between the number of samples and the precision of the estimate of average [Hg] exposure in the colony. The uncertainty of the estimated average [Hg] declined markedly beyond six samples, and the per sample increase in accuracy was sharply reduced beyond 10 samples, suggesting a sample size of at least 10 to characterize Hg exposure in colonies.[51]

2.2 Results

2.2.1 Sub-lethal Effects: Results from Experimental Studies

Dosage and accumulation trials showed that captive great egret nestlings exposed to low (0.5 $mg\,kg^{-1}$ food ww) and high (0.5 $mg\,kg^{-1}$ food ww) dietary MeHg during their rapid growth period showed reduced appetite and subsequent weight loss compared to a control group.[46] All birds were in good condition at the end of the experiment, but lack of food stress under experimental conditions could have been masking effects of MeHg that might be evident in a multistressor field environment. Low-dose birds also showed significant histological and immunological effects, while high-dose birds showed strong neurological effects too, becoming severely ataxic.[52] Further, although the fish hunting efficiency of low-dose birds did not seem to be compromised they were less likely to hunt fish and sought shade more often than control birds.[53] A comparison of results of that experiment with observational data from wild birds suggested that severe effects of MeHg exposure occurred at lower doses in wild birds than in captive birds, possibly because wild birds were exposed to many other environmental stressors that were absent in captive birds.[52]

To assess the effects of exposure to Hg on reproductive success, an experiment was conducted using white ibises. In 2005 nestling white ibises were collected from breeding colonies in the Everglades and randomly assigned to 4 different treatment groups: control, low-dose (0.05 $mg\,kg^{-1}$ ww), medium-dose (0.1 $mg\,kg^{-1}$ ww) and high-dose (0.2 $mg\,kg^{-1}$ ww) dietary MeHg.[47,54] These doses were within exposure estimates in the field and therefore environmentally relevant, although the highest dose was not as high as the high category used with great egrets (discussed earlier). All nestlings were kept on an Hg-free diet until 90 days of age, to delete possible background differences among individuals before experimental dosing. The experiment lasted for 3.5 years (April 2005 to September 2008), overlapping three full breeding seasons. Exposure to Hg at all dose levels resulted in some degree of altered endocrine expression, though there was inter-annual variation and clear dose–response relationships were not always clear. Hg in the higher dose groups induced demasculinized endocrine expression that resulted in reduced courtship behavior, reduced breeding success and male–male pairs.[47,54] In addition, dosed pairs laid smaller clutches compared to control pairs, although the decrease in clutch size was not linearly related to dose.[32] Breeding attempts by dosed birds were also less likely to reach the egg-laying stage, and in consequence exposure to Hg resulted in comparatively smaller breeding populations.[32] Exposure to Hg resulted in fewer fledglings produced by dosed males, even after discarding homosexual breeding attempts from analysis.[47] Levels of [Hg] in nestling feathers correlated strongly with blood [Hg] measured in breeders of the same dosage groups.[32]

2.2.2 Evidence of Mercury Effects in Field Conditions

In 1994 and 1995, 40 and 70 great egret nestlings were radio-tagged to analyze their post-fledgling survival in response to Hg exposure. In 1994, the survival up to 10.5 months of age of 40 nestlings exposed to naturally (non-manipulated) variable Hg concentrations did not differ between high- and low-exposure individuals.[55] In 1995 the survival up to 8 months of age under natural conditions of 30 nestlings orally dosed with a total 3 mg methylmercury chloride (MeHgCl) over 15 days in their nest did not differ from that of 40 control nestlings.[55]

From 2015 to 2018, we non-destructively sampled egg albumen, nestling blood and feathers (discussed earlier). We compared tissue [Hg] in these three tissues against several reproductive endpoints monitored in the same nests. Endpoints assessed included clutch size and hatching success, as well as nestling survival beyond 21 days of age. Albumen [Hg] was not associated with any of the endpoints, either within the entire data set or within colonies.[56] Similarly, there was no association between nestling blood [Hg] and any of the endpoints assessed, not even within a subset of similarly aged nestlings (16 ± 3 days). Nestling feather [Hg], in turn, was associated negatively with the early survival probability of hatchlings, negatively with the probability of hatchlings to survive to fledglings and negatively with the probability of laid eggs resulting in successful fledglings.[56] However, it was noteworthy that most of the failures were hatching failures and none of the Hg indicator tissues assessed were associated with hatching success.

Later we revisited the analyses of albumen [Hg] and nestling feather [Hg] in that data set but accounted for variation in food availability, as we suspected that variation in environmental stressors could behave interactively with Hg effects to influence responses in natural populations.[52] We modeled food availability using proxies of both availability and abundance. We used average fish biomass around each colony as a proxy of local food abundance, and the recession range (the difference between the highest water depth at the start of the season and the lowest depth before the recession) as a proxy for food vulnerability. To include fish biomass we used data from over 100 fish sampling points over the whole study area sampled twice a year.[57–59] We used wet-season fish biomass estimates because they have been shown to be a good predictor of fish biomass in dry-season water bodies where breeding wading birds prey and forage.[19] To model vulnerability, we used daily average water depths measured in 128 water gauging stations of the Everglades Depth Estimation Network (EDEN; https://sofia.usgs.gov/eden/). We assumed that differences between maximum and minimum water depths approximated the amount and duration of the water recession and how much and for how long fish were being concentrated in shallow drying pools where they were highly vulnerable to wading birds.[16,18,60] In both cases, to approximate food availability conditions in each colony we used fish sampling stations and water gages within 20 km of the colony (>95% of foraging flights were within 20 km of breeding colonies).[42] Accounting for food availability revealed associations between albumen and nestling feather [Hg] and every endpoint assessed, and suggested that the main effect of Hg was in an interaction with food availability, particularly its quadratic term, rather than simply by dose.[61]

However, terms including Hg (either as an additive effect or in interaction with components of food availability) accounted for different amounts of variation depending on endpoints. Nestling feather [Hg] explained more of the variation in post-hatching survival or numbers of fledglings. Both of these are close in time to the development of the feather and exposure to Hg captured by that indicator. In contrast, albumen [Hg] accounted for a relevant part of variation in early-mid breeding cycle endpoints, particularly hatching success. However, both indicator tissues were associated well with and explained intermediate amounts of variation in more integrative endpoints like the probability of a laid egg resulting in a successful nestling.[61]

We also used a colony-based approach, in which the exposure was approximated by average [Hg] in feathers in a subset of nests. This allowed us to include a larger sample size of years and food conditions. If food availability was high, there was no apparent effect of Hg-induced hatching failure within the range of Hg exposures recorded in the study. When food availability was low, the hatching failure increased with exposure to Hg and the failure was near total in the highest Hg exposures recorded.[61] The estimated dose–response curve under high food availability[15] fell within dose–response patterns observed in other avian species. However, when food availability was low, the estimated response was much higher than expected from other studies. Our models suggested a 50–100% decrease in hatching success at low food availability, while at high food availability only a 10–20% decrease was predicted.[15,61]

As the dosing experiment with breeding white ibises resulted in comparatively smaller breeding populations in dosed groups, we looked for evidence of reduced breeding population sizes associated with higher exposures to Hg in great egrets in the field. To approximate the population size we used peak counts of annual systematic aerial surveys of nests over 21 years. We accounted for the influence of annual food availability, adverse weather conditions and exposure to Hg. The recession range, water depth at the start of the breeding season and exposure to Hg explained most of the observed variation in nesting numbers.[32] Bigger recessions were associated with higher numbers of breeders in colonies, as did high water at the start of the season. High water represented more flooded surfaces and thus more fish breeding habitats, while large recession represented fish getting concentrated and becoming vulnerable in shallow pools over larger areas or a longer period of time. After controlling for these two factors, inter-annual and geographical variation in Hg in nestling feathers was associated with variation in nesting pairs, suggesting that egrets simply did not attempt to breed or failed prior to the time they might have been counted in surveys in higher Hg exposure years.[32] When controlling for the effect of food availability, our model suggested that at Hg exposure levels where a 10–20% reduction in breeding success would be expected, a decrease in nest numbers of as much as 50% could be expected. Further, we used our model to predict the numbers of egret nests expected if Hg exposure was high every year and in every colony, as it was supposed to be in the late 1970s, the 1980s and the early 1990s. Our model predicted a reduction in numbers of nesting pairs (62.7% ± 22.5) that encompassed and was similar to the observed reduction in nest counts in the late 1970s and early 1980s (59.7%).[32] This suggests that the modeled results were at the least in the range of observed results, and that Hg may have been an important variable in ecosystem-wide trends in numbers of nests.

2.3 General Discussion, Lessons Learnt and Pros and Cons of Our and Alternative Approaches

2.3.1 Evidence and Estimation of Hg Effects in Natural Populations

Several studies have reported evidence of deleterious Hg effects in birds in natural settings, often comparing a contaminated area against a reference one,[62–64] while other studies have found no effect, at least in some of the years or populations analyzed.[55,63,65,66] Our results suggest that in natural conditions Hg effects can be masked and modulated by food availability. Of course, other environmental stressors like weather and predators are likely to play a role too.[63,67] Yet, few studies have assessed the interactive influence of environmental stressors and particularly of food availability on contaminant effects. To the best of our knowledge, a single experimental study addressed interactions between food stress and a contaminant.[68] Ringed turtle doves (*Streptopelia risoria*) were exposed to DEE [1,1,-dichloro-2,2-bis(*p*-chlorophenyl)ethylene], a metabolite of DDT, and food restriction. Although the authors could not test for statistical significance, results suggested a reduction in breeding success in groups experimentally exposed to both DEE and food stress, which was higher than the apparent additive effects of each stressor. Physiological indicators yielded partial evidence and some parameters, like ovary weight, supported interactive effects of DEE and food reduction, while several others did not.[68,69] We know of no other study that accounted for variation in food availability in natural conditions. A study with Tengmalm's owl (*Aegolius funereus*) breeding in a heavy metal gradient surveyed prey abundance and observed concurrent gradients of increases in breeding success and prey abundance with increasing distance to the source of pollution, a smelter.[66] Prey liver heavy metal concentrations increased with proximity to the smelter, but the chick liver values did not for most metals. The authors concluded that direct effects of metals on owl breeding success were unlikely. While contamination could have reduced prey abundance, reduced breeding success seemed a consequence of reduced prey availability.[66] The small mammal prey was subject to cyclic variations in that area and there could be interactions between food availability and heavy metals that might be causing early failures in the breeding cycle, before they were detectable in field conditions.[32] Another study, assessing the breeding success of great tits (*Parus major*) along a heavy metal contamination gradient from a smelter, found that pairs breeding in contaminated areas performed significantly worse in several endpoints.[70] Further, nest box occupancy was slightly lower toward the contaminated area, a pattern observed every year but not statistically significant.[70] The authors argued that some of the effects could be caused by food limitation, probably as a consequence of contamination. Our results suggest that there might have been a combined effect of heavy metals and food stress, and thus these studies might have partially underestimated contaminant effects.

In general, our results suggest that in natural environments contaminant effects are likely to go unnoticed or to be masked or attributed to other factors if environmental stressors and their interactions are not accounted for. Similarly, focusing on the endpoints of the late breeding cycle, often the ones that are easily observable in natural studies, such as hatching success and fledglings produced in a set of monitored nests, might fail to detect early failures. Of course, this might not always be possible to monitor in the field, but researchers should be aware of that blind spot.

The literature on the effect of Hg on reproductive failure in natural populations is unclear. Several studies have looked for effects of Hg in breeding endpoints of natural populations or experimental groups but often with different endpoints (*e.g.*, hatching success, production of at least one young to fledging).[15] This focus on just one part of the breeding cycle has the potential to bias toxicity reference values especially if the effect of Hg is not constant during the breeding cycle. Where studies have observed several stages of the breeding cycle, the endpoint sensitivity to Hg apparently varies.[71] In the case of our study area, we have shown that higher exposure to Hg not only results in smaller annual breeding population size,[32] but also reduces fledgling survival and influences hatching success in a variable way depending on food availability.[51,61] These effects (reduced breeding probability and reduced breeding success) are likely to be additive over the breeding cycle, particularly considering that we reported them over the same study population and period. There could also be further effects on clutch size, post-fledgling survival and breeding performance as adults[72] that we have not tested. Therefore, estimates of net effects remain elusive, and our current estimates of effect sizes are probably underestimates.

2.3.2 Indicator Tissues: Comparative Advantages, Limitations and Uncertainty

Measuring exposure to Hg is often problematic in the field. Ideally MeHg in food would provide a good approximation to exposure and a clear link to experimental studies. However, birds often eat a range of species, within which Hg varies with location, species and individual.[73]

For these reasons, Hg is commonly measured directly from bird tissues. We will focus here on tissues that can be non-destructively sampled, including adult and nestling blood, adult and nestling feathers and more recently albumen. Each indicator has advantages and limitations. Adult blood [Hg] has been widely used; there are studies supporting its use, and there are good toxicity reference curves for it and equations to transform it to concentrations in other tissues, enhancing the comparability of results.[15,71,74–76] In addition, maternal blood [Hg] is related to *in ovo* Hg, thus providing a good approximation of risk to the embryo.[77] However, blood [Hg] can be strongly influenced by foods eaten in the short term, which creates snapshots of recent intake rather than long-term exposure.[78] This could be a potential issue if there were strong differences in [Hg] among prey or size classes of prey, or if there were fast changes in Hg in the system.[73,79] Diet in wintering areas and Hg accumulated outside breeding areas can also influence blood [Hg] in breeding areas and thus misrepresent Hg exposure in breeding grounds.[80] Yet maternal blood [Hg] would still represent *in ovo* exposure regardless of the region of origin of Hg. In our view, the main drawback of adult blood as an indicator of Hg exposure is that capturing adults can be difficult, costly and invasive. Further, adults captured might not be directly related to a nest or breeding area, resulting in uncertainty in the estimate of nestlings' exposure.[81] Nestling blood, in turn, seems to have several problems and few advantages. Blood [Hg] is sequestered by growing feathers, and as nestlings are growing feathers rapidly, feathers are continuously drawing Hg from blood.[46,80] This results in rapid changes in blood [Hg] as chicks mature and uneven increases in blood Hg in response to Hg dose at even slightly different ages.[55,82]

Egg albumen [Hg] can be transformed into approximated whole egg [Hg] and there are well-developed toxicological reference values for it too.[15] It is correlated with maternal blood [Hg], and clearly indicates the exposure of the embryo, a life stage that might be particularly sensitive.[77,83] On the other hand, the variation in egg [Hg] in the same clutch can be considerable and measuring a single egg might not be representative of the whole clutch.[77,84–86] In some cases variation among eggs of the same clutch was small, however.[87,88] On breeding grounds, egg albumen might be influenced by maternal Hg burden from non-nesting areas. Egg albumen also might not accurately reflect nestling exposure, particularly if there are temporal changes in Hg exposure in breeding grounds, birds rely on different foods through the nesting period, or prey consumed by adults differ from that provided to nestlings. As with parental blood, a potential issue with albumen sampling is logistics. Albumen micro-sampling requires some time and needs to be carefully performed to prevent damaging the egg. It can only be performed in a short window of time (fresh eggs <3 days of incubation), and in several cases limited access to colonies or nests might not enable that.

Using feathers of adult birds to approximate Hg exposure has been strongly questioned, as often there is no correlation, or only a comparatively weaker one, between Hg in feathers and other tissues.[81,89,90] Adult birds accumulate Hg in internal tissues while they are not growing any feather and mobilize it to feathers during molting. First-grown feathers may take up higher quantities than later ones, as body Hg is partially depleted as feathers grow.[80,91] This results in strong variability among feathers of the same individual.[90,92–94] Adult feathers may be an indicator of cumulative Hg intake between molting periods or of Hg in blood at the time of feather growth.[95] Birds molt feathers in a particular order that may be peculiar to species or even population and using the same feather of different individuals can help comparisons. However, many factors can influence molt, and this kind of approach should be therefore carefully and critically considered.[96] In certain cases, adult feathers might provide relevant information as to the general degree of exposure on species, Hg in the environment of different resident populations and approximate Hg exposure for the location where feathers were grown[97] or provide insight into long-term changes in environmental Hg using museum specimens.[30,98]

Nestling feathers present a different set of opportunities and constraints as indicators of Hg. These feathers have been occasionally used to approximate Hg exposure but we are not aware of any study, other than ours, which used nestling feathers [Hg] as an indicator for sub-lethal effects.[99–101] Virtually all albumen Hg is excreted in chick down, therefore the only source of Hg for the growing body and feathers of nestlings is food from the breeding area.[102] Thus any effect of maternal body burden is muted. Further, dosage experiments showed that nestling feather Hg correlates strongly with cumulative Hg intake during the nestling phase.[32,46] In captivity where there was little variation in Hg content in food, nestling feather [Hg] also correlated very well ($R^2 > 0.95$) with breeding adult blood [Hg].[32] Further, in our study species nestling scapular feather [Hg] in replicated samples of the same individual showed very little (<2%) variation and variation among siblings was small (<6%). Therefore, feathers from a single individual were representative of all the brood. Logistically, nestling feathers can be quickly and easily collected, and this requires little training and bird manipulation. Thus, this method is also less invasive. Nestling feathers can also be easily preserved, stored and analyzed, and typically show a strong signal of Hg body burden. In study areas like ours, where access and time in colonies are limited to one hour a week and visiting requires traveling and wading, these advantages make a difference. A particular feature of nestling feathers is that while they capture exposure to Hg during feather growth their [Hg] remains unaltered later. Therefore, they represent exposure over a certain window of time but, unlike blood or albumen, they can be collected well after that time passed from carcasses, museum specimens and shed feathers.

On the other hand, nestling feathers have several drawbacks that need to be carefully considered. First, as we have said, feathers capture Hg exposure during a short period of 14–20 days, and then further exposure is unrecorded. If the Hg exposure changes rapidly over time or birds feed on areas where the Hg levels in food are different, feathers would not be indicative of Hg in other stages of the breeding cycle. As feathers have not been widely used, there are no toxicity reference values to estimate risk directly from nestling feather [Hg]. For the same reason, comparison with other studies is not possible unless researchers sample other tissues and develop transformation formulas. We encourage researchers conducting work with birds and Hg to report, in addition to other indicators, Hg in nestling feathers, to help in constructing an interpretative framework. Lastly, nestling feathers can only be collected from successful nests, or nests that fail late in the nesting cycle. If there is uneven exposure to Hg within the breeding area, nests with higher Hg exposure could fail early in larger proportions and nestling feathers would underestimate exposure to Hg in the area. Although this drawback affects every indicator, its effect might be particularly stark with nestling feathers, because they are collected at the end of the breeding process. In our study area, albumen and nestling feather [Hg] from the same individual correlated significantly but poorly ($P < 0.001$; $R^2 = 0.279$), while colony averaged values showed a stronger correlation ($P < 0.001$; $R^2 = 0.706$).[50] This finding suggested that the geographical signature of Hg was strong and persistent in time but there was strong variability in individual exposure. In consequence, nestling feather Hg approximated general Hg exposure in colonies but its use and results with individual nests need to be considered in light of that limitation.

Inter- and intra-individual variation in exposure, or in [Hg] in any tissue, is an interesting phenomenon that has received little attention from researchers. Within our study site there was considerable variation in Hg among nestlings of the same colony that is commonly interpreted as an indication of strong variation in individual exposure.[50,51] However, captivity experiments with *ad libitum* food with a constant and controlled Hg concentration also resulted in high inter-individual variability in Hg in tissues of birds in the same treatment group.[55,74] This suggests that some amount of variability in Hg in tissues might be related to inter-individual differences in physiology, perhaps associated with genetic and familial characteristics.[71,103] To gain insight into the contribution of individual variability to the variance observed, we compared dispersion around the average of Hg metrics in nestling feathers of wild and captive nestlings and in the blood of captive breeders. To analyze variation in nature we used 116 great egret nestling feathers collected in 2018 in three different colonies of the Everglades.[51] To assess variation in captive animals fed controlled food with constant Hg content, we used the blood of breeding adults and nestling feathers collected in the last year of a three year experiment.[32,47] To allow a comparison of deviations around different averages, we divided the difference (individual value – average value in colony or treatment groups) by the average value in the colony or treatment groups, and multiplied the result by 100 and took its absolute value. Thus, for each colony or group we estimated the percentage of deviation from the average, an adimensional metric similar to the coefficient of variation. In great egret nestling feathers from colonies in the Everglades the mean deviation from average was 24.02 ± 16.21 ($N = 116$), while in nestling feathers of captive white ibises it was 12.48 ± 10.49 and that difference was significant (see Figure 2.5). Differences between blood [Hg] levels of individual breeding ibises and the average of their dosed group were 13.28 ± 10.33. There were no significant differences between the variability in ibis blood and nestling feathers. Further, there were no significant trends in deviation from average as the exposure to Hg increased (see Figure 2.5). The absolute value of the difference became bigger in groups or colonies with higher exposure to Hg but their proportional

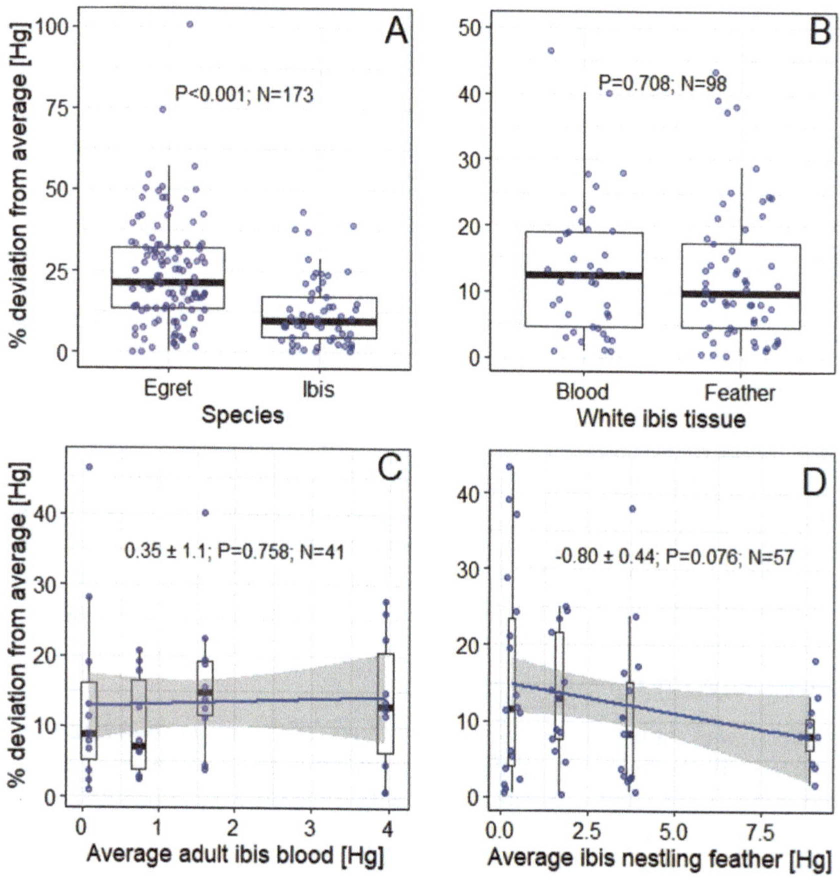

Figure 2.5 Deviation of individual [Hg] from average [Hg] in avian tissues in colonies or dosed groups. (A) Individual deviation from colony/dosed-group average [Hg] in wild great egrets in the Everglades (Egret) and captive white ibises (Ibis). (B) Deviation from average breeding adult blood and nestling feather [Hg] in dosed captive white ibises. (C) Deviation from average breeding white ibis blood [Hg] in relation to average blood [Hg] in the dosed groups. (D) Deviation from average in white ibis nestling feather [Hg] in relation to average nestling feather [Hg] in the dosed groups. In all plots clear blue dots indicate individual values (in A and B we added some horizontal jitter to enhance visibility), boxes encompass the values between 25th and 79th percentiles and the thick central line indicates the median value. We indicate the *P* value of the test assessing differences between groups in A and B, while in C and D we show the slope estimate, standard error and *P* value for the linear trend shown in the plot, with shaded standard error. In all cases we indicate the sample size (*N*).

deviation from average was the same, or even seemed to decrease with higher exposure in the case of ibis nestling feathers (see Figure 2.5). These results suggest that not all the variation observed in colonies arises from variation in exposure but that inter-individual differences in physiology account for a relevant part of it.

2.4 Conclusion and Advice for Similar Cases or Final Remarks

If research studies need to work with Hg in areas where no model species is available and want to develop their own model, we advise to select a species common in the study area. The decision on indicator tissues should be made in concordance with endpoints selected, preferably in tissues developed (or sampled) close in time to the endpoint they wish to analyze, and logistics. We strongly encourage to sample two or more indicators, including adult blood or albumen, at least in the first years, to assess their coherence. If possible, researchers should run dosage trials to establish correlations between Hg in food and Hg in indicators and other tissues. This is a must if adult blood or egg albumen cannot be obtained and researchers wish to compare their results with others or determine Hg risk based on toxicity reference values. Finally, in studies in natural settings we strongly encourage researchers to carefully consider what are the main environmental or ecological stressors in their area, with particular attention to possible fluctuation in food variability, and account for them and their interaction with Hg.

Acknowledgements

This work was supported by the US Army Corps of Engineers (W912HZ-12-02-0007) and by grants of equipment from the University of Florida Institute of Food and Agricultural Sciences. Egret monitoring was performed under University of Florida IACUC permit 201708650. Nestling ibises were collected from the field under Florida Fish and Wildlife Conservation Commission permit WX03527, which also allowed for their maintenance in captivity, and, with a modification, re-release into the wild. Ibises were held in captivity under Institutional Animal Care and Use Committee permit D424-2006. Pictures of the Everglades in Figure 2.1, white ibises in Figure 2.2, breeding great egrets in Figure 2.3 and a wood stork in Figure 2.4 were taken by J. Zabala. The image of the Florida peninsula used in Figure 2.1 was modified from Google Earth (Data SIO, NOAA, US Navy, NGA, GEBCO, Image Landsat/Copernicus).

References

1. D. P. Krabbenhoft and E. M. Sunderland, Global change and mercury, *Science*, 2013, **341**, 1457–1458.
2. J. O. Nriagu, A history of global metal pollution, *Science*, 1996, **272**, 223–224.

3. S. Hong, J. Candelone, C. C. Patterson and C. F. Boutront, History of ancient copper smelting pollution during Roman and Medieval times recorded in Greenland ice, *Science*, 1996, **272**, 246–249.
4. J. Hoguet, J. M. Keller, J. L. Reiner, J. R. Kucklick, C. E. Bryan, A. J. Moors, R. S. Pugh and P. R. Becker, Spatial and temporal trends of persistent organic pollutants and mercury in beluga whales (*Delphinapterus leucas*) from Alaska, *Sci. Total Environ.*, 2013, **449**, 285–294.
5. A. Goutte, C. Barbraud, D. Herzke, P. Bustamante, F. Angelier, S. Tartu, C. Clément-Chastel, B. Moe, C. Bech, G. W. Gabrielsen, J. O. Bustnes and O. Chastel, Survival rate and breeding outputs in a high Arctic seabird exposed to legacy persistent organic pollutants and mercury, *Environ. Pollut.*, 2015, **200**, 1–9.
6. R. Bargagli, F. Monaci, J. C. Sanchez-Hernandez and D. Cateni, Biomagnification of mercury in an Antarctic marine coastal food web, *Mar. Ecol.: Prog. Ser.*, 1998, **169**, 65–76.
7. A. Goutte, P. Bustamante, C. Barbraud, K. Delord, H. Weimerskirch and O. Chastel, Demographic responses to mercury exposure in two closely-related Antarctic top predators, *Ecology*, 2014, **95**, 131007010809000.
8. O. Lindqvist, Atmospheric mercury – a review, *Tellus B*, 1985, **37B**, 136–159.
9. H. M. Amos, D. J. Jacob, D. G. Streets and E. M. Sunderland, Legacy impacts of all-time anthropogenic emissions on the global mercury cycle, *Global Biogeochem. Cycles*, 2013, **27**, 410–421.
10. J. W. M. Rudd, R. A. Bodaly, N. S. Fisher, C. A. Kelly, D. Kopec and C. Whipple, Fifty years after its discharge, methylation of legacy mercury trapped in the Penobscot Estuary sustains high mercury in biota, *Sci. Total Environ.*, 2018, **642**, 1340–1352.
11. R. J. Letcher, J. O. Bustnes, R. Dietz, B. M. Jenssen, E. H. Jørgensen, C. Sonne, J. Verreault, M. M. Vijayan and G. W. Gabrielsen, Exposure and effects assessment of persistent organohalogen contaminants in arctic wildlife and fish, *Sci. Total Environ.*, 2010, **408**, 2995–3043.
12. K. C. Jones and P. de Voogt, Persistent organic pollutants (POPs): state of the science, *Environ. Pollut.*, 1999, **100**, 209–221.
13. M. F. Wolfe, S. Schwarzbach and R. A. Sulaiman, Effects of mercury on wildlife: A comprehensive review, *Environ. Toxicol. Chem.*, 1998, **17**, 146–160.
14. A. M. Scheuhammer, M. W. Meyer, M. B. Sandheinrich and M. W. Murray, Effects of environmental methylmercury on the health of wild birds, mammals, and fish, *Ambio*, 2007, **36**, 12–18.
15. P. C. Fuchsman, L. E. Brown, M. H. Henning, M. J. Bock and V. S. Magar, Toxicity reference values for methylmercury effects on avian reproduction: Critical review and analysis, *Environ. Toxicol. Chem.*, 2017, **36**, 294–319.
16. P. C. Frederick and M. W. Collopy, Nesting success of five ciconiiform species in relation to water conditions in the Florida Everglades, *Auk*, 1989, **106**, 625–634.

17. J. M. Beerens, P. C. Frederick, E. G. Noonburg and D. E. Gawlik, Determining habitat quality for species that demonstrate dynamic habitat selection, *Ecol. Evol.*, 2015, **5**, 5685–5697.
18. D. E. Gawlik, The effects of prey availability on the numerical response of wading birds, *Ecol. Monogr.*, 2002, **72**, 329–346.
19. B. A. Botson, D. E. Gawlik and J. C. Trexler, Mechanisms that generate resource pulses in a fluctuating wetland, *PLoS One*, 2016, **11**, 1–20.
20. W. F. Loftus and A.-M. Eklund, in *Everglades: the ecosystem and its restoration*, ed. S. Davis and J. C. Ogden, St. Lucie Press, Delray Beach, FL, 1994, pp. 461–483.
21. L. A. Nell, P. C. Frederick, F. J. Mazzotti, K. A. Vliet and L. A. Brandt, Presence of breeding birds improves body condition for a crocodilian nest protector, *PLoS One*, 2016, **11**, 1–16.
22. L. A. Nell and P. C. Frederick, Fallen nestlings and regurgitant as mechanisms of nutrient transfer from nesting wading birds to crocodilians, *Wetlands*, 2015, **35**, 723–732.
23. W. Gabel, P. Frederick and J. Zabala, Nestling carcasses from colonially breeding wading birds: patterns of access and energetic relevance for a vertebrate scavenger community, *Sci. Rep.*, 2019, **9**, 14512.
24. L. B. Cleckner, C. C. Gilmour, J. P. Hurley and D. P. Krabbenhoft, Mercury methylation in periphyton of the Florida Everglades, *Limnol. Oceanogr.*, 1999, **44**, 1815–1825.
25. M. C. Gabriel, N. Howard and T. Z. Osborne, Fish mercury and surface water sulfate relationships in the Everglades Protection Area, *Environ. Manage.*, 2014, **53**, 583–593.
26. S. J. Klapstein and N. J. O'Driscoll, Methylmercury biogeochemistry in freshwater ecosystems: A review focusing on DOM and photodemethylation, *Bull. Environ. Contam. Toxicol.*, 2018, **100**, 14–25.
27. D. Jordan, Mercury Contamination: Another threat to the Florida panther, *Endanger. Species Tech. Bull.*, 1990, **15**, 1–6.
28. S. F. Sundlof, M. G. Spalding, J. D. Wentworth and C. K. Steible, Mercury in livers of wading birds (Ciconiiformes) in Southern Florida, *Arch. Environ. Contam. Toxicol.*, 1994, **27**, 299–305.
29. M. G. Spalding, R. D. Bjork, G. V. N. Powell and S. F. Sundlof, Mercury and cause of death in great white herons, *J. Wildl. Manage.*, 1994, **58**, 735–739.
30. P. C. Frederick, B. Hylton, J. A. Heath and M. G. Spalding, A historical record of mercury contamination in Southern Florida (USA) as inferred from avian feather tissue, *Environ. Toxicol. Chem.*, 2004, **23**, 1474–1478.
31. P. C. Frederick, M. G. Spalding and R. Dusek, Wading birds as bioindicators of mercury contamination in Florida, USA: Annual and geographic variation, *Environ. Toxicol. Chem.*, 2002, **21**, 163–167.
32. J. Zabala, J. C. Trexler, N. Jayasena and P. Frederick, Early breeding failure in birds due to environmental toxins: A potentially powerful but hidden effect of contamination, *Environ. Sci. Amp. Technol.*, 2020, **54**, 13786–13796.

33. D. J. Forrester and M. G. Spalding, *Parasites and diseases of wild birds in Florida*, University Press of Florida, Gainesville, 2003.
34. K. A. Williams, P. C. Frederick and J. D. Nichols, Use of the super-population approach to estimate breeding population size: An example in asynchronously breeding birds, *Ecology*, 2011, **92**, 821–828.
35. V. L. B. Jaspers, Selecting the right bird model in experimental studies on endocrine disrupting chemicals, *Front. Environ. Sci.*, 2015, **3**, 1–7.
36. J. A. Heath, P. C. Frederick, J. A. Kushlan and K. L. Bildstei, White Ibis, https://birdsna.org/Species-Account/bna/species/whiibi/introduction (accessed 11 December 2019).
37. D. McCrimmon, J. C. Ogden, G. T. Bancroft, A. Martínez-Vilalta, A. Motis, G. M. Kirwan and P. F. D. Boesman, in *Birds of the World*, ed. S. M. Billerman, Cornell Lab of Ornithology, Ithaca, New York, 2020.
38. J. M. Beerens, D. E. Gawlik, G. Herring and M. I. Cook, Dynamic habitat selection by two wading bird species with divergent foraging strategies in a seasonally fluctuating wetland, *Auk*, 2011, **128**, 651–662.
39. P. C. Frederick, K. L. Bildstein, B. Fleury and J. Ogden, Conservation of large, nomadic populations of white ibises (*Eudocimus albus*) in the United States, *Conserv. Biol.*, 1996, **10**, 203–216.
40. P. C. Frederick, M. G. Spalding, M. S. Sepúlveda, G. E. Williams, L. Nico and R. Robins, Exposure of great egret (*Ardea albus*) nestlings to mercury through diet in the everglades ecosystem, *Environ. Toxicol. Chem.*, 1999, **18**, 1940–1947.
41. R. A. Boyle, N. J. Dorn and M. I. Cook, Nestling diet of three sympatrically nesting wading bird species in the Florida Everglades, *Waterbirds*, 2012, **35**, 154–159.
42. G. T. Bancroft, A. M. Strong, R. J. Sawicki, W. Hoffman and S. D. Jewell, in *Everglades: the ecosystem and its restoration*, ed. S. Davis and J. C. Ogden, St. Lucie Press, Delray Beach, FL, 1994, pp. 615–657.
43. P. C. Frederick, D. E. Gawlik, J. C. Ogden, M. I. Cook and M. Lusk, The white ibis and wood stork as indicators for restoration of the everglades ecosystem, *Ecol. Indic.*, 2009, **9**, 83–95.
44. P. C. Frederick and J. Ogden, in *Monitoring ecosystems: interdisciplinary approaches for evaluating ecoregional initiatives*, ed. D. Bush and J. C. Trexler, Island Press, Washington, DC, 2003, pp. 321–340.
45. G. Herring, C. A. Eagles-Smith, J. T. Ackerman, D. E. Gawlik and J. M. Beerens, Landscape factors and hydrology influence mercury concentrations in wading birds breeding in the Florida Everglades, USA, *Sci. Total Environ.*, 2013, **458–460**, 637–646.
46. M. G. Spalding, P. C. Frederick, H. C. McGill, S. N. Bouton and L. R. McDowell, Methylmercury accumulation in tissues and its effects on growth and appetite in captive great egrets, *J. Wildl. Dis.*, 2000, **36**, 411–422.
47. P. C. Frederick and N. Jayasena, Altered pairing behaviour and reproductive success in white ibises exposed to environmentally relevant concentrations of methylmercury, *Proc. Biol. Sci.*, 2011, **278**, 1851–1857.

48. M. S. Sepúlveda, P. C. Frederick, M. G. Spalding and G. E. Williams, Mercury contamination in free-ranging great egret nestlings (*Ardea albus*) from Southern Florida, USA, *Environ. Toxicol. Chem.*, 1999, **18**, 985–992.
49. K. R. Stebbins, J. D. Klimstra, C. A. Eagles-Smith, J. T. Ackerman and G. H. Heinz, A nonlethal microsampling technique to monitor the effects of mercury on wild bird eggs, *Environ. Toxicol. Chem.*, 2009, **28**, 465–470.
50. J. Zabala, I. A. Rodriguez-Jorquera, S. C. Orzechowski and P. Frederick, Non-lethal sampling of avian indicators reveals reliable geographic signals of mercury despite highly dynamic patterns of exposure in a large wetland, *Environ. Pollut.*, 2019, **24**, 577–585.
51. J. Zabala, A. M. Meade and P. Frederick, Variation in nestling feather mercury concentrations at individual, brood, and breeding colony levels: Implications for sampling mercury in birds, *Sci. Total Environ.*, 2019, **671**, 617–621.
52. M. G. Spalding, P. C. Frederick, H. C. McGill, S. N. Bouton, L. J. Richey, I. M. Schumacher, C. G. M. Blackmore and J. Harrison, Histologic, neurologic, and immunologic effects of methylmercury in captive great egrets, *J. Wildl. Dis.*, 2000, **36**, 423–435.
53. S. N. Bouton, P. C. Frederick, M. G. Spalding and H. McGill, Effects of chronic, low concentrations of dietary methylmercury on the behavior of juvenile great egrets, *Environ. Toxicol. Chem.*, 1999, **18**, 1934–1939.
54. N. Jayasena, P. C. Frederick and I. L. V Larkin, Endocrine disruption in white ibises (*Eudocimus albus*) caused by exposure to environmentally relevant levels of methylmercury, *Aquat. Toxicol.*, 2011, **105**, 321–327.
55. M. S. Sepúlveda, G. E. Williams, P. C. Frederick and M. G. Spalding, Effects of mercury on health and first-year survival of free-ranging great egrets (*Ardea albus*) from southern Florida, *Arch. Environ. Contam. Toxicol.*, 1999, **37**, 369–376.
56. J. Zabala, I. A. Rodriguez-Jorquera, S. C. Orzechowski and P. Frederick, Mercury concentration in nestling feathers better predicts individual reproductive success than egg or nestling blood in a piscivorous bird, *Environ. Sci. Technol.*, 2019, **53**, 1150–1156.
57. F. Jordan, S. Coyne and J. C. Trexler, Sampling fishes in vegetated habitats: Effects of habitat structure on sampling characteristics of the 1-m 2 throw trap, *Trans. Am. Fish. Soc.*, 1997, **126**, 1012–1020.
58. J. V. Gatto and J. C. Trexler, Seasonality of fish recruitment in a pulsed floodplain ecosystem: Estimation and hydrological controls, *Environ. Biol. Fishes*, 2019, **102**, 595–613.
59. D. L. DeAngelis, J. C. Trexler, C. Cosner, A. Obaza and F. Jopp, Fish population dynamics in a seasonally varying wetland, *Ecol. Modell.*, 2010, **221**, 1131–1137.
60. S. M. Lantz, D. E. Gawlik and M. I. Cook, The effects of water depth and emergent vegetation on foraging success and habitat selection of wading birds in the Everglades, *Waterbirds*, 2012, **34**, 439–447.

61. J. Zabala, I. Rodríguez-Jorquera, J. C. Trexler, S. Orzechowski, L. Garner and P. Frederick, Accounting for food availability reveals contaminant-induced breeding impairment, food-modulated contaminant effects, and endpoint-specificity of exposure indicators in free ranging avian populations, *Sci. Total Environ.*, 2021, **791**, 148322.
62. R. L. Brasso and D. A. Cristol, Effects of mercury exposure on the reproductive success of tree swallows (*Tachycineta bicolor*), *Ecotoxicology*, 2008, **17**, 133–141.
63. K. K. Hallinger and D. A. Cristol, The role of weather in mediating the effect of mercury exposure on reproductive success in tree swallows, *Ecotoxicology*, 2011, **20**, 1368–1377.
64. M. W. Meyer, D. C. Evers, J. J. Hartigan and P. S. Rasmussen, Patterns of common loon (*Gavia immer*) mercury exposure, reproduction, and survival in Wisconsin, USA, *Environ. Toxicol. Chem.*, 1998, **17**, 184.
65. G. Herring, D. E. Gawlik and D. G. Rumbold, Feather mercury concentrations and physiological condition of great egret and white ibis nestlings in the Florida Everglades, *Sci. Total Environ.*, 2009, **407**, 2641–2649.
66. B. Hornfeldt and N. E. I. Nyholmt, Breeding performance of Tengmalm's owl in a heavy metal pollution gradient, *J. Appl. Ecol.*, 1996, **33**, 377–386.
67. A. Sih, A. M. Bell and J. L. Kerby, Two stressors are far deadlier than one, *Trends Ecol. Evol.*, 2004, **19**, 274–276.
68. J. O. Keith and C. A. Mitchell, Effects of DDE and food stress on reproduction and body condition of ringed turtle doves, *Arch. Environ. Contam. Toxicol.*, 1993, **25**, 192–203.
69. J. O. Keith, Synergistics effects of DDE and food stress on reproduction in Brown Pelicans and Ringdoves, PhD thesis, The Ohio State University, 1978.
70. E. Janssens, T. Dauwe, R. Pinxten and M. Eens, Breeding performance of great tits (*Parus major*) along a gradient of heavy metal pollution, *Environ. Toxicol. Chem.*, 2003, **22**, 1140–1145.
71. C. W. Varian-Ramos, J. P. Swaddle and D. A. Cristol, Mercury reduces avian reproductive success and imposes selection: An experimental study with adult- or lifetime-exposure in zebra finch, *PLoS One*, 2014, **9**, 1–9.
72. O. J. Paris, J. P. Swaddle and D. A. Cristol, Exposure to dietary methylmercury solely during embryonic and juvenile development halves subsequent reproductive success in adult zebra finches, *Environ. Sci. Technol.*, 2018, **52**, 3117–3124.
73. W. F. Loftus, Accumulation and fate of mercury in an Everglades aquatic food web, PhD thesis, Florida International University, 2000.
74. K. A. Buck, C. W. Varian-Ramos, D. A. Cristol and J. P. Swaddle, Blood mercury levels of zebra finches are heritable: Implications for the evolution of mercury resistance, *PLoS One*, 2016, **11**, 1–14.

75. C. A. Eagles-Smith, J. T. Ackerman, T. L. Adelsbach, J. Y. Takekawa, A. K. Miles and R. A. Keister, Mercury correlations among six tissues for four waterbird species breeding in San Francisco Bay, California, USA, *Environ. Toxicol. Chem.*, 2008, **27**, 2136–2153.
76. J. T. Ackerman, C. A. Eagles-Smith, M. P. Herzog, C. A. Hartman, S. H. Peterson, D. C. Evers, A. K. Jackson, J. E. Elliott, S. S. Vander Pol and C. E. Bryan, Avian mercury exposure and toxicological risk across western North America: A synthesis, *Sci. Total Environ.*, 2016, **568**, 749–769.
77. J. T. Ackerman, C. A. Eagles-Smith, M. P. Herzog and C. A. Hartman, Maternal transfer of contaminants in birds: Mercury and selenium concentrations in parents and their eggs, *Environ. Pollut.*, 2016, **210**, 145–154.
78. C. A. Caldwell, M. A. Arnold and W. R. Gould, Mercury distribution in blood, tissues, and feathers of double-crested cormorant nestlings from arid-lands reservoirs in south central New Mexico, *Arch. Environ. Contam. Toxicol.*, 1999, **36**, 456–461.
79. C. A. Eagles-Smith and J. T. Ackerman, Rapid changes in small fish mercury concentrations in estuarine wetlands: Implications for wildlife risk and monitoring programs, *Environ. Sci. Technol.*, 2009, **43**, 8658–8664.
80. R. A. Lavoie, C. J. Baird, L. E. King, T. K. Kyser, V. L. Friesen and L. M. Campbell, Contamination of mercury during the wintering period influences concentrations at breeding sites in two migratory piscivorous birds, *Environ. Sci. Technol.*, 2014, **48**, 13694–13702.
81. C. A. Hartman, J. T. Ackerman, G. Herring, J. Isanhart and M. Herzog, Marsh wrens As bioindicators of mercury in wetlands of great salt lake: Do blood and feathers reflect site-specific exposure risk to bird reproduction?, *Environ. Sci. Technol.*, 2013, **47**, 6597–6605.
82. J. T. Ackerman, C. A. Eagles-Smith and M. P. Herzog, Bird mercury concentrations change rapidly as chicks age: Toxicological risk is highest at hatching and fledging, *Environ. Sci. Technol.*, 2011, **45**, 5418–5425.
83. G. H. Heinz, D. J. Hoffman, J. D. Klimstra, K. R. Stebbins, S. L. Kondrad and C. A. Erwin, Species differences in the sensitivity of avian embryos to methylmercury, *Arch. Environ. Contam. Toxicol.*, 2009, **56**, 129–138.
84. J. A. Akearok, C. E. Hebert, B. M. Braune and M. L. Mallory, Inter- and intraclutch variation in egg mercury levels in marine bird species from the Canadian Arctic, *Sci. Total Environ.*, 2010, **408**, 836–840.
85. M. Morera, C. Sanpera, S. Crespo, L. Jover and X. Ruiz, Inter- and intraclutch variability in heavy metals and selenium levels in Audouin's gull eggs from the Ebro Delta, Spain, *Environ. Contam. Toxicol.*, 1997, **75**, 71–75.
86. L. Ou, C. W. Varian-Ramos and D. A. Cristol, Effect of laying sequence on egg mercury in captive zebra finches: An interpretation considering individual variation, *Environ. Toxicol. Chem.*, 2015, **34**, 1787–1792.

87. R. L. Brasso, M. K. Abdel Latif and D. A. Cristol, Relationship between laying sequence and mercury concentration in tree swallow eggs, *Environ. Toxicol. Chem.*, 2010, **29**, 1155–1159.
88. G. H. Heinz, D. J. Hoffman, J. D. Klimstra and K. R. Stebbins, Reproduction in mallards exposed to dietary concentrations of methylmercury, *Ecotoxicology*, 2010, **19**, 977–982.
89. J. T. Ackerman, C. A. Eagles-Smith, J. Y. Takekawa, J. D. Bluso and T. L. Adelsbach, Mercury concentrations in blood and feathers of prebreeding Forster's terns in relation to space use of San Francisco Bay, California, USA, habitats, *Environ. Toxicol. Chem.*, 2008, **27**, 897.
90. K. E. Low, D. K. Ramsden, A. K. Jackson, C. Emery, W. D. Robinson, J. Randolph and C. A. Eagles-Smith, Songbird feathers as indicators of mercury exposure: high variability and low predictive power suggest limitations, *Ecotoxicology*, 2020, **29**, 1281–1292.
91. P. Granadeiro, D. Lopes, M. Claire, R. Furtado, E. Pereira and P. Catry, Untangling causes of variation in mercury concentration between flight feathers, *Environ. Pollut.*, 2021, **269**, 116105.
92. Z. Varela, R. García-Seoane, J. A. Fernández, A. Carballeira and J. R. Aboal, Study of temporal trends in mercury concentrations in the primary flight feathers of *Strix aluco*, *Ecotoxicol. Environ. Saf.*, 2016, **130**, 199–206.
93. A. Martínez, D. Crespo, J. Á. Fernández, J. R. Aboal and A. Carballeira, Selection of flight feathers from *Buteo buteo* and *Accipiter gentilis* for use in biomonitoring heavy metal contamination, *Sci. Total Environ.*, 2012, **425**, 254–261.
94. I. Roque, R. Lourenço, A. Marques, J. P. Coelho, C. Coelho, E. Pereira, J. E. Rabaça and A. Roulin, Barn owl feathers as biomonitors of mercury: sources of variation in sampling procedures, *Ecotoxicology*, 2016, **25**, 469–480.
95. C. L. J. Bottini, S. A. Macdougall-shackleton, B. A. Bran and K. A. Hobson, Feathers accurately reflect blood mercury at time of feather growth in a songbird, *Sci. Total Environ.*, 2021, **775**, 145739.
96. I. Zuberogoitia, J. Zabala and J. E. Martínez, Moult in birds of prey: A review of current knowledge and future challenges for research, *Ardeola*, 2018, **65**, 183–207.
97. I. J. Stenhouse, E. M. Adams, L. M. Phillips, S. Weidensaul and C. L. McIntyre, A preliminary assessment of mercury in the feathers of migratory songbirds breeding in the North American subarctic, *Ecotoxicology*, 2020, **29**, 1221–1228.
98. M. Perkins, O. P. Lane, D. C. Evers, A. Sauer, E. M. Adams, N. J. O'Driscoll, S. T. Edmunds, A. K. Jackson, J. C. Hagelin, J. Trimble and E. M. Sunderland, Historical patterns in mercury exposure for North American songbirds, *Ecotoxicology*, 2020, **29**, 1161–1173.
99. D. E. Kramar, B. Carstensen, S. Prisley and J. Campbell, Mercury concentrations in blood and feathers of nestling bald eagles in coastal and inland Virginia, *Avian Res.*, 2019, 1–7.

100. S. Espín, E. Martínez-López, M. León-Ortega, J. F. Calvo and A. J. García-Fernández, Factors that influence mercury concentrations in nestling eagle owls (*Bubo bubo*), *Sci. Total Environ.*, 2014, **470–471**, 1132–1139.
101. I. Rubio, M. Martinez-Madrid, L. Méndez-Fernández, A. Galarza and P. Rodriguez, Heavy metal concentration in feathers of little egret (*Egretta garzetta*) nestlings in three coastal breeding colonies in Spain, *Ecotoxicology*, 2016, **25**, 30–40.
102. J. T. Ackerman and C. A. Eagles-Smith, Integrating toxicity risk in bird eggs and chicks: Using chick down feathers to estimate mercury concentrations in eggs, *Environ. Sci. Technol.*, 2009, **43**, 2166–2172.
103. C. W. Varian-Ramos, J. P. Swaddle and D. A. Cristol, Familial differences in the effects of mercury on reproduction in zebra finches, *Environ. Pollut.*, 2013, **182**, 316–323.

CHAPTER 3

The Importance of Ecological Traits in Assessing Seabird Vulnerability to Environmental Risks

CAN ZHOU,*[a] JOAN A. BROWDER[b] AND YAN JIAO[a]

[a] Department of Fish and Wildlife Conservation, Virginia Tech, Blacksburg, Virginia, USA; [b] NOAA National Marine Fisheries Service, Miami, Florida, USA
*Email: eidotog@gmail.com

3.1 Introduction

It is generally accepted that seabirds are one of the most threatened groups of birds.[1] Due to their high trophic level in the marine food web and wide distribution, connecting coastal zones and high seas, seabirds play a key role in ecosystem functioning, and they are generally regarded as a useful indicator of its health status.[2–4] The latest global assessment of seabird populations found that fishery bycatch, invasive species and climate change are the top three threats in terms of the number of seabirds impacted.[5] In particular, at least 160 000 seabirds are killed annually by longline fisheries alone, not accounting for potentially substantial losses due to other fishing gears, *e.g.* illegal, unreported and unregulated (IUU) fisheries as well as pre-haul drop-off.[6–8] In this chapter, drawing from our research experience in assessing the impact of a particular risk – fishery bycatch on seabird populations, we explore different approaches to assess the vulnerability of seabirds

Issues in Toxicology No. 45
Bird and Reptile Species in Environmental Risk Assessment Strategies
Edited by Guillermo Eli Liwszyc and Marcelo L. Larramendy

Published by the Royal Society of Chemistry, www.rsc.org

to environmental risks, discuss current challenges, propose a trait-based solution to these challenges and explore additional complicating factors in using non-conventional model species in environmental risk assessments.

3.2 Failings of the Standard Approach

Quantifying the hazard of a risk factor to biological receptors of interest is one of the key steps in an ecological risk assessment process.[9] The standard approach is to calculate a reference risk, *e.g.* toxicity reference values, based on extensive animal testing with model/standard species. These test animals can be easily procured and maintained under laboratory conditions. Data based on model species are then extrapolated to native receptors of interest in native species or to humans.[10] This extrapolation may fail because of the inherent limitations of the standard approach for the following reasons: (1) the use of mortality as the only endpoint for measuring toxicity in acute studies may not be relevant for certain risk factors, (2) the toxicokinetic and toxicodynamic processes of a contaminant are mostly ignored, (3) the early life stage of the biotic matrix employed as a model species used in such studies may not be relevant for other biological processes, and (4) the life history and ecological traits of native species are mostly ignored.[11] Researchers previously assumed that animals from the same taxon would respond similarly to environmental risks.[12] However, more recent evidence has shown that the hazard of a risk factor to a phylogenetically related non-model species can be substantially under- or overestimated, undercutting the usefulness of using model species to predict the hazard level for native non-model species (for a review, see Allard *et al.*[13] and Vardy *et al.*[14]). The lack of predictive performance to extrapolate toxicity results to native non-model species is a major source of uncertainty in environmental risk assessment, calling for the use of non-conventional model species in this research area.

3.3 Vulnerability to Anthropogenic and Natural Risks

The particular vulnerability of seabirds to environmental risks comes from their relatively slow life history, the combination of a high maturation age and a low reproductive output, the attraction of seabirds to fishing vessels and the significant overlapping between the distribution of seabirds and the footprint of fisheries.[15–17] A sustainable fishing pressure for the target fish species may not be ecologically sustainable for some of the associated bycatch species, which may have a much lower reproductive potential, such as seabirds, and a constant removal of seabirds in the form of fishery bycatch will be detrimental until efficient mitigation. For example, for multiple seabird populations in the South Atlantic Ocean, including the grey-headed albatross (*Thalassarche chrysostoma*), the black-browed albatross (*Thalassarche melanophris*), the wandering albatross (*Diomedea exulans*) and the white-chinned petrel (*Procellaria aequinoctialis*), longline bycatch has been identified as the principal factor driving their global population decline.[18]

Besides the vulnerability of seabirds to anthropogenic and natural perturbations, seabirds are sentinels of the state of marine ecosystems, serving as bio-monitors of ecosystem-scale changes as well as quantitative indicators of ecosystem components.[19] Their widespread distribution at sea enables seabirds to be an efficient indicator of multiple risk factors over vast spatial scales, such as organic pollutants, heavy metals and fishery bycatch.[20–23] Arguably, it is more cost effective and ecologically relevant to collect stomach content samples from roosting seabirds on land than conducting at-sea ship-based surveys. As top predators, seabirds prey on lower trophic species and changes in the composition of their diet reflect changes in the marine community.[24,25] Again, in these studies, seabirds serve as efficient samplers offering near real-time information about community structure and ecosystem functioning. In comparison, a traditional ecological survey is limited by the number of sampling stations that a program can support per year; for example, the groundfish survey of the Southeast Area Monitoring and Assessment Program (SEAMAP) in the Gulf of Mexico is conducted only twice every year.[26] In sum, seabirds offer an alternative approach to traditional surveys for investigating environmental risks in marine ecosystems.

In the following section, we discuss the exposure stage of seabirds to environmental risks. Many of the examples to be given will be related to one particular type of risk – fishing operations. Besides, it is important to know that the direct impact of fishery interactions of seabirds will be physical, *e.g.* due to collisions with cable wires or drowning.[27] However, the exposure stage of seabirds to bycatch risk is common among other environmental risk factors, such as heavy metals and/or organic compounds. It is the process following risk exposure that differs. Even for heavy metals, the mechanistic actions of internal interactions with biomolecules differ.[28] Therefore, the particular type of risk considered in this chapter is also relevant for other environmental risk factors. Note that we do not consider internal physiological processes here. Among species, variations in internal physiological processes, such as toxicokinetic and toxicodynamic processes, add another layer of complexity to the analysis, and the novel development and tools in cross-species extrapolation of internal physiological processes has already been thoroughly reviewed elsewhere.[10]

3.4 Challenges for an Observational Study

One issue arising from using seabird-based environmental risk assessment is the multitude of seabird species involved in such studies. The same environmental conditions may pose different levels of risk to different seabird species due to differences in life history and behavior.[11,29] For example, in the Southern Ocean, black-browed albatrosses (*T. melanophris*) are commonly observed to follow fishing vessels and to engage in bait-taking interactions with longline fishing vessels, but storm petrels in the Oceanitidae family rarely interact with fishing operations in this region.[30] In this case, black-browed albatrosses are more exposed to bycatch risks than storm petrels. In the Southern Ocean, between black-browed albatrosses and storm petrels, there is a wide spectrum of seabird species that are exposed to bycatch risks to varying degrees.[31] Then, it is not clear if a specific species should be chosen to represent the risk, or an ecological community-based approach is more appropriate.

One may think that a specific seabird species that has a global distribution, *e.g.* the black-browed albatross (see Figures 3.1 and 3.2), may serve as a reference species for comparing risks among different regions, but the actual situation is more complicated. For example, in a long-term study conducted in the Southern Ocean, none of the seven species of storm petrels observed interacted with fishing operations of a Japanese longline fleet, and thus, there was zero bycatch risk.[30] To the contrary, in the North Atlantic, an average of 98 to a maximum of 190 storm petrels were projected to have been bycaught between 1992 and 2016 by the USA pelagic longline fleet.[32] The sharp contrast may be due to not only differences in fishing practices, such as the bait type and bycatch mitigation measures, but also differences in seabird communities. The same individual may be exposed to different levels of risk depending on the competitive environment and the presence of other seabird species at the time of the interaction. It turns out that different individuals either from the same or different species are not independent in their bait-taking interactions: some species are more capable of detecting and locating bait items, and they facilitate the bait-taking interactions of larger species that will fight aggressively for feeding opportunities.[33] Due to the presence of density dependences and interspecies interactions, it is difficult to pick a reference seabird species for standardization purposes, and an ecological community-based approach seems to be more appropriate.

Figure 3.1 An aggregation of mostly black-browed albatrosses (*Thalassarche melanophris*) and some possibly sooty shearwaters (*Ardenna grisea*, all-black birds at a distance to the vessel) near a bottom-fishing factory trawler. The trawler was conducting a post-season survey for *Doryteuthis gahi* squid on the shelf break east of the Falkland/Malvinas Islands in September 2009. Image credit: Dr Andreas Winter; Falkland Islands Fisheries Department.

Figure 3.2 Black-browed albatrosses (*Thalassarche melanophris*) and a few giant petrels (*Macronectes* spp.) feeding at the stern of a commercial bottom trawler operating within the Falkland/Malvinas Islands maritime zone in July 2013. Image credit: Dr Ross James; Government of South Georgia & the South Sandwich Islands.

3.5 A Trait-based Approach

An ecological community is composed of various individual species that coexist and interrelate in the same biotope or habitat. If assessing the environmental risk to a single species is already hard, a community-based approach does not make the problem any easier. As an analogy, fishery management faces a similar problem transitioning from a single species-based approach to an ecosystem-based approach.[34,35] Despite these difficulties, it is clear that a community-based holistic approach is more conducive to maintaining ecosystem quality and sustaining associated benefits.[36] For the problem of environmental risk assessment using seabirds, to make the community-based approach more tractable one can factor in the risk to each species based on ecological traits. In a sense, we decompose each species into its constituting parts, such as distribution patterns, feeding behavior and generation time, analyze the associated risks due to these parts using statistical regression and reconstruct/extrapolate the risk to a new species based on its traits. Studies by Zhou *et al.*[23] and Richards *et al.*[37] are examples of such an approach.

3.5.1 An Ecological Dimension Reduction Technique

Before diving into more details, it is worthwhile to understand how and why such an approach works, and how it simplifies the community-based (multi-species) approach if it works. The major difficulties associated with a multi-species approach are that each additional species represents an independent dimension in the analysis, incurs a new set of unknowns that need to be estimated and may further increase the number of unknown parameters

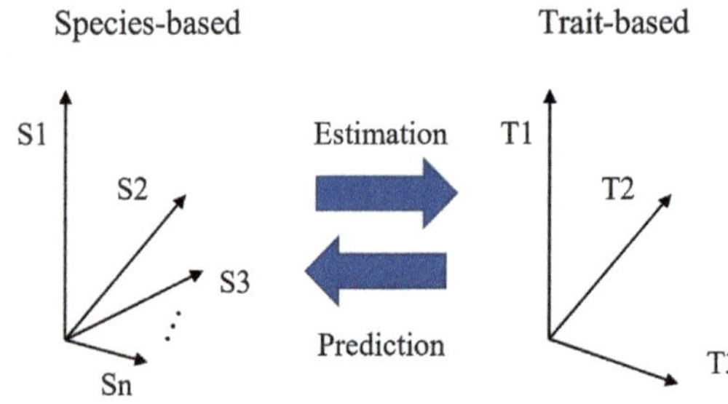

Figure 3.3 The trait-based approach is an ecological dimension reduction technique to effectively reduce the dimension of model parameter space. In a species-based approach, each of the species, such as species 1 (S1), 2 (S2), 2 (S3), up to n (Sn), is treated as an independent dimension, and the dimension of the parameter space grows to the second power of the number of species. In a trait-based approach, the parameter space is fixed given a predefined set of traits (3 as shown here), and its dimension does not grow with the number of species in the model.

because of more species interactions. For example, say we need to evaluate the vulnerability of species S1 to a risk factor; there is just a single parameter to estimate. But if species S1 interacts with species S2 and the interaction affects the exposure of these species to the risk factor, we have four parameters to estimate, one for each species plus two interaction terms. Similarly, with n coexisting species, the number of unknowns grows to n^2 (see Figure 3.3). The innovation of the trait-based approach is to recognize the commonality among different species, and the fact that their vulnerability to different risk factors is due to their specific feeding behavior, spatial and temporal distribution patterns, intra- and inter-species interactions, life history traits, *etc.*, rather than species identity, which is just a label we attach to them. The trait-based approach can substantially reduce the number of parameters to be estimated. The dimension of the model only depends on the number of key ecological traits chosen for a particular study, and it does not grow with each additional species in the model, and hence is not affected by the curse of dimensionality. If these traits are carefully chosen, this approach can reveal more meaningful biological information than the traditional approach.

The trait-based approach is an ecological dimension reduction technique to make sense of datasets with a high species dimension. In contrast, the principal component analysis (PCA) is a statistical dimension reduction technique that computes a small number, typically one or two, of PCs to represent the original dataset. These PCs are linear combinations of the original variables, and the biological/ecological meaning of these PCs is not usually clear. In a trait-based approach, new data usually originate from outside of the risk monitoring program and constitute additional input to the analysis. The construction of the trait space is entirely motivated by consideration of the physiological and ecological processes underlying the exposure of seabirds to environmental risks, and thus the estimated parameters have readily interpretable biological meanings. The statistical significance of each trait to risk exposure can be determined afterward through the significance of regression parameters. In a sense, the trait-based approach is an ecological hypothesis driven (ecology first and statistics second) approach instead of the other way around for the PCA for example.

The trait-based approach is not only useful for environmental risk assessments, but also enjoys wide usage in ecological studies.[38–42] A recent study by Richards *et al.*[37] showed that anthropogenic impacts are selectively removing large, long-lived, pelagic surface feeders with narrow habitat breadths. They found that seabirds with narrow habitat breadths, fast reproductive speeds and varied diets are more prone to risk from habitat modifying processes, such as pollution and natural variation. Threatened and non-threatened species occupy a unique ecological niche in the trait space, suggesting the ecological value of protecting both groups to preserve the function and stability of the ecosystem.

3.5.2 Useful Traits for Ecological Risk Evaluation

An important part of a seabird species' traits profile is geographical traits which include the range, migration patterns, breeding colony locations, feeding locations and seasonality. These are often applied at the population, rather than species, level and, in fact, used to define populations of a species. Fishing operations are one of the several risk factors that influence seabird populations variously according to specific geographical traits. Traits that make seabird species/populations vulnerable to fisheries must, first, bring them to locations that coincide with fishing operations. Similarly, traits that expose them to predator invasions or habitat loss in breeding colonies also have a geographical basis.

Various seabird species may apply many different migration strategies to wind up at the same place, often at the same time. This likely is a feeding location such as those found in the Middle Atlantic Bight of the USA off Cape Hatteras and the Outer Banks, which supports a diversity of foraging seabirds. Migration strategies bringing them together there include bi-hemispheric post-breeding flights of the great shearwater (*Ardenna gravis*) and the sooty shearwater (*Ardenna grisea*) from nesting colonies on oceanic islands in the southern South Atlantic Ocean. These flights can extend even further north because they also have been documented in the Bay of Fundy (Rob Ronconi, www.seaturtle.org/tracking). The northern gannet (*Morus bassanus*) migrates south from its breeding grounds in Canada and feeds in the Middle Atlantic Bight as well as in the Gulf of Mexico. Gulls and terns in feeding flocks off Cape Hatteras and the Outer Banks may come from colonies on the nearby coast. The black-capped petrel (*Pterodroma hasitata*), which nests on cliffs on the island of Hispaniola, also feeds in the Cape Hatteras–Outer Banks coastal ocean area. The Bermuda petrel (*Pterodroma cahow*) makes foraging flights to the same area from nesting colonies in Bermuda. Two other rare gadfly petrels, the Trindade petrel (*Pterodroma arminjoniana*) or called the Herald petrel before the split of the species, and the Maderian or Fea's petrel (*Pterodroma feae*), are also found there.[43] The black-capped petrel is also found in the Gulf of Mexico and has been reported in Canadian waters. Common loons (*Gavia immer*) summer across the middle latitudes of North America, where they nest on freshwater lakes and ponds, and summer along the eastern and southern coasts of North America, including the Middle Atlantic Bight, where they face the risk of capture as bycatch in small-scale nearshore shallow water gillnet fisheries. Foraging in the Middle Atlantic Bight and Northeast Coastal fishery regions [statistical regions demarcated by the National Oceanic and Atmospheric Administration (NOAA) National Marine Fisheries Service (NMFS) for the use of reporting and scientific research] makes oceanic seabirds vulnerable to several USA Atlantic fisheries.[8,44] The great shearwater and the northern gannet are especially prominent in the seabird bycatch of deep-water gillnet fisheries.[8] The great shearwater and herring gull (*Larus argentatus smithsonianus*) are estimated to be the most numerous seabird bycatch species in the pelagic longline fishery in this region.[44]

Seabird feeding ecological traits, such as the trophic level and diet composition, are another important trait dimension to consider. Using stable isotope analysis, Lenzi *et al.*[45] found that the consumption of urban refuse, mainly composed of beef and chicken product, correlated negatively with young kelp gull (*Larus dominicanus*) nestling success, and on the other hand, that of a fish-based diet correlated positively with young nestling success. In the eastern North Atlantic, Thompson *et al.*[46] reported the Hg concentration in multiple seabird species and found that the mercury methylation in mesopelagic water is responsible for the greatest increases in seabird Hg concentration, and the presence of mesopelagic prey is more important than the trophic level in determining its concentration in seabirds. In the South Atlantic, by examining the Hg concentration of 13 seabird species, Becker *et al.*[47] reported the importance of the trophic level in determining feather Hg levels; in another study also in the South Atlantic, Anderson *et al.*[48] confirmed that the trophic level was an important factor in determining Hg concentrations among species. These studies highlight the presence of regional variation in risk exposure to the same risk factor. In particular, Anderson *et al.*[48] also emphasized the importance of a community-based approach in providing a more complete picture of the complex interactions among risk factors and foraging ecology of seabirds than a single species approach.

The adult body size of a species is another important trait dimension.[49–51] Virtually, an organism's size affects all aspects of its physiology and ecology.[52] For birds, in particular, body size affects their trophic and non-trophic interactions, speciation and diversity, and extinction risk.[37,50,53,54] Larger seabirds usually have a longer generation time, slow maturation and a smaller clutch size, and such traits make them more vulnerable to both natural and anthropogenic perturbations. For fishery bycatch risks, seabirds with a larger body size tend to compete more aggressively for baits, and severe competition environments tend to produce deadly outcomes.[7,30]

Occasionally, we may find ourselves in a situation where there are more traits than the number of species for a particular study, where the dimension of the trait space is larger than that of the species space. In such situations, one should feel lucky about the multitude of independent variables available for prediction, although dependences among these traits can cause statistical issues due to multicollinearity. Ecological traits are defined based on ecological significance and not statistical independence, and many of these traits will be correlated. For example, a larger body size is generally correlated with a larger skull due to allometry and similar body shapes for groups of seabirds. To remove any statistical dependence among traits, one may use a statistical dimension reduction technique such as the PCA as mentioned earlier, for example, see the work of Baird and Van den Brink,[11] but now the biological significance of the trait space may become blurred. Alternatively, one can remove traits that are highly correlated with others or engineer new traits that are relatively independent. Take the body and skull sizes for example; one may use body size and the ratio of skull length to body length instead. It is not clear how this transformation can be conducted automatically, and such an *ad hoc* transformation may be developed case by case based on ecological insights.

Web-based data-sharing has been found to be essential in facilitating the development of new trait-based tools.[40] For seabirds, web resources, such as the BirdLife International species factsheets (http://datazone.birdlife.org/species/factsheet/) and the ACAP species assessments (https://www.acap.aq/acap-species), provide up-to-date and relevant information for assimilation into risk assessments on a global scale. There are also regional bird phenology databases, for example, a Colombian Tropical Andes bird species database,[55] the North American Bird Phenology Program[56] and the USGS Patuxent Wildlife Research Center Seabirds Compendium of seabirds in the North Atlantic,[57] useful for risk assessments on a regional scale. These public domain datasets provide invaluable species information for risk assessments.

3.5.3 Trait-based Prediction

Another advantage of the trait-based approach is the ability to predict the environmental risk to a new species, which may not be well represented in the dataset due to the limitations of the data collection protocol. One inherent difficulty with environmental risk assessments is a shortage of quality data, and also the high cost of additional data collection. To establish reasonably precise seabird bycatch estimates, a coverage of 20% is recommended.[58] With a sufficient sample size, a design-based approach can be applied. In reality, this level of coverage is rarely achieved in national and regional observer programs with notable exceptions.[17,59,60] It is far more common to obtain a smaller sample, and some researchers may call seabird bycatch a "rare event". To be clear, such events appear to be rare because (1) the observer programs are generally not designed to monitor bycatch risks but to monitor the catch of target species, such that high bycatch risk areas may not substantially overlap with observer coverage and (2) substantial amounts of seabird bycatch are lost pre-haul and are substantially unobserved in traditional observer programs.[6,7,32,61] To get a complete picture, a model-based approach is preferred, which can further take into account the underlying biological/ecological connections by conditioning on either climatic variables or species traits.

A recent regional study by Zhou *et al.*[23] used an external biodiversity dataset (USGS Patuxent Wildlife Research Center Seabirds Compendium, https://www1.usgs.gov/obis-usa/ipt/resource?r=usgs_pwrc_seabirdscompendium) to infer the species identity of unidentified seabird bycatch by the pelagic longline fishery in the North Atlantic based on ecological traits and spatial and temporal information. Only a proportion of the observed seabird carcasses were identified to either the species or genus level during the early part of the monitoring program due to the lack of sufficient observer training. It is important to predict the species identity of these unidentified carcasses because (1) the extinction risks are species specific, and a high bycatch mortality may be sustainable for an abundant species but can be detrimental for a species already low in abundance or rapidly declining, and (2) some globally threatened seabird species depend on the North Atlantic for their survival or migration path, *e.g.* Bermuda petrel and the Atlantic puffin (*Fratercula arctica*), and it is unknown if any of these species were among those unidentified carcasses. Based on a trait analysis, the authors found that the Bermuda petrel and the Atlantic puffin were not vulnerable to bycatch risks in this fishery.

On a global scale, Richards *et al.*[37] reported a vulnerability framework that integrates exposure, sensitivity and adaptive capacity of 341 seabird species to fishery bycatch. They combined global seabird distribution data from BirdLife International (http://datazone.birdlife.org/species/requestdis), fishery footprint from Global Fishing Watch (https://globalfishingwatch.org/) and the ecological trait data to predict the potential bycatch risk to each seabird species. The findings revealed that 46 species that have high exposure to fishing activities are not identified as vulnerable to bycatch in the International Union for Conservation of Nature (IUCN) database. This trait-based approach is particularly attractive because it is unaffected by (1) the spotted coverage of fishery observer programs in developing countries, where the loss of biodiversity may be most severe due to management oversight, and (2) the insufficient coverage and unobserved cryptic mortality in most current observer programs in developed countries. However, Richards *et al.*[37] did not incorporate fishery observer data, which is a weakness of their study. As the quality and quantity of fishery observer data continue to improve over time, we may be able to validate and update their results soon.

3.6 Other Challenges

It is difficult to experiment on seabirds in a standard ecotoxicological approach due to ethical and logistical considerations. As a result, most data will be observational rather than experimental, and this may cause additional practical and/or analytical problems when assessing the impact of environmental risks on native non-conventional species. In the following sections, we discuss common situations that may need additional care.

3.6.1 Correlation *vs.* Causation

Without conducting a controlled experiment as in the standard approach, we lose the ability to make strong inferences on the cause-and-effect relationship between the risk factor and the biological process of interest, both of which are observed with error. One issue that may arise is called the difficulty of "inverse inference": using dependent variables to estimate the independent variable.[62] The difficulty is due to the inconsistency of the ordinary least squares (OLS) estimator. Various bias correction techniques and the standardized major axis regression may be used instead of the OLS method.[63] Even if a regression coefficient has been consistently estimated, its statistical significance still does not indicate causation due to the potential presence of an unknown third factor affecting both variables. In these situations, hypothesizing one or more biological/ecological pathways linking the risk factor and the biological process is as important as showing a statistically significant result.

Another statistical problem that may arise is the "spurious correlation" between non-stationary variables. Say two time series data have been obtained, one for the intensity of a risk factor, such as parent blood Hg concentration, and the other for the value of an ecological process of interest, such as hatching success. If both time series are non-stationary, such as random walks, the p-value from Pearson's correlation test is biased downward, meaning that it is more likely that the p-value is smaller than the test significance level indicates. OLS estimates are also affected. The standard approach is to test for non-stationarity first, such as using the Dickey–Fuller test,[64] and then take the first difference if any of the time series exhibits non-stationarity. Alternatively, one can use the co-integration test, which is widely used in economic studies where non-stationarity is commonly observed.[65] See the study of Zhou *et al.*[66] for an example applying the co-integration test in an ecological context.

3.6.2 The Default of No Risk

Another issue that may arise from the reliance on observational data is the default of no effect in a statistical model. In a standard statistical test, the procedure is to accept the null of no effect when no significant effect is found. In a risk assessment setting, it means that when the sample size is small, the statistical test would default to no risk. In the standard approach, this is not problematic as the researcher fully controls the sample size, and the sample size can be increased to a level through a power analysis such that the probability of committing such errors is acceptable, but this is not the case for an observational study, especially for endangered species with a low abundance. In the latter case, it is important to report the associated statistical power and mark the result as inconclusive if the error of failing to reject the null of no effect is high when there is an ecologically significant effect, also called a type II error or false negative, instead of accepting the null of no risk. The inability to control sample size is an inherent limitation of observational studies, and it may entail prolonged observations if the accumulation of usable data is slow. In this case, a trait-based prediction/extrapolation combined with observational data may offer some relief. Meanwhile, it is important to emphasize that these trait-based predictions are subject to the validity of model assumptions, which may or may not hold true, and ultimately the researcher needs to collect real-world data to validate/invalidate these assumptions.

3.6.3 Ecological Regulation

Native seabirds do not live in a test tube, and one generally does not have the luxury of controlling all the auxiliary factors to single out the risk factor of interest in an observational study. Native seabirds, or other non-conventional model species, have a unique ecological niche in the local ecological community, and provide ecological functions and services by interacting with other species. A biological process of interest (*P*), such as nestling success, is subject to various time-varying factors, such as climatic conditions, density dependence, prey abundance, invasive species and human activities, in addition to the risk factor of interest (*R*). Again, the statistical analysis of *P* and *R* is subject to many failings discussed earlier if the statistical procedure is chosen poorly. In another example, fishing is a risk factor of interest and seabird population dynamics is the biological process of interest. A higher availability of lower trophic level prey fish due to the effect of fishing down the food web can compensate for to some degree or even reverse the detrimental effect of fishery bycatch of seabirds.[67] The point is that the observed ecological response of any risk factors is the end-response from the ecological community, and this response has been modulated by intra- and inter-species interactions within the community.[68] The multi-species interaction network may be a useful framework for organizing indirect effects in a local community and it deserves further investigation.[69]

3.6.4 Climate Change and Variability

Dependences on certain feeding locations and times make seabird populations vulnerable to changes in oceanographic conditions, whether episodic, cyclic or long term, as in climate change. Nesting locations can make them especially vulnerable to sea-level rise. An example is given by the mortality events of common murre (*Uria aalge*) in Alaska, recurrent at least since recorded in 1970, which reached a zenith in the 2015–2016 event.[70,71] Starvation was recorded as the probable cause of death in several events, and widespread breeding colony abandonment and reproductive failure were indicated.[72] Mortality events from 2014 through 2016 coincided with the presence of a warm anomaly in 2014 in the North-eastern Pacific, later referred to as the 2014/15 North Pacific marine heatwave and also "The Blob".[73–75] The unusually warm water for the area set off a series of food-web-related malfunctions believed to have suppressed the availability of fish and other organisms that usually supported the murres in the Gulf of Alaska and along the West Alaskan coast, leaving them with nothing to eat.[72,76] Conspicuous flights inland, away from the coast, were associated with these events, suggesting that the birds were searching for food. Problems for common murres extended down the USA Pacific coast, where long-term monitoring of colonies at the Yaquina Head Outstanding Natural Area in Oregon revealed reduced reproductive success for seven years from 2011 to 2017 with reproductive failure in several of these years.[77] Since 1980, this murre colony has been one of the most rapidly growing and productive murre colonies on the Oregon coast. Besides starvation conditions, predation on both chicks and adults, principally by the bald eagle (*Haliaeetus leucocephalus*), became a factor. Increased predation in seabird colonies occurs with signs of feeding stress and promotes reproductive failure.[78] This is an example in which a colony, by being detrimentally influenced by one risk factor (shortage of prey), becomes more vulnerable to another risk factor (colony predation).

Whether the North Pacific Heat Wave is a manifestation of climate change or an extreme representation of the El Niño–Southern Oscillation (ENSO) may not be settled, but, regardless, it is a startling example of the devastation to seabirds that can be caused by conditions that interrupt their food supply, especially at the time and place of nesting, when use of alternative sites is most limited. Other climate patterns affecting oceanographic conditions, such as the Pacific Decadal Oscillation (PDO) and North Atlantic Oscillation (NAO) (singly, in combination with each other or with ENSO), also provide opportunities to evaluate effects on seabirds of changes in conditions. Discussing such climate patterns, several studies examining indicators of seabird conditions have focused on the Northeast Atlantic. A 0.6 °C increase in sea surface temperature in the mid-1990s was interpreted as an underlying controlling factor for observed northward shifts of plankton and prey fish species supporting the range expansion of the critically endangered Balearic shearwater (*Puffinus mauretanicus*).[79]

Testing for bottom-up effects across four trophic levels (phytoplankton, zooplankton, fish larvae and seabirds) over a 17 year period (1991–2007) uncovered little evidence of bottom-up regulation in three ecosystems of the Northeast Atlantic where sea surface temperature had increased substantially in recent decades: Irish Sea, Celtic Sea and English Channel.[80]

The authors only found one significant positive relationship: between kittiwake species (*Rissa tridactyla* and *Rissa brevirostris*) productivity and abundance of fish larvae at one colony in the Irish Sea. On the other hand, trophic changes can sometimes occur without any apparent connection to oceanographic change or any change in condition of the feeding seabird. Cory's shearwaters (*Calonectris borealis*) feeding in the Canary Current switched from a diet heavily dependent on chub mackerel (*Scomber japonicus*) to one dependent on longspine snipefish (*Macroramphosus scolopax*) in concert with a change in dominance of the two species in their ecosystem, as indicated by fishery catches.[25] The dietary shift was not accompanied by any change in the bird's breeding success, nor was it correlated with any apparent change in oceanographic variables. These two studies indicate resiliency in the ecosystems that support seabirds as well as in the birds themselves.

Despite the signs of resiliency, seabird conservationists and resource managers should brace for future signs of stress on Atlantic seabirds because the expected negative (cooling) phase of the Atlantic Multidecadal Oscillation (AMO) may not be coming and so Atlantic temperatures may warm faster than expected. Recent results suggest that "the AMO is not a regular internal source of variability but has been driven by both volcanism and sulfate aerosol emissions that have influenced temperature negatively and a period of greenhouse gas accumulation causing temperatures to be higher than normal".[81]

One more reason to be concerned about seabird populations dependent on the North Atlantic is shown by an analysis of seabird colony reproductive success in the Northern and Southern Hemispheres, which suggests that the reproductive success is lower in the Northern Hemisphere.[82] The reason can be partly explained by differences in the representation of feeding traits (surface *vs.* subsurface feeding and planktonic *vs.* piscivorous *vs.* omnivorous diets) in the two hemispheres among bird species participating in the colonies. The analysis was based on 122 compiled time series of annual breeding productivity (proportionate change from the long-term mean) for 66 seabird species during the 1964–2018 period. These data were used to analyze seabird breeding success in relation to hemispheric asymmetry in ocean warming. Data were distinguished by decade. Trends in seabird breeding productivity varied by trophic level; within each trophic level, they varied by hemisphere. There was no overall effect of foraging depth (type). Omnivorous species, many of which provided their young with small pelagic fish, showed the most substantial changes in both hemispheres, with larger decreasing trends in normalized breeding productivity in the Northern Hemisphere. The breeding productivity of piscivorous species declined in the north but increased in the south. Planktivorous seabirds had increasing productivity over time in the north and stable productivity in the south. Investigators proposed that northern breeding seabird populations needed management intervention to improve the availability of appropriate food within the foraging distance of their breeding colonies, possibly by restricting fishing on those resources when colonies could be affected by shortages (*e.g.* use management to reduce one risk factor related to fishing), whereas Southern Hemisphere species needed less direct intervention, possibly by creation of marine reserves.

3.6.5 Other Approaches

The focus of this chapter has been on the analysis of the external exposure stage of risk factors in environmental risk assessments. The approach taken is a type of *in silico* approach, referring to any computer-based methodologies on the planning, analysis, evaluation or prediction of risk factors.[83] Other non-conventional experimental approaches are available, including (1) *in vitro* approaches, using subcellular components, cell lines or tissue cultures, *e.g.* that of Kennedy *et al.*,[84] and (2) early-stage testing including *in ovo* testings, *e.g.* those of Yao *et al.*[85] and Bosveld *et al.*,[86] among others. These experimental approaches investigate the internal response of the specific biological receptor of interest to risk factors given a prescribed level of exposure, and thus complements the *in silico* approach discussed here that models processes regulating the level of exposure. When combined, these approaches should provide a more complete picture of the impact of a risk factor on the local species and ecological community and integrating non-standard species in environmental risk assessments is one important step toward this goal.

Disclaimer

The scientific results and conclusions, as well as any views or opinions expressed herein, are those of the authors and do not necessarily reflect those of NOAA or the Department of Commerce.

Acknowledgements

In preparing this manuscript, the authors benefited from years of accumulated experience and knowledge from working together in association with a multi-year seabird project at the Southeast Fisheries Science Centre, NOAA-NMFS, Miami, Florida, and from the support of the NMFS National Seabird Program sponsored by the NMFS Bycatch Reduction Engineering Program. The authors also benefited from the encouragement of the SEFSC Pelagic Observer Program.

References

1. J. P. Croxall, S. H. Butchart, B. Lascelles, A. J. Stattersfield, B. Sullivan, A. Symes and P. Taylor, Seabird conservation status, threats and priority actions: a global assessment, *Bird Conserv. Int.*, 2012, **22**, 1–34.

2. R. Moreno, L. Jover, C. Diez, F. Sardà and C. Sanpera, Ten Years after the Prestige Oil Spill: Seabird Trophic Ecology as Indicator of Long-Term Effects on the Coastal Marine Ecosystem, *PLoS One*, 2013, **8**, e77360.
3. W. F. Mills, P. Bustamante, R. A. R. McGill, O. R. J. Anderson, S. Bearhop, Y. Cherel, S. C. Votier and R. A. Phillips, Mercury exposure in an endangered seabird: long-term changes and relationships with trophic ecology and breeding success, *Proc. Biol. Sci.*, 2020, **287**, 20202683.
4. R. W. Furness and K. Camphuysen, Seabirds as monitors of the marine environment, *ICES J. Mar. Sci.*, 1997, **54**, 726–737.
5. M. P. Dias, R. Martin, E. J. Pearmain, I. J. Burfield, C. Small, R. A. Phillips, O. Yates, B. Lascelles, P. G. Borboroglu and J. P. Croxall, Threats to seabirds: a global assessment, *Biol. Conserv.*, 2019, **237**, 525–537.
6. N. Brothers, A. R. Duckworth, C. Safina and E. L. Gilman, Seabird bycatch in pelagic longline fisheries is grossly underestimated when using only haul data, *PLoS One*, 2010, **5**, e12491.
7. C. Zhou, N. Brothers, J. Browder and Y. Jiao, Seabird bycatch loss rate variability in pelagic longline fisheries, *Biol. Conserv.*, 2020, **247**, 108590.
8. J. M. Hatch, Comprehensive estimates of seabird–fishery interactions for the US Northeast and mid-Atlantic, *Aquat. Conserv.*, 2018, **28**, 182–193.
9. G. W. Suter II, *Ecological risk assessment*, CRC Press, 2016.
10. M. Hecker, in *A Systems Biology Approach to Advancing Adverse Outcome Pathways for Risk Assessment*, ed. N. Garcia-Reyero and C. A. Murphy, Springer, 2018, pp. 107–132.
11. D. J. Baird and P. J. Van den Brink, Using biological traits to predict species sensitivity to toxic substances, *Ecotoxicol. Environ. Saf.*, 2007, **67**, 296–301.
12. J. P. Baker, in *Ecological Assessment of Hazardous Waste Sites: A Field and Laboratory Reference*, ed. W. Warren-Hicks, B. R. Parkhurst and J. S. S. Baker, Environmental Research Laboratory, Office of Research and Development, US Environmental Protection Agency, Corvallis, OR, 1991.
13. P. Allard, A. Fairbrother, B. K. Hope, R. N. Hull, M. S. Johnson, L. Kapustka, G. Mann, B. McDonald and B. E. Sample, Recommendations for the development and application of wildlife toxicity reference values, *Integr. Environ. Assess. Manage.*, 2010, **6**, 28–37.
14. D. W. Vardy, J. Oellers, J. A. Doering, H. Hollert, J. P. Giesy and M. Hecker, Sensitivity of early life stages of white sturgeon, rainbow trout, and fathead minnow to copper, *Ecotoxicology*, 2013, **22**, 139–147.
15. R. L. Lewison and L. B. Crowder, Estimating fishery bycatch and effects on a vulnerable seabird population, *Ecol. Appl.*, 2003, **13**, 743–753.
16. T. A. Clay, C. Small, G. N. Tuck, D. Pardo, A. P. B. Carneiro, A. G. Wood, J. P. Croxall, G. T. Crossin and R. A. Phillips, A comprehensive large-scale assessment of fisheries bycatch risk to threatened seabird populations, *J. Appl. Ecol.*, 2019, **56**, 1882–1893.

17. O. R. Anderson, C. J. Small, J. P. Croxall, E. K. Dunn, B. J. Sullivan, O. Yates and A. Black, Global seabird bycatch in longline fisheries, *Endanger. Species Res.*, 2011, **14**, 91–106.
18. R. Gales, N. Brothers and T. Reid, Seabird mortality in the Japanese tuna longline fishery around Australia, 1988–1995, *Biol. Conserv.*, 1998, **86**, 37–56.
19. E. Velarde, D. W. Anderson and E. Ezcurra, Seabird clues to ecosystem health, *Science*, 2019, **365**, 116–117.
20. S. Avery-Gomm, J. F. Provencher, M. Liboiron, F. E. Poon and P. A. Smith, Plastic pollution in the Labrador Sea: an assessment using the seabird northern fulmar Fulmarus glacialis as a biological monitoring species, *Mar. Pollut. Bull.*, 2018, **127**, 817–822.
21. A. E. Cartraud, M. Le Corre, J. Turquet and J. Tourmetz, Plastic ingestion in seabirds of the western Indian Ocean, *Mar. Pollut. Bull.*, 2019, **140**, 308–314.
22. T. Berr, J. Naudet, C. Lagourgue, K. Vuibert, K. Bourgeois and E. Vidal, Plastic ingestion by seabirds in New Caledonia, South Pacific, *Mar. Pollut. Bull.*, 2020, **152**, 110925.
23. C. Zhou, Y. Jiao and J. Browder, Seabird bycatch vulnerability to pelagic longline fisheries: ecological traits matter, *Aquat. Conserv.*, 2019, **29**, 1324–1335.
24. J. Provencher, A. Gaston, P. O. Hara and H. Gilchrist, Seabird diet indicates changing Arctic marine communities in eastern Canada, *Mar. Ecol.: Prog. Ser.*, 2012, **454**, 171–182.
25. J. Romero, P. Catry, H. Alonso and J. P. Granadeiro, Seabird diet analysis suggests sudden shift in the pelagic communities of the subtropical Northeast Atlantic, *Mar. Environ. Res.*, 2021, **165**, 105232.
26. P. A. Thompson and N. Bane, *SEAMAP environmental and biological atlas of the Gulf of Mexico, 1984*, Gulf States Marine Fisheries Commission, 1986.
27. B. Watkins, S. Petersen and P. Ryan, Interactions between seabirds and deep-water hake trawl gear: an assessment of impacts in South African waters, *Anim. Conserv.*, 2008, **11**, 247–254.
28. M. Balali-Mood, K. Naseri, Z. Tahergorabi, M. R. Khazdair and M. Sadeghi, Toxic Mechanisms of Five Heavy Metals: Mercury, Lead, Chromium, Cadmium, and Arsenic, *Front. Pharmacol.*, 2021, **12**, 643972.
29. P. Calow, R. M. Sibly and V. Forbes, Risk assessment on the basis of simplified life-history scenarios, *Environ. Toxicol. Chem.*, 1997, **16**, 1983–1989.
30. N. Brothers, *How accurate are observer reported kills of albatrosses on longlines?* Duke University and Blue Ocean Institute, 2008.
31. C. Zhou and N. Brothers, Seabird bycatch vulnerability in pelagic longline fisheries based on modelling of a long-term dataset, *Bird Conserv. Int.*, 2021, 1–16.
32. C. Zhou, Y. Jiao and J. Browder, How much do we know about seabird bycatch in pelagic longline fisheries? A simulation study on the potential bias caused by the usually unobserved portion of seabird bycatch, *PLoS One*, 2019, **14**, e0220797.

33. C. Zhou and N. Brothers, Interaction frequency of seabirds with longline fisheries: risk factors and implications for management, *ICES J. Mar. Sci.*, 2021, **78**, 1278–1287.
34. E. K. Pikitch, C. Santora, E. A. Babcock, A. Bakun, R. Bonfil, D. O. Conover, P. Dayton, P. Doukakis, D. Fluharty and B. Heneman, Ecosystem-based fishery management, *Science*, 2004, **305**, 346–347.
35. M. J. Fogarty, The art of ecosystem-based fishery management, *Can. J. Fish. Aquat. Sci.*, 2014, **71**, 479–490.
36. J. Brodziak and J. Link, Ecosystem-based fishery management: what is it and how can we do it?, *Bull. Mar. Sci.*, 2002, **70**, 589–611.
37. C. Richards, R. S. Cooke and A. E. Bates, Biological traits of seabirds predict extinction risk and vulnerability to anthropogenic threats, *Glob. Ecol. Biogeogr.*, 2021, **30**, 973–986.
38. M. Schleuning, E. L. Neuschulz, J. Albrecht, I. M. Bender, D. E. Bowler, D. M. Dehling, S. A. Fritz, C. Hof, T. Mueller and L. Nowak, Trait-based assessments of climate-change impacts on interacting species, *Trends Ecol. Evol.*, 2020, **35**, 319–328.
39. J. Chan, Y. Zeng and D. C. Yeo, Invasive species trait-based risk assessment for non-native freshwater fishes in a tropical city basin in Southeast Asia, *PLoS One*, 2021, **16**, e0248480.
40. D. J. Baird, M. N. Rubach and P. J. Van, den Brinkt, Trait-based ecological risk assessment (TERA): The new frontier?, *Integr. Environ. Assess. Manage.*, 2008, **4**, 2–3.
41. C. Violle and L. Jiang, Towards a trait-based quantification of species niche, *J. Plant Ecol.*, 2009, **2**, 87–93.
42. D. D. Ackerly and W. K. Cornwell, A trait-based approach to community assembly: partitioning of species trait values into within-and among-community components, *Ecol. Lett.*, 2007, **10**, 135–145.
43. D. S. Lee, Pelagic seabirds and the proposed exploration for fossil fuels off North Carolina: a test for conservation efforts of a vulnerable international resource, *J. Elisha Mitchell Sci. Soc.*, 1999, 294–315.
44. C. Zhou, Y. Jiao and J. Browder, Seabird bycatch vulnerability to pelagic longline fisheries: Ecological traits matter, *Aquatic. Conserv.: Mar. Freshwater Ecosyst.*, 2019, **29**, 1324–1335.
45. J. Lenzi, I. González-Bergonzoni, E. Flaherty, D. Hernández, E. Machín and B. Pijanowski, The relationship between urban refuse with fecundity and nestlings' success of a generalist seabird in the Río de la Plata Estuary – Uruguay, *Mar. Pollut. Bull.*, 2021, **173**, 113000.
46. D. R. Thompson, R. W. Furness and L. R. Monteiro, Seabirds as biomonitors of mercury inputs to epipelagic and mesopelagic marine food chains, *Sci. Total Environ.*, 1998, **213**, 299–305.
47. P. H. Becker, J. González-Solís, B. Behrends and J. Croxall, Feather mercury levels in seabirds at South Georgia: influence of trophic position, sex and age, *Mar. Ecol.: Prog. Ser.*, 2002, **243**, 261–269.

48. O. Anderson, R. Phillips, R. McDonald, R. Shore, R. McGill and S. Bearhop, Influence of trophic position and foraging range on mercury levels within a seabird community, *Mar. Ecol.: Prog. Ser.*, 2009, **375**, 277–288.
49. M. LaBarbera, Analyzing body size as a factor in ecology and evolution, *Annu. Rev. Ecol. Syst.*, 1989, **20**, 97–117.
50. V. A. Olson, R. G. Davies, C. D. L. Orme, G. H. Thomas, S. Meiri, T. M. Blackburn, K. J. Gaston, I. P. Owens and P. M. Bennett, Global biogeography and ecology of body size in birds, *Ecol. Lett.*, 2009, **12**, 249–259.
51. N. J. Isaac, K. E. Jones, J. L. Gittleman and A. Purvis, Correlates of species richness in mammals: body size, life history, and ecology, *Am. Nat.*, 2005, **165**, 600–607.
52. M. LaBarbera, in *Patterns and Processes in the History of Life*, Springer, 1986, pp. 69–98.
53. P. L. Mancini, K. A. Hobson and L. Bugoni, Role of body size in shaping the trophic structure of tropical seabird communities, *Mar. Ecol.: Prog. Ser.*, 2014, **497**, 243–257.
54. P. Almaraz and D. Oro, Size-mediated non-trophic interactions and stochastic predation drive assembly and dynamics in a seabird community, *Ecology*, 2011, **92**, 1948–1958.
55. L. J. Cardona-Salazar, Y. A. Benavides-Ossa, Á. M. Vargas-Daza, J. F. Betancurt-Grisales, M. Bohada-Murillo, E. T. Martínez-Sánchez, M. Cardona-Romero, A. Busi, W. D. Tobón-Escobar and M. Ortíz-Giraldo, A morphological, reproductive, and molt phenology database for 379 bird species from the Colombian Tropical Andes, *Ecology*, 2020, **101**, e03016.
56. S. Droege, D. Masaki and E. Sellers, USGS PWRC – Bird Phenology Program, United States Geological Survey, Occurrence dataset, 2021.
57. USGS, USGS Patuxent Wildlife Research Center Seabirds Compendium, 2021.
58. I. Debski, J. Pierre and K. Knowles, Observer coverage to monitor seabird captures in pelagic longline fisheries, 2016.
59. T. Sippel, N. Nasby-Lucas and S. Kohin, Description of the Hawaii longline observer program, *Int. Sci. Comm. Tuna Tuna-Species North Pac. ISC/14/SHARKWG-3/01*, 2014.
60. M. A. Collins, P. R. Hollyman, J. Clark, M. Soeffker, O. Yates and R. A. Phillips, Mitigating the impact of longline fisheries on seabirds: Lessons learned from the South Georgia Patagonian toothfish fishery (CCAMLR Subarea 48.3), *Mar. Policy*, 2021, **131**, 104618.
61. R. Bi, Y. Jiao, H. Bakka and J. A. Browder, Long-term climate ocean oscillations inform seabird bycatch from pelagic longline fishery, *ICES J. Mar. Sci.*, 2020, **77**, 668–679.
62. J. M. Durant, D. Ø. Hjermann, M. Frederiksen, J.-B. Charrassin, Y. Le Maho, P. S. Sabarros, R. J. Crawford and N. C. Stenseth, Pros and cons of using seabirds as ecological indicators, *Clim. Res.*, 2009, **39**, 115–129.
63. R. J. Carroll, D. Ruppert, L. A. Stefanski and C. M. Crainiceanu, *Measurement error in nonlinear models: a modern perspective*, Chapman and Hall/CRC, 2006.

64. D. A. Dickey and W. A. Fuller, Distribution of the estimators for autoregressive time series with a unit root, *J. Am. Stat. Assoc.*, 1979, **74**, 427–431.
65. S. Johansen, Estimation and hypothesis testing of cointegration vectors in Gaussian vector autoregressive models, *Econometrica*, 1991, 1551–1580.
66. C. Zhou, M. Fujiwara and W. E. Grant, Finding regulation among seemingly unregulated populations: a practical framework for analyzing multivariate population time series for their interactions, *Environ. Ecol. Stat.*, 2016, **23**, 181–204.
67. D. Pauly, V. Christensen, J. Dalsgaard, R. Froese and F. Torres, Fishing down marine food webs, *Science*, 1998, **279**, 860–863.
68. V. E. Forbes and P. Calow, Species Sensitivity Distributions Revisited: A Critical Appraisal, *Hum. Ecol. Risk Assess.*, 2002, **8**, 473–492.
69. J. R. Rohr, J. L. Kerby and A. Sih, Community ecology as a framework for predicting contaminant effects, *Trends Ecol. Evol.*, 2006, **21**, 606–613.
70. R. M. Corcoran, Common Murre (Uria aalge) die-off in the Kodiak Archipelago, Alaska, April 2015–April 2016, *Wilson J. Ornithol.*, 2021, **133**, 136–145.
71. H. Renner, K. Kuletz, D. Irons, R. Kaater, L. Labunski, J. Parrish, H. Burgess, B. Bodenstein, J. Piatt, S. Schoen and M. Renner, *The Murre Die Off*, 2016.
72. J. F. Piatt, J. K. Parrish, H. M. Renner, S. K. Schoen, T. T. Jones, M. L. Arimitsu, K. J. Kuletz, B. Bodenstein, M. García-Reyes and R. S. Duerr, Extreme mortality and reproductive failure of common murres resulting from the northeast Pacific marine heatwave of 2014-2016, *PLoS One*, 2020, **15**, e0226087.
73. N. A. Bond, M. F. Cronin, H. Freeland and N. Mantua, Causes and impacts of the 2014 warm anomaly in the NE Pacific, *Geophys. Res. Lett.*, 2015, **42**, 3414–3420.
74. E. Di Lorenzo and N. Mantua, Multi-year persistence of the 2014/15 North Pacific marine heatwave, *Nat. Clim. Change*, 2016, **6**, 1042–1047.
75. W. Peterson, N. Bond and M. Robert, The blob (part three): Going, going, gone?, *PICES Press*, 2016, **24**, 46.
76. M. L. Arimitsu, J. F. Piatt, S. Hatch, R. M. Suryan, S. Batten, M. A. Bishop, R. W. Campbell, H. Coletti, D. Cushing and K. Gorman, Heatwave-induced synchrony within forage fish portfolio disrupts energy flow to top pelagic predators, *Glob. Change. Biol*, 2021, **27**, 1859.
77. R. Suryan, S. Loredo, J. Dolliver, A. Roman, J. Porquez, R. Orben and I. Justiniano, *Yaquina Head Seabird Colony Monitoring 2017 Season Summary. End of season summary report to the U.S. Fish and Wildlife Service and Bureau of Land Management*, Oregon State University, Newport, Oregon, 2017.
78. R. M. Suryan, D. B. Irons, E. D. Brown, P. G. Jodice and D. D. Roby, Site-specific effects on productivity of an upper trophic-level marine predator: bottom-up, top-down, and mismatch effects on reproduction in a colonial seabird, *Prog. Oceanogr.*, 2006, **68**, 303–328.

79. R. B. Wynn, S. A. Josey, A. P. Martin, D. G. Johns and P. Yésou, Climate-driven range expansion of a critically endangered top predator in northeast Atlantic waters, *Biol. Lett.*, 2007, **3**, 529–532.
80. V. Lauria, M. J. Attrill, A. Brown, M. Edwards and S. C. Votier, Regional variation in the impact of climate change: evidence that bottom-up regulation from plankton to seabirds is weak in parts of the Northeast Atlantic, *Mar. Ecol.: Prog. Ser.*, 2013, **488**, 11–22.
81. G. Beaugrand, R. Faillettaz and R. Kirby, Absence of an internal multidecadal oscillation in the North Atlantic has consequences for anticipating the future of marine ecosystems, *Clim. Res.*, 2021, **85**, 107–111.
82. W. J. Sydeman, D. S. Schoeman, S. A. Thompson, B. A. Hoover, M. García-Reyes, F. Daunt, P. Agnew, T. Anker-Nilssen, C. Barbraud and R. Barrett, Hemispheric asymmetry in ocean change and the productivity of ecosystem sentinels, *Science*, 2021, **372**, 980–983.
83. T. Hartung and S. Hoffmann, Food for thought… on *in silico* methods in toxicology, *ALTEX*, 2009, **26**, 155–166.
84. S. Kennedy, A. Lorenzen, S. Jones, M. Hahn and J. Stegeman, Cytochrome P4501 a induction in avian hepatocyte cultures: A promising approach for predicting the sensitivity of avian species to toxic effects of halogenated aromatic hydrocarbons, *Toxicol. Appl. Pharmacol.*, 1996, **141**, 214–230.
85. C. Yao, B. Panigrahy and S. Safe, Utilization of cultured chick embryo hepatocytes as in vitro bioassays for polychlorinated biphenyls (PCBs): quantitative structure-induction relationships, *Chemosphere*, 1990, **21**, 1007–1016.
86. B. A. Bosveld, M. van den Berg and R. M. Theelen, Assessment of the EROD inducing potency of eleven 2,3,7,8-substituted PCDD/Fs and three coplanar PCBs in the chick embryo, *Chemosphere*, 1992, **25**, 911–916.

CHAPTER 4

A Review of the Levels and Distribution Patterns of Organochlorine Pesticides in the Eggs of Wild Birds in India

DHANANJAYAN VENUGOPAL,*[a] JAYAKUMAR SAMIDURAI,[b] JAYANTHI PALANIYAPPAN,[c] JAYAKUMAR RAJAMANI[d] AND MURALIDHARAN SUBRAMANIAN[e]

[a] ICMR – Regional Occupational Health Centre (S), NIOH, Bangalore 562 110, India; [b] A.V.C. College (Autonomous), Mayiladuthurai – 609 305, Tamil Nadu, India; [c] Department of Environmental Science, Periyar University, Salem – 636 011, Tamil Nadu, India; [d] Vivekanandha College of Arts and Science for Women, Tiruchengode – 637 205, Tamil Nadu, India; [e] Sálim Ali Centre for Ornithology and Natural History, Anaikatty, Coimbatore 641108, Tamil Nadu, India

*Email: dhananjayan_v@yahoo.com

4.1 Introduction

4.1.1 Organochlorine Pesticides

Distribution and accumulation of a wide range of pesticide residues have been found to occur in various matrices of the environment, including soil, water, air and biota. This has resulted from the indiscriminate use of large quantities of chemical pesticides for agriculture and public health purposes.[1]

Issues in Toxicology No. 45
Bird and Reptile Species in Environmental Risk Assessment Strategies
Edited by Guillermo Eli Liwszyc and Marcelo L. Larramendy

Published by the Royal Society of Chemistry, www.rsc.org

Organochlorine pesticides (OCPs) are a group of chemicals used as pesticides for agriculture and related activities in order to protect crops and increase agricultural production. Since pesticides have a multitude of applications, and distribution and transport patterns, it is very difficult to assess their risks and effects. Despite their omnipresence in normal terrestrial and aquatic environments, the occurrence and persistent nature of OCPs have led to their detection in remote and pristine regions due to physical phenomena such as atmospheric drift and long-range transport. Under the purview of the Green Revolution in the past decades, meeting the food demand for the exploding world population, large quantities of pesticides have been used across the world in order to increase global agricultural productivity.[2] Among the pesticides used indiscriminately for the above cause, OCPs have been predominant due to their broad-spectrum pesticide properties. Several studies have documented the occurrence of OCP residues in almost all environmental matrices including fishes.[3,4] Hence OCP residues are ubiquitously present in the environment due to their chemical persistence and bioaccumulation potential.[5]

OCPs also have a higher tendency to get accumulated in non-target wildlife through the food chain in the natural environment.[5–7] Furthermore, upon accumulation, they result in severe reproductive disorders in most of the sensitive wildlife species, especially birds.[8] Several studies have highlighted the threat of chemical pesticides on the survival of bird species across the world. Most of the wild birds occupy the higher position in trophic levels of the food chain. Their higher exposure level to toxic environmental contaminants has, in turn, created adverse toxic effects in avian species.[9,10] Due to this factor, bird species serve as sentinel and indicator organisms for chemical contaminant monitoring and risk assessment studies across the world, underpinning the fact that these contaminants have high bioaccumulation potentials in the tissues of birds, thus producing severe toxicity.[8,11,12] Bearing in mind this context, the eggs of birds serve as an ideal tool for reflecting the toxicity of chemical contaminants, particularly OCPs.[13]

Numerous studies have advocated the toxic effects of OCPs on wildlife and humans as a concerning issue.[14,15] Besides, it has been proved that reproductive toxic effects in birds like hatching failure, egg-laying abnormalities and thinning of eggshells,[16] leading to their population decline,[17] are due to the accumulation of OCPs therein. Moreover, it is documented that the OCP residue concentration exceeding 500 $ng\,g^{-1}$ in the eggs of wild birds resulted in adverse long-term toxic effects, leading to a decline in their population.[17,19–21]

In India, even the most common species, *e.g.*, the house sparrow (*Passer domesticus*), has become rare to sight, and infrequent sighting or the absence of such birds was reported.[22] Many such species of birds were recently added to the red list for their global conservation by the Royal Society for the Protection of Birds (RSPB), UK.[22] Several research works on avian species in India reported the occurrence and accumulation patterns of OCPs.[23,24] Hence in this review an attempt has been made to systematically document the occurrence of OCP residues in eggs and their toxic effects on common and wild birds of India.

4.1.2 Organochlorine Pesticides – Marketing and Consumption

The consumption of pesticides has increased severalfold in recent years, which, in turn, has caused severe environmental issues.[8,13,25] The excess quantity of fertilizers remaining after absorption by plants gets further transferred to other environmental components such as air and water through volatilization and runoff processes, respectively.[8] Increased amounts of nitrates and phosphates accumulate in the irrigated agricultural land of arid and semiarid regions due to the evaporation of the irrigated water from the soil. This type of accumulation varies significantly in relationship to the depth of the soil. Nitrogenous fertilizers get converted to nitrates through nitrification by soil microorganisms. Therefore, these studies required the implementation of environmental monitoring programmes and protective measures to human health.[26]

4.2 Pesticides – Indian Scenario

Pesticides have been in use across the globe for several purposes since the early 1940s and the magnitude of their use has increased severalfold in recent decades.[27] The Food and Agriculture Organization (FAO) reported that the use of pesticides doubled within 20 years, as it was about 4 million tons in 2017 against 2 million tons in 1990. Out of this global use, India accounts for approximately 1% of global pesticide consumption.[28] In terms of manufacture, India ranked as the fourth largest manufacturer and consumer of agrochemicals next to the USA, Japan and China.[29]

Initially, DDT and HCH were imported to India in 1948 for malaria and locust control respectively.[30] However, DDT was banned in later years.[30] In spite of this, its use in India was reported until the early 2000s.[31] It was also extended to agricultural activities in 1949, and its use in India steadily increased to 50 583 metric tons during the 1960s from a mere 5000 metric tons in 1950 (www.indiastat.com). Later, it was estimated that around 25 000 metric tons of chlorinated pesticides were consumed in India annually, out of which DDT alone contributes to 40% consumption.[32] Among the total quantum of pesticides consumed in India, insecticides contribute their major share of 76%.[33] The residues of the OCPs used across the country have raised toxicological issues for both biotic and abiotic factors of the environment in recent decades.

In India, though several types of pesticides are being applied, OCPs have been found to be dominantly occurring in non-target organisms like fishes, amphibians and mammals and also in humans.[5,34–36] In addition, traditional application methods without any awareness of their toxicity have also accounted for their extensive presence in the environment, which in turn affects human health.[37,38]

It has been widely reported that OCPs tend to bioaccumulate and biomagnify in the lipid tissues of all biota, mainly in birds.[39] Birds serve as the best indicators for any chemical contaminants prevailing in the environment due to the high possibility for their bioaccumulation and biomagnification properties, as they occupy the higher trophic levels in any ecological food chain. Birds are known for their varied ecosystem services like pollination, predation of pathogens and the linking factor for ecosystem balance, among others.[40] Despite the ecosystem services they provide, they are vulnerable to inadvertent pesticide poisoning due to their feeding and foraging habits. Direct feeding of granulated pesticides as their food, consuming seeds coated with pesticides and exposure to pesticides during their application and feeding on pesticide-contaminated food resources were reported as the common routes of exposure leading to the mortality of birds to pesticides.[41] In this context, the United States Environmental Protection Agency (US EPA) listed around 50 commonly used pesticides causing toxicity to commonly occurring songbirds, water birds and even raptors.[42]

In India, the OCP residue accumulation in various tissues of terrestrial and water birds and acetylcholinesterase inhibition due to pesticide residues have been reported,[43–46] and the same have also been reported across the world.[7] Besides the unintentional pesticide accumulation, OCP residues were reported in terrestrial birds[45,47] and colonial nesting birds.[48] However, despite these studies, there is a need for systematic, comparable biomonitoring studies using birds and their eggs as a tool for evidence-reporting of the magnitude of OCP toxicity to non-target environmental components.

4.3 Impact of Pesticides in Eggs of Wild Birds in India

Birds serve as a bioindicator as they are omnipresent in all ecosystems of the world. They are used as a tool for biomonitoring studies, due to their vulnerability to bioaccumulate organochlorine chemicals through the food chain.[12] In addition, they are the sentinel species in many agroecosystems. Their eggs serve as a non-invasive tool for assessing the toxic effects of chemical contaminants.[14,28] Parameters like thinning of eggshells, hatching failure and reproductive impairments are directly correlated with the toxic effects of organochlorine chemicals.[49,50] OCP residue concentrations exceeding 500 $ng\,g^{-1}$ ww in the eggs of wild birds were reported to result in adverse long-term toxic effects, leading to their population decline.[51] Globally, the reductions in the survival rates of birds at different stages of their lifecycle have been reported for commonly occurring terrestrial birds (Figure 4.1).[20,21]

4.3.1 OCP Residues in Birds' Eggs

Rather than birds, their eggs provide us vital information on the presence of chemical contaminants in the environment. As OCPs are highly lipophilic, they get highly accumulated in egg fat, and hence, the burden of residues accumulated in the living system of birds rests in the eggs that are laid by them. Hence eggs are conveniently used as an easily available non-invasive tool for assessing the toxic effects of chemical contaminants, especially OCP residues. In India, the research and experimental studies oriented towards

Figure 4.1 Common wild birds in their natural habitats.

this aspect are very few and sporadic. However, some studies and reviews were carried out in India to assess the OCP residues in the eggs of common wild birds of India (see Table 4.1).[25,44,47,52,53]

Table 4.1 Comparison of pesticide residues recorded in eggs of birds in India.[a]

Species	Scientific name	Category	Location	Unit	α-HCH	β-HCH	γ-HCH	δ-HCH	∑HCHs	p,p′-DDE	p,p′-DDD	p,p′-DDT	TDDT	Aldrin	Total endosulfan	HE	Dieldrin	Source
House sparrow	*Passer domesticus*	T	Namakkal, Tamil Nadu, India	$\mu g\,g^{-1}$, wet wt	BDL	0.37	BDL	BDL	0.37	0.64	0.38	0.27	1.29		0.33	0.19	0.13	Dhananjayan *et al.*, 2011[39]
House sparrow	*Passer domesticus*	T	Kotagiri, Tamil Nadu, India	$\mu g\,g^{-1}$, wet wt	0.07	0.28	0.79	BDL	1.14	0.07	0.01	0.06	0.14		BDL	0.07	BDL	
House sparrow	*Passer domesticus*	T	Mettur, Tamil Nadu, India	$\mu g\,g^{-1}$, wet wt	BDL	BDL	0.89	0.46	1.35	0.59	BDL	0.03	0.62		BDL	0.23	BDL	
House sparrow	*Passer domesticus*	T	Anaikatty, Tamil Nadu, India	$\mu g\,g^{-1}$, wet wt	BDL	0.01	BDL	BDL	0.01	BDL	BDL	0.02	0.02		BDL	BDL	BDL	
White-rumped vulture	*Gyps bengalensis*	R & S	Rajasthan, India	$\mu g\,g^{-1}$, ww						2.86	2.15	1	6.01				0.009	Muralidharan *et al.*, 2008[42]
House crow	*Corvus splendens*	R & S	Southern India	$ng\,g^{-1}$ lw					49 000				3400					Senthilkumar *et al.*, 2001[48]
Red junglefowl	*Gallus gallus*	T	Southern India	$ng\,g^{-1}$ lw					4100				430					
Spotted dove	*Spilopelia chinensis*	T	Southern India	$ng\,g^{-1}$ lw					1800				740					
Collared dove	*Streptopelia decaocto*	T	Southern India	$ng\,g^{-1}$ lw					3900				780					
Blue rock pigeon	*Columba livia*	T	Southern India	$ng\,g^{-1}$ lw					3000				110					
Baya weaver	*Ploceus philippinus*	T	Southern India	$ng\,g^{-1}$ lw					1400				660					
House sparrow	*Passer domesticus*	T	Southern India	$ng\,g^{-1}$ lw					9700				7900					
White wagtail	*Motacilla alba*	T	Southern India	$ng\,g^{-1}$ lw					2100				880					
Asian koel	*Eudynamys scolopaceus*	T	Southern India	$ng\,g^{-1}$ lw					ND	410	ND	10	420			11	ND	Dhananjayan *et al.*, 2020[23]
Baya weaver	*Ploceus philippinus*	T	Southern India	$ng\,g^{-1}$ lw					350	49	3	25	75			55	ND	

Table 4.1 (*Continued*)

Species	Scientific name	Category	Location	Unit	α-HCH	β-HCH	γ-HCH	δ-HCH	∑HCHs	*p,p′*-DDE	*p,p′*-DDD	*p,p′*-DDT	TDDT	Aldrin	Total endosulfan	HE	Dieldrin	Source
Eurasian blackbird	*Turdus merula*	T	Southern India	ng g^{-1} lw					5	ND	ND	3	3			ND	ND	
Small button quail	*Turnix sylvaticus*	T	Southern India	ng g^{-1} lw					60	13	ND	49	62			68	7	
Common tailorbird	*Orthotomus sutorius*	T	Southern India	ng g^{-1} lw					ND	ND	ND	100	100			310	ND	
Indian peafowl	*Pavo cristatus*	T	Southern India	ng g^{-1} lw					18	630	51	16	690			16	14	
Indian robin	*Copsychus fulicatus*	T	Southern India	ng g^{-1} lw					860	160	20	60	240			135	40	
Jungle bush quail	*Perdicula asiatica*	T	Southern India	ng g^{-1} lw					ND	11	ND	ND	11			ND	2	Dhananjayan *et al.*, 2020[23]
Long-tailed shrike	*Lanius schach*	T	Southern India	ng g^{-1} lw					1370	ND	ND	46	46			300	ND	
Nilgiri flycatcher	*Eumyias albicaudatus*	T	Southern India	ng g^{-1} lw					350	130	57	54	240			22	8	
Nilgiri pipit	*Anthus nilghiriensis*	T	Southern India	ng g^{-1} lw					40	ND	ND	300	300			ND	ND	
Paddy field pipit	*Anthus rufulus*	T	Southern India	ng g^{-1} lw					140	76	ND	54	130			16	ND	
Pied bushchat	*Saxicola caprata*	T	Southern India	ng g^{-1} lw					1560	130	65	24	220			125	41	
Plain prinia	*Prinia inornata*	T	Southern India	ng g^{-1} lw					1680	26	5	24	54			118	14	
Purple sunbird	*Cinnyris asiaticus*	T	Southern India	ng g^{-1} lw					290	680	ND	26	710			37	4	
Brown-cheeked fulvetta	*Alcippe poioicephala*	T	Southern India	ng g^{-1} lw					310	220	ND	ND	220			170	ND	
Grey junglefowl	*Gallus sonneratii*	T	Southern India	ng g^{-1} lw					2000	500	250	250	1000			700	200	
Red-vented bulbul	*Pycnonotus cafer*	T	Southern India	ng g^{-1} lw					1120	220	81	120	420			600	85	

Red-whiskered bulbul	*Pycnonotus jocosus*	T	Southern India	ng g^{-1} lw					1700	280	84	550	900			370	240	
Scaly-breasted munia	*Lonchura punctulata*	T	Southern India	ng g^{-1} lw					2800	320	20	53	390			660	35	
Nilgiri blue robin	*Sholicola major*	T	Southern India	ng g^{-1} lw					290	77	ND	30	107			240	ND	
Yellow-billed babbler	*Argya affinis*	T	Southern India	ng g^{-1} lw					1160	160	39	170	370			440	55	
Painted stork	*Mycteria leucocephala*	WB	Southern India	ng g^{-1} wet mass					17.9				8.1		2.17	0.61	0.34	Muralidharan *et al.*, 2012[61]
Black-headed ibis	*Threskiornis melanocephalus*	WB	Southern India	ng g^{-1} wet mass					27.1				14.8		2.31	0.38	0.5	
Spot-billed pelican	*Pelecanus philippensis*	WB	Southern India	ng g^{-1} wet mass					28.35				7.36		4.27	0.74	2.09	
Farm hen	*Gallus gallus*	OT	Uttar Pradesh, India	µg			1.6		4.6	5.9	1.9	6.4		0.9				Siddiqui and Saxena, 1983[47]
Domestic hen	*Gallus domesticus*	OT	Uttar Pradesh, India	µg			1.2		1.4	3.1	0.41	3.57		0.4				
Poultry chicken	*Gallus gallus*	OT	Punjab, India	µg kg^{-1}	0.04	0.16	0.06		0.26	0.055	0.13	0.07	0.51		31.2	0.11		Aulakh *et al.*, 2006[51]
Commercial chicken	*Gallus gallus*	OT	Tamil Nadu, India	ng g^{-1}											25			Vanitha *et al.*, 2014[52]
Farm hen	*Gallus gallus*	OT	Uttar Pradesh, India	µg			1.6		4.6	5.9	1.9	6.4		0.9				Siddiqui and Saxena, 1983[47]
Desi chicken	*Gallus domesticus*	OT	Coimbatore, Tamil Nadu, India	ng g^{-1}	BDL	0.5	1.1	0.2	1.81	0.11	BDL	0.1	0.2		0.4	BDL	0.2	Vanitha *et al.*, 2014[52]
Turkey	*Meleagris gallopavo*	OT	Southern India	ng g^{-1} lw					350				34					Senthilkumar *et al.*, 2001[48]

[a]Note: T – terrestrial birds; R & S – raptorial and scavenging birds; WB – water birds; OT – other birds.

A few studies have reported that the decline in population of insectivorous birds is associated with high pesticide concentrations.[17,54] A recent study conducted by Dhananjayan *et al.*[25] reported the OCP residues in the eggs from 22 species of common Indian birds. A total of 107 egg samples were analysed for the presence of OCP residues, and it was reported that all the samples contained the residues of at least one pesticide. In this paper, the residues of ΣOCPs (>5000 $ng\,g^{-1}$ ww) were recorded in the eggs of the red-whiskered bulbul (*Pycnonotus jocosus*), scaly-breasted munia (*Lonchura punctulata*) and grey junglefowl (*Gallus sonneratii*). However, the toxicity effects of OCPs may depend on the chemical nature of the OCPs, and hence, the following sections review the different types of OCP residues found in the eggs of wild birds.

4.3.1.1 Residues of HCH and Its Isomers

HCH is a broad-spectrum pesticide used from the historic past until recent times. Though it was banned for use in the early 2000s, it is more resistant to natural degradation and its residues remain in the environment for several years resulting in toxic effects. HCH residues present in the environment exist in their isomeric forms, namely, α-HCH, β-HCH and γ-HCH. In general, these residues undergo transformation and get metabolized into their isomeric forms. When HCH residues accumulate in birds' tissues, they metabolize into α-, β- and γ-HCH, among which β-HCH is more recalcitrant in nature and biomagnifies through the food web.[55] Confirming this fact, β-HCH residues were found to occur predominantly in egg samples.[25,56,57] Moreover, higher levels of γ-HCH residues were also observed in the aforementioned studies, confirming the transformation process and their higher bioaccumulation potential. Technically speaking, HCH was banned for use across the world; however, its production has continued in India since 2011. Further use of HCH has been restricted in India since 2013. Hence the presence of HCH residues until the recent past indicated their historical and recent use as well as discharges across the world.[58] The distributions of HCH isomers in birds' eggs were also in line with this, detecting high residual levels of β- and γ-HCH.[59]

The study on the OCP residues in the eggs of wild birds conducted by Dhananjayan *et al.*[24] reported high concentrations of ΣHCHs in the eggs of the red-whiskered bulbul, plain prinia (*Prinia inornata*), scaly-breasted munia and grey junglefowl, suggesting that this may be due to the consumption of pesticide-contaminated food. The ΣHCH residues documented in this study were comparatively much higher than the HCH residues detected in the eggs of common birds of Southern India,[53] white-rumped vultures (*Gyps bengalensis*)[47] and house sparrows (*P. domesticus*).[44] In support of these findings, high levels of HCH residues were also reported in soil samples across India at the same time.[60] Moreover, Muñoz Cifuentes[61] stated that HCH residues are responsible for the reduced hatching and reproduction of common gulls (*Larus canus*) at the Elbe Estuary. He also added that this may be due to the synergistic effect of the HCH isomers, leading to chronic toxicity. Hence, the presence of HCH residues in the eggs of common birds should be viewed with concern due to their widespread occurrence, in spite of the fact that the compounds was banned several years ago.

4.3.1.2 Residues of DDT and Its Metabolites

DDT residues get metabolized into DDE and DDD residues in environmental matrices.[30] Monitoring the DDT residues in birds' eggs will help us estimate the trend of input of these pesticides into the environment and food chain. Dhananjayan *et al.*[24] documented higher ΣDDT residues in the eggs of the grey junglefowl (1035 ng g^{-1}), red-whiskered bulbul (917 ± 872 ng g^{-1}), purple sunbird (*Cinnyris asiaticus*) (707 ± 434 ng g^{-1}) and Indian peafowl (*Pavo cristatus*) (693 ± 474 ng g^{-1}). This study suggested that a higher ratio of *p,p′*-DDE (72% of the ΣDDT) accumulation in eggs (ND to 621 ng g^{-1}), and the DDE levels were above the threshold level of 500 ng g^{-1} in 4% of the eggs sampled. This indicates the continued exposure of these species to DDT metabolites in India. Besides 79% of the egg samples contained *p,p′*-DDE residues, confirming the persistent nature and metabolization of DDT to DDE in the environment.

The concentration of *p,p′*-DDE residues in the eggs of birds in the range of 0.5 to 6 µg g^{-1} is correlated with impaired reproduction[17,18] and also significant eggshell thinning, resulting in hatching failure.[62,63] The study conducted by Senthilkumar *et al.*[53] in the eggs of common birds in Southern India recorded the ΣDDT residues ranging from 600 ng g^{-1} lw to 3400 ng g^{-1} lw, indicating DDT contamination of the birds' habitat. All these studies point to the long-term persistent nature of DDT compounds and their consumption for public health activities for mosquito control until the recent past. However, use of DDT for agricultural activity could not be confirmed.

4.3.1.3 Residues of Cyclodiene Compounds

The cyclodiene group of compounds collectively includes aldrin, endosulfan, heptachlor epoxide (HCE) and dieldrin residues. Aldrin and endosulfan residues were not detected at considerable levels in the reviewed studies. However, the more toxic compound among the cyclodiene group is dieldrin. The dieldrin concentration of 1 µg g^{-1} was represented as "total dieldrin equivalents"[64] and as the threshold level for bird eggs.[65] About 38% of the egg samples contained dieldrin residues in a study conducted by Dhananjayan *et al.*[24] In contrast to dieldrin residues, HCE residues (>1.5 ppm) were present in 5% of the eggs sampled. Dhananjayan *et al.*[24] reported that this level of residues was correlated with reproductive abnormalities as observed by Henny *et al.*[49] However, these residues were high in the scaly-breasted munia (664 ± 461 ng g^{-1}). The HCE levels detected in the red-whiskered bulbul and grey junglefowl (224 ppb) of this study were higher than the LC_{50} values (93 ppb) reported for the ring-necked pheasant (*Phasianus colchicus*) and Japanese quail (*Coturnix japonica*).[66] These levels were similar to the HE concentration in the eggs of house sparrows of the same region.[44]

4.3.2 Variation in Residue Levels Based on Species and Food Habits

Based on the available literature, a comparison was made in order to understand the variation in residue levels based on species and food habits. Under this context, the bird species whose eggs were subjected to OCP residue analysis were categorized as terrestrial birds, raptorial and scavenging birds, water birds and other birds (see Table 4.1) and are discussed in the following sections.

4.3.2.1 *Terrestrial Birds*

The birds commonly occurring in the agricultural ecosystem across India are categorized as terrestrial birds. The reviewed studies document the presence of OCP residues in the eggs of such birds (see Table 4.1). The occurrence of OCP residues in the eggs of house sparrows sampled from four various locations in Tamil Nadu was reported by Dhananjayan *et al.*[44] The sparrows' eggs sampled from Kotagiri and Mettur recorded the maximum ΣHCHs, whereas eggs from Namakkal recorded the highest ΣDDT, HCE and dieldrin concentrations. Similarly, Senthilkumar *et al.*[53] reported the levels of ΣHCHs (9700 ng g^{-1} lw) and ΣDDTs (7900 ng g^{-1} lw) in the eggs of the house sparrow (*P. domesticus*). Besides, the concentration of ΣHCHs in the eggs of the red junglefowl, spotted dove (*Spilopelia chinensis*), collared dove (*Streptopelia decaocto*), blue rock pigeon (*Columba livia*), baya weaver (*Ploceus philippinus*) and white wagtail (*Motacilla alba*) were about 4100 ng g^{-1}, 1800 ng g^{-1}, 3900 ng g^{-1}, 3000 ng g^{-1}, 1400 ng g^{-1} and 2100 ng g^{-1}, respectively. On the other hand, the levels of ΣDDTs in the eggs of red junglefowl, spotted dove, collared dove, blue rock pigeon, baya weaver and white wagtail were about 430 ng g^{-1}, 740 ng g^{-1}, 780 ng g^{-1}, 110 ng g^{-1}, 660 ng g^{-1} and 880 ng g^{-1}, respectively (see Table 4.1).

The study on the OCP residues in the eggs of terrestrial birds conducted by Dhananjayan *et al.*[24] reported higher concentrations of ΣHCHs in the eggs of the pied bushchat (*Saxicola caprata*) (1560 ng g^{-1}), red-whiskered bulbul (1700 ng g^{-1}), plain prinia (1680 ± 1290 ng g^{-1}), scaly-breasted munia (2800 ng g^{-1}) and grey junglefowl (2000 ng g^{-1}). The presence of these residues was attributed to their feeding habits in pesticide-sprayed regions. Similarly, higher ΣDDT residues were found in the eggs of the grey junglefowl (1035 ng g^{-1}), red-whiskered bulbul (917 ± 872 ng g^{-1}), purple sunbird (*C. asiaticus*) (707 ± 434 ng g^{-1}) and Indian peafowl (*P. cristatus*) (693 ± 474 ng g^{-1}). This study suggested that a higher ratio of *p,p′*-DDE (72% of the ΣDDT) accumulation in eggs (ND to 621 ng g^{-1}) and the DDE levels were above the threshold level of 500 ng g^{-1} in 4% of the eggs of the common terrestrial birds sampled. This indicates that these commonly occurring terrestrial birds are continuously exposed to DDT metabolites in India. Besides, 79% of the egg samples contained *p,p′*-DDE residues, which, in turn, confirm the environmental resistance nature and metabolization of DDT residues in the environment.

About 38% of the egg samples of terrestrial birds reported in this study contained dieldrin residues. However, none of the samples exceeded the threshold level of 1 $\mu g\,g^{-1}$. Similarly, HCE residues (>1.5 ppm) were present in 5% of the eggs of the terrestrial birds sampled and these levels of residues were reported to cause reproductive failure.[49] These residues were high in scaly-breasted munia (664 ± 461 $ng\,g^{-1}$). Similar levels were observed in the HCE concentration in the eggs of house sparrows of the same region.[44] Hence, the reviewed studies clearly indicate that common terrestrial birds are highly vulnerable and susceptible to OCP contamination in the environment.

4.3.2.2 Raptorial and Scavenging Birds

Studies on the occurrence of OCP residues in the eggs of raptorial and scavenging birds are very few in India. Muralidharan *et al.*[47] reported concentrations of *p,p′*-DDE (2.86 $\mu g\,g^{-1}$), *p,p′*-DDD (2.15 $\mu g\,g^{-1}$), *p,p′*-DDT (1.0 $\mu g\,g^{-1}$) and ΣDDTs (6.01 $\mu g\,g^{-1}$) in the eggs of white-rumped vultures (*G. bengalensis*) sampled from Rajasthan, India (see Table 4.1). The study also reported the levels of dieldrin (0.006 $\mu g\,g^{-1}$) in the eggs of this species. Similarly, Senthilkumar *et al.*[53] described the occurrence of ΣHCHs (49 000 $ng\,g^{-1}$) and ΣDDTs (3400 $ng\,g^{-1}$) in the eggs of house crows (*Corvus splendens*) in Southern India. Besides these, there were no studies on raptors and scavenger birds to discuss extensively, and this is considered to be a considerable limitation for this review.

4.3.2.3 Water Birds

The levels of OCP residues in three species of water birds, *i.e.*, the black-headed ibis (*Threskiornis melanocephalus*), the spot-billed pelican (*Pelecanus philippensis*) and the painted stork (*Mycteria leucocephala*), were reported from Southern India;[67] all species are listed under the category of near threatened species by the International Union for Conservation of Nature (IUCN) Red List. Among them, the maximum total concentration of HCH was reported in the black-headed ibis (28.35 $ng\,g^{-1}$) followed by the spot-billed pelican (27.1 $ng\,g^{-1}$). However, the painted stork had the lowest concentration of ΣHCHs (17.9 $ng\,g^{-1}$). The ΣDDT concentration was 14.8 $ng\,g^{-1}$ in the spot-billed pelican (*P. philippensis*), whereas the eggs of the painted stork (*M. leucocephala*) and the black-headed ibis (*T. melanocephalus*) contained 8.1 $ng\,g^{-1}$ and 7.36 $ng\,g^{-1}$, respectively. The concentrations of Σendosulfan in the eggs were reported as 4.27 $ng\,g^{-1}$ for the black-headed ibis, 2.31 $ng\,g^{-1}$ for the spot-billed pelican and 2.17 $ng\,g^{-1}$ for the painted stork. Among the three species of water birds, the maximum concentration of HCE was recorded in the eggs of the black-headed ibis (0.74 $ng\,g^{-1}$). The minimum concentration was noted in the eggs of the spot-billed pelican. Similarly, the highest concentration of dieldrin was reported in black-headed ibis eggs, whereas the lowest was seen in painted stork eggs (see Table 4.1). Though varying levels of OCP residues were detected in the eggs of these water birds, they did not pose any toxicity risk; however, continuous exposure of these birds to the OCP residues cannot be ignored, which may lead to chronic reproductive failure.

4.3.3 Eggshell Thinning and Reproductive Impairment

Thinning of eggshells was not the only risk associated with an increased pollutant burden in birds. Subtle changes in behaviour have also been seen and associated with pollution in many terrestrial birds too. No such studies have been done on Indian birds, so far. Exposure to higher concentrations of pesticide residues was reported to alter the behaviour of birds in terms of their incubation nest attention, courtship activities and feeding habits.[68] It has been reported that nests of great black-backed gulls with higher blood levels of polychlorinated biphenyls (PCBs), DDE, β-HCH and oxy-chlordane were more prone to be predated by ravens and crows, when compared to birds that had lower levels. These birds also had smaller eggs.[69] Abnormal reproductive behaviour, skewed sex ratios, differential gender-related mortalities and feminization of male embryos are other potential effects related to pesticide exposure (see Figure 4.2).[45]

4.4 Conclusions and Further Recommendations

There are only a very few studies documenting the presence of OCP residues in the eggs of avifauna with reference to the Indian context; nevertheless, these studies adequately report the toxic effects of OCP residues on birds that get accumulated in their eggs. The occurrence of higher concentrations of HCH, DDT and cyclodiene residues clearly indicates that the habitats of

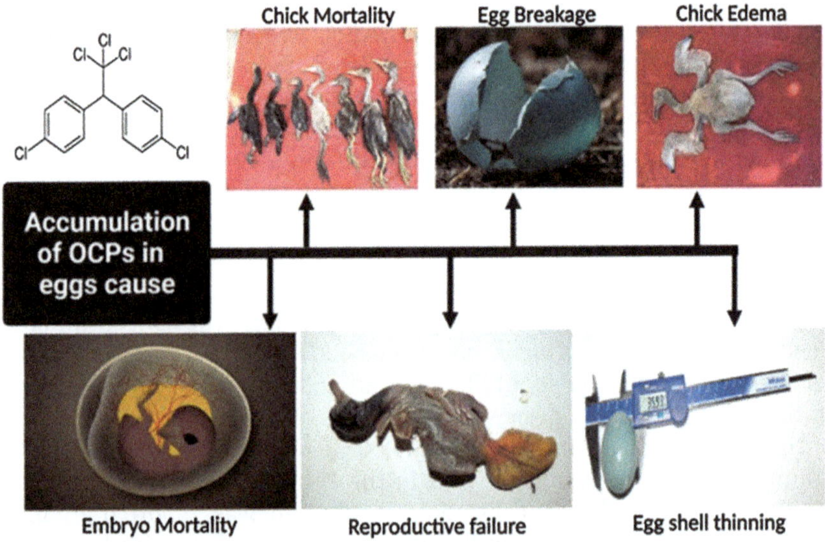

Figure 4.2 Impact of organochlorine pesticides on eggs of birds.

the birds reviewed were contaminated with pesticide residues irrespective of legal bans and restrictions. Moreover, several minor reproductive effects and behavioural abnormalities in birds go unnoticed on a larger scale. Despite this, the threshold levels for OCP residues were also periodically evaluated based on various comparable regional metrics. In order to overcome these limitations, further research efforts should be made with a planned systematic sampling including spatiotemporal attributes. Biomonitoring programmes with higher orders of sampling along with ecological observations and habitat studies, among others, are highly appreciated and recommended to document the environmental contaminants posing risk to the survival of birds.

References

1. M. F. Jallow, D. G. Awadh, M. S. Albaho, V. Y. Devi and B. M. Thomas, Pesticide knowledge and safety practices among farm workers in Kuwait: results of a survey, *Int. J. Environ. Res. Public Health*, 2017, **14**, 340.
2. P. Abhilash and N. Singh, Pesticide use and application: an Indian scenario, *J. Hazard. Mater.*, 2009, **165**, 1–12.
3. V. Dhananjayan and S. Muralidharan, Organochlorine pesticide residues in inland wetland fishes of Karnataka, India and their implications on human dietary intake, *Bull. Environ. Contam. Toxicol.*, 2010, **85**, 619–623.
4. J. Samidurai, M. Subramanian and D. Venugopal, Levels of organochlorine pesticide residues in fresh water fishes of three bird sanctuaries in Tamil Nadu, India, *Environ. Sci. Pollut. Res. Int.*, 2019, **26**, 1983–1993.
5. A. Ashesh, S. Singh, N. L. Devi and I. C. Yadav, Organochlorine pesticides in multi-environmental matrices of India: A comprehensive review on characteristics, occurrence, and analytical methods, *Microchem. J.*, 2022, 107306.
6. V. Dhananjayan and B. Ravichandran, Organochlorine pesticide residues in foodstuffs, fish, wildlife, and human tissues from India: Historical trend and contamination status, in *Environmental Deterioration and Human Health*, ed. A. Malik, E. Grohmann and R. Akhtar, Springer, Dordrecht, 2014, pp. 229–262.
7. P. N. Smith, M. Afzal, R. Al-Hasan, H. Bouwman, L. E. Castillo, M. H. Depledge, M. Subramanian, V. Dhananjayan, C. Fossi and M. Kitulagodage, Global perspectives on wildlife toxicology emerging issues, in *Wildlife Toxicology: Emerging Contaminant and Biodiversity Issues*, ed. R. J. Kendall, T. E. Lacher, G. C. Cobb and S. B. Cox, CRC Press, Boca Raton, 2010, p. 197.
8. V. Dhananjayan, P. Jayanthi, S. Jayakumar and B. Ravichandran, Agrochemicals impact on ecosystem and bio-monitoring, in *Resources Use Efficiency in Agriculture*, ed. S. Kumar, R. S. Meena and M. K. Jhariya, Springer Nature Singapore Pte Ltd, 2020, pp. 349–388.
9. T. Custer, C. M. Custer, R. Hines and D. Sparks, Trace elements, organochlorines, polycyclic aromatic hydrocarbons, dioxins, and furans

in lesser scaup wintering on the Indiana Harbor Canal, *Environ. Pollut.*, 2000, **110**, 469–482.
10. K. C. Parsons, P. Mineau and R. B. Renfrew, Effects of pesticide use in rice fields on birds, *Waterbirds*, 2010, **33**, 193–218.
11. X.-L. Zhang, X.-J. Luo, J. Liu, Y. Luo, S.-J. Chen and B.-X. Mai, Polychlorinated biphenyls and organochlorinated pesticides in birds from a contaminated region in South China: association with trophic level, tissue distribution and risk assessment, *Environ. Sci. Pollut. Res.*, 2011, **18**, 556–565.
12. M. Eens, V. L. Jaspers, E. Van den Steen, M. Bateson, C. Carere, P. Clergeau, D. Costantini, Z. Dolenec, J. E. Elliott and J. Flux, Can starling eggs be useful as a biomonitoring tool to study organohalogenated contaminants on a worldwide scale?, *Environ. Int.*, 2013, **51**, 141–149.
13. V. Dhananjayan, S. Jayakumar and B. Ravichandran, Conventional methods of pesticide application in agricultural field and fate of the pesticides in the environment and human health, in *Controlled Release of Pesticides for Sustainable Agriculture*, ed. K. R. Rakhimol, S. Thomas, T. Volova and K. Jayachandran, Springer, Cham., 2020, pp. 1–39.
14. E. Van den Steen, M. Poisbleau, L. Demongin, A. Covaci, A. C. Dirtu, R. Pinxten, H. J. van Noordwijk, P. Quillfeldt and M. Eens, Organohalogenated contaminants in eggs of rockhopper penguins (*Eudyptes chrysocome*) and imperial shags (*Phalacrocorax atriceps*) from the Falkland Islands, *Sci. Total Environ.*, 2011, **409**, 2838–2844.
15. P. Jayanthi, V. Dhananjayan, D. Elango and B. Ravichandran, Pesticides and human health implications, in *Pesticides Remediation Technologies from Water and Wastewater*, ed. M. H. Dehghani, R. R. Karri and I. Anastopoulos, Elsevier B.V., 2022, pp. 3–21.
16. D. Venugopal, P. Karunamoorthy, R. Beerappa, D. Sharma, M. Aambikapathy, K. Rajasekar, A. Gaikwad and S. Kondhalkar, Evaluation of work place pesticide concentration and health complaints among women workers in tea plantation, Southern India, *J. Expo. Sci. Environ. Epidemiol.*, 2021, **31**, 560–570.
17. L. Castillo, E. Thybaud, T. Caquet and F. Ramade, Organochlorine contaminants in common tern (*Sterna hirundo*) eggs and young from the river Rhine area (France), *Bull. Environ. Contam. Toxicol.*, 1994, **53**, 759–764.
18. C. Lundholm, DDE induced eggshell thinning in birds: effects of *p,p′*-DDE on the calcium and prostaglandin metabolism of the eggshell gland, *Comp. Biochem. Physiol. C Pharmacol. Toxicol.*, 1997, **118**, 113–128.
19. L. Lotter and H. Bouwman, Organochlorine pesticide residues in Helmeted Guineafowl (*Numida meleagris*), South Africa, *Bull. Environ. Contam. Toxicol.*, 2001, **66**, 162–170.
20. W. Krol, T. Arsenault and M. Mattina, Persistent organochlorine pesticide contamination of birds collected in Connecticut during the year 2000, *Bull. Environ. Contam. Toxicol.*, 2002, **69**, 452–458.
21. R. Mateo, J. Carrillo and R. Guitart, *p′*-DDE residues in eggs of European kestrel *Falco tinnunculus* from Tenerife, Canary islands, Spain, *Bull. Environ. Contam. Toxicol.*, 2000, **65**, 780–785.

22. V. S. Vijayan, Where all the sparrows have are gone?, *Down Earth*, 2014, **31**, 50–51.
23. V. Dhananjayan, *Levels of organic contaminants and select biomarkers in birds of Gujarat and Tamil Nadu, India*, PhD thesis, Bharathiar University, Coimbatore, India, 2009.
24. V. Dhananjayan, S. Muralidharan and P. Jayanthi, Distribution of persistent organochlorine chemical residues in blood plasma of three species of vultures from India, *Environ. Monit. Assess.*, 2011, **173**, 803–811.
25. D. Venugopal, M. Subramanian, J. Rajamani, J. Palaniyappan, J. Samidurai and A. Arumugam, Levels and distribution pattern of organochlorine pesticide residues in eggs of 22 terrestrial birds from Tamil Nadu, India, *Environ. Sci. Pollut. Res. Int.*, 2020, **27**, 39253–39264.
26. D. Venugopal, P. Karunamoorthy, R. Beerappa, D. Sharma, M. Aambikapathy, K. Rajasekar, A. Gaikwad and S. Kondhalkar, Evaluation of work place pesticide concentration and health complaints among women workers in tea plantation, Southern India, *J. Expo. Sci. Environ. Epidemiol.*, 2021, **31**, 560–570.
27. M. Grung, Y. Lin, H. Zhang, A. O. Steen, J. Huang, G. Zhang and T. Larssen, Pesticide levels and environmental risk in aquatic environments in China – A review, *Environ. Int.*, 2015, **81**, 87–97.
28. I. C. Yadav, N. L. Devi, J. H. Syed, Z. Cheng, J. Li, G. Zhang and K. C. Jones, Current status of persistent organic pesticides residues in air, water, and soil, and their possible effect on neighbouring countries: a comprehensive review of India, *Sci. Total Environ.*, 2015, **511**, 123–137.
29. Government of India, Department of chemical of petrochemicals, https://chemicals.nic.in/chemical (accessed on 12th June 2019).
30. I. C. Yadav, N. L. Devi, J. H. Syed, Z. Cheng, J. Li, G. Zhang and K. C. Jones, in Pesticide Residues and International Regulations, in *Sustainable Management of Potato Pests and Diseases*, ed. S. K. Chakrabarti, S. Sharma and M. A. Shah, Springer, 2022, pp. 353–367.
31. P. K. Gupta, Pesticide exposure – Indian scene, *Toxicology*, 2004, **198**, 83–90.
32. S. Mathur, Pesticide industries in India, *Pesticide Inf.*, 1993, **19**, 7–15.
33. S. Mathur and S. Tannan, Future of Indian pesticides industry in next millennium, *Pesticide Inf.*, 1999, **24**, 9–23.
34. V. Dhananjayan, B. Ravichandran, N. Anitha and H. Rajmohan, Assessment of acetylcholinesterase and butyrylcholinesterase activities in blood plasma of agriculture workers, *Indian J. Occup. Environ. Med.*, 2012, **16**, 127–130.
35. P. Chakraborty, S. N. Khuman, S. Selvaraj, S. Sampath, N. L. Devi, J. J. Bang and A. Katsoyiannis, Polychlorinated biphenyls and organochlorine pesticides in River Brahmaputra from the outer Himalayan Range and River Hooghly emptying into the Bay of Bengal: Occurrence, sources and ecotoxicological risk assessment, *Environ. Pollut.*, 2016, **219**, 998–1006.
36. V. Dhananjayan, B. Ravichandran and H. Rajmohan, Organochlorine pesticide residues in blood samples of agriculture and sheep wool

workers in Bangalore (rural), India, *Bull. Environ. Contam. Toxicol.*, 2012, **88**, 497–500.
37. V. Dhananjayan and B. Ravichandran, Occupational health risk of farmers exposed to pesticides in agricultural activities, *Curr. Opin. Environ. Sci. Health*, 2018, **4**, 31–37.
38. V. Dhananjayan, B. Ravichandran, K. Panjakumar, K. Kalaiselvi, K. Rajasekar, A. Mala, G. Avinash, K. Shridhar, A. Manju and R. Wilson, Assessment of genotoxicity and cholinesterase activity among women workers occupationally exposed to pesticides in tea garden, *Mutat. Res.*, 2019, **841**, 1–7.
39. T. W. Custer, C. M. Custer, P. M. Dummer, E. Bigorgne, E. M. Oziolor, N. Karouna-Renier, S. Schultz, R. A. Erickson, K. Aagaard and C. W. Matson, EROD activity, chromosomal damage, and oxidative stress in response to contaminants exposure in tree swallow (*Tachycineta bicolor*) nestlings from Great Lakes areas of concern, *Ecotoxicology*, 2017, **26**, 1392–1407.
40. O. M. Smith, J. M. Taylor, A. Echeverri, T. Northfield, K. A. Cornell, M. S. Jones, C. E. Latimer, J. P. Owen, W. E. Snyder and C. M. Kennedy, Big wheel keep on turnin': Linking grower attitudes, farm management, and delivery of avian ecosystem services, *Biol. Conserv.*, 2021, **254**, 108970.
41. F. M. Fishel, Pesticides effects on nontarget organisms. PI-85. Pesticide Information Office, Florida Cooperative Extension Service, IFAS, University of Florida, Gainesville, FL, USA, 2011 (http://edis.ifas.ufl.edu/pi122) (accessed 14 October 2013).
42. Birdlife International, Species factsheet (http://www.birdlife.org/datazone), 2014 (accessed 10 June 2019).
43. V. Dhananjayan and S. Muralidharan, Levels of polycyclic aromatic hydrocarbons, polychlorinated biphenyls, and organochlorine pesticides in various tissues of white-backed vulture in India, *BioMed Res. Int.*, 2013, 190353.
44. V. Dhananjayan, S. Muralidharan and S. Ranapratap, Organochlorine pesticide residues in eggs and tissues of house sparrow, *Passer domesticus*, from Tamil Nadu, India, *Bull. Environ. Contam. Toxicol.*, 2011, **87**, 684–688.
45. V. Dhananjayan and S. Muralidharan, Levels of organochlorine pesticide residues in blood plasma of various species of birds from India, *Bull. Environ. Contam. Toxicol.*, 2010, **85**, 129–136.
46. V. Dhananjayan, Accumulation pattern of persistent organochlorine pesticides in liver tissues of various species of birds from India, *Environ. Sci. Pollut. Res. Int.*, 2013, **20**, 3149–3156.
47. S. Muralidharan, V. Dhananjayan, R. Risebrough, V. Prakash, R. Jayakumar and P. H. Bloom, Persistent organochlorine pesticide residues in tissues and eggs of white-backed vulture, *Gyps bengalensis* from different locations in India, *Bull. Environ. Contam. Toxicol.*, 2008, **81**, 561–565.

48. S. Jayakumar, S. Muralidharan and V. Dhananjayan, Organochlorine pesticide residues among colonial nesting birds in Tamil Nadu, India: A maiden assessment from their breeding grounds, *Arch. Environ. Contam. Toxicol.*, 2020, **78**, 555–567.
49. C. J. Henny, L. J. Blus and C. J. Stafford, Effects of heptachlor on American kestrels in the Columbian Basin, Oregon, *J. Wildlife Management.*, 1983, **47**, 1080–1087.
50. L. J. Blus and C. J. Henny, Field studies on pesticides and birds: Unexpected and unique relations, *Ecol. Appl.*, 1997, **7**, 1125–1132.
51. I. C. Nisbet and L. M. Reynolds, Organochlorine residues in common terns and associated estuarine organisms, Massachusetts, USA, 1971–81, *Mar. Environ. Res.*, 1984, **11**, 33–66.
52. M. Siddiqui and M. Saxena, Biological monitoring of environmental contaminants: Chlorinated hydrocarbons in eggs of hens, *Sci. Total Environ.*, 1983, **32**, 29–34.
53. K. Senthilkumar, K. Kannan, A. Subramanian and S. Tanabe, Accumulation of organochlorine pesticides and polychlorinated biphenyls in sediments, aquatic organisms, birds, bird eggs and bat collected from south India, *Environ. Sci. Pollut. Res. Int.*, 2001, **8**, 35–47.
54. C. Hallmann, R. Foppen and C. Turnhout, Declines in insectivorous birds are associated with high neonicotinoid concentrations, *Nature*, 2014, **511**, 341–344.
55. H. Hop, K. Borgå, G. W. Gabrielsen, L. Kleivane and J. U. Skaare, Food web magnification of persistent organic pollutants in poikolotherms and homeotherms, from the Barents Sea, *Environ. Sci. Technol.*, 2002, **36**, 2589–2597.
56. R. S. Aulakh, J. P. S. Gill, J. S. Bedi, J. K. Sharma, B. S. Joia and H. W. Ockerman, Organochlorine pesticide residues in poultry feed, chicken muscle and eggs at a poultry farm in Punjab, *India, J. Sci. Food. Agric.*, 2006, **86**, 741–744.
57. P. Vanitha, G. S. Chandra, R. Asha Rajini, M. Sivachandran and K. Venukopalan, Estimation of organochlorine pesticide residues in poultry meat and egg, *Indian J. Anim. Sci. Res.*, 2014, **10**, 121–130.
58. Y. Zhou, G. Yin, L. Asplund, Y. Qiu, A. Bignert, Z. Zhu, J. Zhao and Å. Bergman, A novel pollution pattern: Highly chlorinated biphenyls retained in black-crowned night heron (*Nycticorax nycticorax*) and whiskered tern (*Chlidonias hybrida*) from the Yangtze River Delta, *Chemosphere*, 2016, **150**, 491–498.
59. B. M. Braune, S. R. Jacobs and A. J. Gaston, Variation in organochlorine and mercury levels in first and replacement eggs of a single-egg clutch breeder, the thick-billed murre, at a breeding colony in the Canadian Arctic, *Sci. Total Environ.*, 2018, **610**, 462–468.
60. N. L. Devi, P. Chakraborty, Q. Shihua and G. Zhang, Selected organochlorine pesticides (OCPs) in surface soils from three major states from the northeastern part of India, *Environ. Monit. Assess.*, 2013, **185**, 6667–6676.

61. J. Muñoz Cifuentes, Seabirds at risk? Effects of environmental chemicals on reproductive success and mass growth of seabirds breeding at the Wadden Sea in the mid 1990s. Common Wadden Sea Secretariat, Trilateral Monitoring and Assessment Group, Wil-helmshaven, 2004, vol. 18, pp. 27–51.
62. D. Herzke, R. Kallenborn and T. Nygård, Organochlorines in egg samples from Norwegian birds of prey: congener-, isomer- and enantiomer specific considerations, *Sci. Total Environ.*, 2002, **291**, 59–71.
63. K. E. Clark, Y. Zhao and C. M. Kane, Organochlorine pesticides, PCBs, dioxins, and metals in postterm peregrine falcon (*Falco peregrines*) eggs from the Mid-Atlantic States, *Arch. Environ. Contam. Toxicol.*, 2009, **57**, 174–184.
64. J. P. Meador, *Environmental contaminants in wildlife-interpreting tissue concentrations*, CRC Press, New York, USA, Boca Raton, Florida, 2011.
65. K. A. King, E. L. Flickinger and H. Hildebrand, Shell thinning and pesticide residues in Texas aquatic birds eggs, *Pestic. Monitor. J.*, 1978, **12**, 16–21.
66. R. G. Heath, J. W. Spann, E. F. Hill and J. F. Kreitzer, Comparative dietary toxicities of pesticides to birds, U.S. Fish and Wildlife Service, Special Scientific Report, 1983, 152, 152, https://pubs.er.usgs.gov/publication/5230135 (accessed August, 2021).
67. S. Muralidharan, S. Jayakumar and V. Dhananjayan, Status of pesticide contamination in birds in India with special reference to threatened species, threatened birds in India: their conservation and requirement, in *Threatened Birds in India: Their Conservation and Requirement*, ed. A. R. Rahmani, Oxford University Press, BNHS, Mumbai, 2012, pp. 73–91.
68. J. O. Bustnes, G. Bangjord and N. G. Yoccoz, Variation in concentrations of organochlorines and brominated flame retardants among eggs in abandoned clutches of a terrestrial raptor, *Chemosphere*, 2015, **118**, 357–360.
69. M. Helberg, J. O. Bustnes, K. E. Erikstad, K. O. Kristiansen and J. U. Skaare, Relationships between reproductive performance and organochlorine contaminants in great black-backed gulls (*Larus marinus*), *Environ. Pollut.*, 2005, **134**, 475–483.

CHAPTER 5

Impacts of Agricultural Intensification on Farmland Birds and Risk Assessment of Pesticide Seed Treatments

JULIE CÉLINE BRODEUR*[a,b] AND MARIA BELÉN POLISERPI[a]

[a] Instituto de Recursos Biológicos, Centro de Investigaciones en Recursos Naturales, Instituto Nacional de Tecnología Agropecuaria, Buenos Aires, Argentina; [b] Consejo Nacional de Investigaciones Científicas y Técnicas (CONICET), Argentina
*Email: julbrodeur@hotmail.com

5.1 The Intensification of Agriculture

In the mid-20th century, an agricultural transformation called the "Green Revolution" prompted farms to expand into large commercial enterprises. During this period, worldwide food production from agriculture sharply increased due to the greater irrigation of lands, newly developed genetically modified crop varieties, and the increase in use of inorganic fertilizers and synthetic pesticides.[1–3] However, this agricultural intensification was linked to considerable harm to the environment, making agricultural systems less efficient by degrading ecosystem services through pollution,[2,4] and loss of biological diversity[5,6] and beneficial insects.[7,8]

Agriculture is nowadays the largest driver of global environmental change, with modern agricultural systems being both a leading cause of water pollution[3]

Issues in Toxicology No. 45
Bird and Reptile Species in Environmental Risk Assessment Strategies
Edited by Guillermo Eli Liwszyc and Marcelo L. Larramendy

Published by the Royal Society of Chemistry, www.rsc.org

Figure 5.1 Pesticides can be applied to crops with a boom sprayer or a plane.

and a primary driver of biodiversity loss.[9,10] The global demand for agricultural crops is increasing, and may continue to do so for decades, propelled by a 2.3 billion person increase in global population.[4] Today, farmlands dominate 38% of the global land surface.[11] The pace of farmland production growth is unlikely to continue, but the demand for agricultural commodities is projected to increase inexorably.[6] A global food revolution based on a new paradigm for agricultural development is urgently required to attain the twin objectives of feeding humanity and living within boundaries of biophysical processes that define the safe operating space of a stable and resilient Earth system.[12]

Modern agricultural systems are highly dependent on pesticides, with an estimated 3 to 4 million tons of pesticides being applied annually to agricultural lands on a global scale (see Figure 5.1).[13–15] Such an intensive use of pesticides has resulted in widespread contamination of air, water, and soil worldwide.[16–20] Recent estimates show that 64% of global agricultural lands are at risk of pesticide pollution by more than one active ingredient, whereas 31% are at high risk.[21] The widespread presence of synthetic pesticides in the environment implies an intrinsic threat to the biodiversity and functioning of both terrestrial[22–25] and aquatic ecosystems.[26–28]

5.2 Agricultural Intensification and Bird Declines

Birds are declining globally, with at least 40% of bird species presenting downward population trends.[29] Although multiple factors have been proposed as causes of this situation, agricultural intensification and expansion are recognized drivers of bird population declines, and severe losses of farmland birds are well documented in Europe and elsewhere.[30–37] Agricultural intensification is a multifactorial process that acts at both the field and landscape levels.[38,39] At the landscape level, the reduction in heterogeneity through the loss and fragmentation of non-cultivated elements (*e.g.*, margins, hedges, fallow land, and wasteland) has critical negative effects on bird populations (see Figure 5.2).[39–43] For example, the reduction of marginal habitats limits the breeding habitat for tree and shrub species, whereas the loss of fallow land affects bird species through the loss of food resources and habitats for bird breeding (see Figure 5.3).[32,34,37,42]

On a field scale, yield and revenue maximization have led to an increase in the use of agriculture-related chemicals such as pesticides, which negatively influence a variety of bird species.[44–49] Given their intrinsic toxic properties, pesticides can directly alter the biology and physiology of birds, and may eventually cause their death. The bird lethality linked to the direct effect of

Figure 5.2 A typical homogenous landscape of soybean monoculture from South America.

Figure 5.3 The reduction of marginal habitats limits the breeding habitat for birds.

pesticides has been estimated to be around 2.7 and 67 to 72 million birds each year in Canada and the USA, respectively.[44,47] Carbamate and organophosphate insecticides such as monocrotophos, terbufos, chlorpyrifos, and carbofuran are well known for their acute toxicity to birds and have been linked to large mortality events in the past.[50–53] Although the use of carbamates and organophosphates is now banned in a majority of countries, their main replacements, neoniconicotinoid insecticides such as imidacloprid and thiamethoxam, are also generating concerns as reports of wild bird poisoning,[54–57] and evidence of negative effects on bird populations accumulates.[24,25,58] Finally, rodenticides and legacy organochlorine pesticides are also known to pose a problem, especially in apex predators and scavenger birds such as vultures and condors,[48,49] although this problem extends beyond agricultural uses and encompasses urban settings (see Figure 5.4).[59]

Figure 5.4 Rodenticides and legacy organochlorine pesticides are known to pose a problem in apex predators such as the Harris's hawk (*Parabuteo unicinctus*).

5.3 Impact of Pesticides on Birds: Direct *vs.* Indirect Effects

5.3.1 Direct Effects

Reports of wild bird poisonings are undeniable evidence of direct acute adverse effects of pesticides on birds. Nevertheless, it is difficult to identify poisoning incidents because carcasses may be quickly scavenged, inconspicuous, or not localized. This is especially true in wide extensions such as agricultural lands, where small birds or birds dying in dense crops are unlikely to be visible. Also, birds receiving a life-threatening exposure in open habitats are likely to seek cover before they die. For these reasons, the frequency of reported incidents generally greatly underestimates the frequency of mortality, and, whereas incidents can confirm a high predicted risk, the absence of incidents does not necessarily indicate a low risk.[60]

Pesticide toxicity may also not be immediately lethal and can be expressed in terms of sublethal effects on behavior and physiology, which can eventually impair bird survival and reproduction and, consequently, population dynamics.[35] Alterations of thermoregulation,[61] enzyme gene expression and activity,[62–64] hematology,[65] hormonal regulation,[66,67] and immune response[62,68] are sublethal effects induced by pesticide exposure. Alterations of the animal bioenergetics that reduce body condition, body mass, or fat accumulation such as reduced feeding[68,69] or increased energy demands[70] may also impair birds and their populations. A number of insecticides are known to alter bird behavior by acting on the neurological system. Reported effects range from the obvious and clearly debilitating alterations seen at high doses (ataxia, breathing difficulties, immobility, tremors)[71] to subtler modifications present at low doses, including reduced exploratory behavior,[72] altered flight orientation,[73,74] and parental care.[75] Finally, direct impacts on reproductive parameters such as egg and brood size, clutch size, fertilization rate, fecundity, and embryonic development[62,68,76–78] are sublethal effects directly linked to population-level effects.

With regard to direct effects, it is important to realize that, whereas laboratory studies aiming at evaluating pesticide toxicity are generally performed on single products, wild birds are normally exposed to a cocktail of substances including insecticide, fungicide, and herbicide products. This is not only because a variety of crop types are normally intermingled in an agricultural landscape, but also because several active ingredients are commonly applied simultaneously to crops to favor complementary effects, increase the scope of action, and reduce application costs.[79,80] Although rarely studied, the exposure of birds to pesticide mixtures and their effects may be greater than expected based on toxicity to single compounds.[35]

5.3.2 Indirect Effects

Birds, like all living organisms, depend on a number of interactions with other species for their survival; either for food or to satisfy their need for vegetative cover for shelter, nesting, and the protection of their offspring. The indirect effects of pesticides are those that occur through these interactions.[81] For example, although herbicides are not normally acutely toxic to birds, the removal of plant species caused by herbicides can impact bird populations by destroying the plant cover they use for habitat, and by reducing the amount of seeds and plants available to herbivorous and granivorous species.[82] For example, Gibbons *et al.*[83] showed that herbicide use decreased the availability of weed seeds that were found to be included in the diet of more than 25 species of birds currently experiencing population declines. Herbicides can also exert indirect negative effects on insectivorous species by reducing insect abundance through the elimination of insect habitats and food plants.[84]

Similarly, the use of insecticides can impact the abundance and diversity of invertebrates that serve as food for insectivorous birds, as well as for the chicks of many granivorous birds, as a number of them feed exclusively on invertebrates during their first few weeks of life.[45,85] For example, the decrease in food resources, which was linked to the application of insecticides, had a negative impact on the body condition of chicks and caused an increase in the frequency of feeding events of the parent in the cerulean bunting (*Emberiza citrinella*).[86] A recent example of indirect effects on bird population is the intrinsic combination of ecological factors that were identified by Breinlinger *et al.*[87] as the causes of bald eagle deaths in North America. The colonization of waterways by an invasive, introduced plant provides a substrate for the growth of a previously unidentified cyanobacterium, which, upon exposure to bromides of anthropogenic origin, results in the production of a neurotoxin that both causes neuropathy in animals that prey on plants and bioaccumulates to kill predators such as bald eagles.

5.4 Seed Treatment With Pesticides: Impacts on Birds

5.4.1 Agricultural Intensification Through Seed Treatment

The practice of treating and coating seeds is a typical component of agricultural intensification strategies, as it promotes growth under adverse conditions and protects seedlings against pathogens, insects, or other pests.[88,89] Seed treatment is the application of chemical ingredients and biological organisms to crop seeds before they are planted. Among the technologies applied to seeds are inoculants, pesticides, micronutrients, seed regulators, insect pests growth regulators, polymers, and colorants.[90]

Since the turn of the century, pesticides have been increasingly used for seed treatment and a majority of corn, soybean, cotton, and other seeds are, nowadays, being treated (see Figure 5.5).[90] Seed treatments have various combinations of active ingredients, and the composition of these products evolves rapidly. A wide variety of pesticide products are currently applied for seed treatment, and up to four or five active ingredients are often combined into a single seed treatment product.[90] Fungicidal and insecticidal active ingredients are the two main categories of pesticides commonly applied for seed treatment. Among fungicides, strobilurins (*e.g.*, azoxystrobin, pyraclostrobin), and triazoles (*e.g.*, cyproconazole, difenoconazole, prothioconazole, tebuconazole) are most frequently employed, together with metalaxyl, fludioxonil, thiram, and iprodione. With regard to insecticides, neonicotinoids (imidacloprid, thiamethoxam, clothianidin), pyrethroids (tefluthrin, lambda-cyhalothrin), and diamides (chlorantraniliprole) are commonly used.

Figure 5.5 Pesticide-treated seeds.

It is often stated that seed treatment reduces active ingredient loading and minimizes the impacts on non-target organisms because it implies lower dosages compared to foliar or soil applications.[91] Nevertheless, research has shown that only 1.6 to 20% of the chemicals put on the seed are incorporated by the plant as it grows,[92,93] implying that the bulk of the active ingredient remains in the soil, where it is available for leaching and runoff movements into surface- and groundwater.[94,95] In the case of neonicotinoid insecticides, this represents a significant risk to surface waters and the aquatic fauna these ecosystems support,[96] as these molecules are persistent, have a high leaching and runoff potential, and are highly toxic to a wide range of invertebrates.[97] Furthermore, the widespread prophylactic use of pesticides for seed treatment has led to the abandonment of the long-established principles of integrated pest management. The philosophy of pest management is based on minimizing the use of chemical pesticides *via* monitoring of pest populations and applying chemical pesticides only when needed, thereby minimizing pesticide use and avoiding impacts on non-target organisms.[98,99]

5.4.2 Bird Exposure to Pesticide-treated Seeds

Birds can be directly exposed to treated seeds when these are left available on the soil due to inadequate burial during sowing or spillage during transport, reloading, and refilling of machinery (see Figure 5.6). The percentage of seeds that remain available on the soil is typically between 0.5 and 9.2%, and depends on the type of crop, the soil conditions, the sowing technique, the season, and the location on the field.[100–102] Many studies have now demonstrated that birds are attracted by spills and consume treated seeds in the field,[101,103–108] or pesticide-containing seedlings emerging from the treated seeds.[109]

Figure 5.6 A spill of wheat seeds that remained in the field for over three weeks is surrounded by seedlings emerging from implanted seeds.

Exposure of birds to pesticide-treated seeds is furthermore demonstrated by mounting evidence that wild birds are exposed to neonicotinoids and fungicides used for seed coating,[108–116] as well as reports of mortality linked to ingestion of pesticide-coated seeds.[54–57,117–120] The dose ingested by birds in a single feeding bout on treated seeds can be sufficient to cause lethal and sublethal effects.[121,122] Because seed coating is done with pesticide active ingredients that are systemic in crop plants, they are translocated to pollen and can thus affect both insect and bird pollinators.[123–125] Evidence also shows that neonicotinoids can reach significant concentrations in top predators such as owls and buzzards once they enter the food chain through either birds or mammals feeding on treated seeds.[126–128]

5.5 Assessment of Risks of Pesticide-treated Seeds to Birds

In most countries, the use and commercialization of pesticides are regulated, and governments require risk quantification to be performed for birds, mammals, and invertebrates before a pesticide product can be registered and sold. Environmental risk assessment is a tiered approach where the initial risk is assessed based on conservative assumptions regarding expected exposure and effects on non-target organisms. If the initial assessment indicates a potential risk, a more refined "higher tier" risk assessment is performed based on more realistic assumptions regarding the exposure and/or the effects.

5.5.1 Regulatory Environmental Toxicity Testing

Risk assessment is based on available toxicity data that define thresholds for mortality, growth, reproduction, and other sublethal effects. As these toxicity data originate from a suite of standardized toxicity tests, it is important to understand the nature and limitations of those tests, to fully appreciate the significance and limits of the risk assessment process. Basic avian testing includes two acute tests: a standardized acute oral toxicity test where a single dose is administered orally by either capsule or gavage,[129,130] and a subacute dietary toxicity test where graded concentrations of a chemical are presented *ad libitum* in the feed for 5 days.[131] In Europe and the United States, a one-generation reproduction test is also mandatory.[132,133] Recommended bird species for these tests are normally a game bird, such as the northern bobwhite (*Colinus virginianus*) in the United States, or the Japanese quail (*Coturnix japonica*) in Europe, and a waterfowl, generally the mallard duck (*Anas platyrhynchos*). A two-generation toxicity test protocol using the Japanese quail (*C. japonica*) has also been released by the United States Environmental Protection Agency (US EPA) to evaluate endocrine effects.[134] Since October 2007, the US EPA has furthermore mandated that avian acute oral toxicity data include a passerine species. The reason for testing passerines originates from the fact that songbirds are commonly more sensitive than other groups of birds, and represent approximately

60% of all living bird species, and many of the non-target species exposed to pesticides are passerine songbirds (see Figure 5.7).[135–137]

In this context, the availability of precise acute toxicity data in a number of bird species is critical for risk assessment, as hazard may be underestimated if too few species are tested.[138] The lack of testing in native species is a general weakness of the regulatory processes in South America and other parts of the world, as most model test species originate from North America and Europe.[139] The use of native species is normally recommended as it provides greater environmental realism and ensures a more conservative endpoint for protecting ecosystems.[135,140,141] Ideally, tested species should represent a broad range of sensitivities, habitats, and trophic groups, and be economically, ecologically, and/or recreationally important.[140,142] To attend the issue of the lack of a native passerine model, Brodeur and Poliserpi[143] proposed the use of cowbirds as an avian model for environmental toxicity testing in South America (see Figure 5.8).

Figure 5.7 Many of the non-target species exposed to pesticides are passerine songbirds.

Figure 5.8 The grayish baywing (*Agelaioides badius*) is proposed as an avian passerine model for environmental toxicity testing in South America.

5.5.2 Tier I Risk Assessment

The Guidance Document on Risk Assessment for Birds and Mammals published by the European Food Safety Authority (EFSA)[60] is one of the most detailed schemes currently available to evaluate the risk of pesticides to birds. The document includes guidelines for performing the risk assessment of spray foliar applications, granular applications, and seed treatment strategies. Regarding exposure to treated seeds, Tier I risk assessment is conducted by calculating the toxicity exposure ratio (TER) for both acute and long-term toxicity using, respectively, the median lethal dose (LD_{50}) and the no-observed adverse effect level (NOAEL) as toxicity values. The seed treatment is considered to present a risk to birds when the obtained TERs pass a critical pre-defined trigger value. These calculations can be performed manually in a spreadsheet or with the software T-REX.[144]

In Tier 1, it is assumed that granivorous birds feed entirely on readily available, freshly treated seeds. This exposure represents a worst-case scenario and provides a conservative level of protection. To cover a range of diets and body sizes, the assessment is performed on at least two different body size categories: small granivorous birds that feed on small seeds, and larger, medium-size birds that feed on larger seeds. Generic bird species are used, whose feed intake rate is pre-defined and obtained from the guidelines.[60,144]

5.5.3 Refinements and Weight-of-evidence Risk Assessment

In cases where Tier 1 analyses conclude that treated seeds present a risk to birds, the risk evaluations must proceed to a second step (*i.e.*, Tier 2), where one or more refinements are incorporated into the analysis. A weight-of-evidence risk assessment is then performed, in which an overall conclusion on the characterization of the risk is formed, giving appropriate weight to each of the lines of evidence available, based on their degree of certainty.[60] Readers can refer to the studies of Lopez-Antia *et al.*[103] and Poliserpi *et al.*[122] for examples of weight-of-evidence risk assessments that examined the risk of seed treatments to birds. Possible refinement options as stated by the EFSA are outlined in the following.[60]

5.5.3.1 Focal Species

Refinements can be performed using the food intake rate and body weight data of actual focal species inhabiting the crop/region under assessment rather than generic table values. Bonneris *et al.*[145] presented a set of field studies that were performed to define focal species for winter cereal fields in France.

5.5.3.2 Availability of Non-treated Seeds

In the first-tier assessment, a diet consisting of 100% treated seeds is considered. It is possible to refine this assumption by considering a mixed diet and adjusting the percentage of treated seeds in the diet.

5.5.3.3 Dehusking Behavior

Many birds are known to dehusk seeds prior to consumption, but the extent of dehusking may vary among bird species and types of seed. The actual intake of a pesticide may be less than estimated from the nominal treatment rate if dehusking occurs. Including experimental data on dehusking can, therefore, serve to refine the risk assessment.

5.5.3.4 Foraging Area

A refinement of the risk assessment may be achieved by estimating the area which needs to be foraged by a bird to obtain a lethal dose. This can be achieved by obtaining information on the density of seeds available on the soil surface after application. This refinement strategy was employed by Poliserpi *et al.*[122] when evaluating the acute risk of imidacloprid seed coating to the grayish baywing (*Agelaioides badius*).

5.5.3.5 Meal Size Approach

Information on the risk of mortality can be obtained by comparing the number of seeds needed to attain a lethal dose with data regarding the typical numbers of seeds a bird can ingest in a single feeding bout. Data on feeding bout are commonly obtained from Prosser.[146]

5.5.3.6 Food Item Preference and Avoidance

Birds may show a preference for either treated or untreated seeds in their diet, which ultimately impacts the treated-seed consumption rate. Information on avoidance behavior can therefore be used to refine the risk assessment. Nevertheless, not all avoidance behaviors effectively prevent death. For example, because aversion to imidacloprid-treated seeds is a conditioned response to post-ingestion distress,[147] birds that feed rapidly may ingest a lethal dose before the onset of the response and mortality can still occur in an acute exposure scenario.[148]

5.5.3.7 Metabolism and Body Burden Modeling

The rates of absorption, distribution, metabolism, and elimination of substances in the gastrointestinal tract of birds and mammals influence the toxicity of the pesticide. These factors may be different for dietary uptake of products from seed treatment than in the gavage experiments that are normally used in first-tier assessments. Therefore, metabolism and body burden models can be used as a potential refinement step at higher tiers. The EFSA opinion on pirimicarb gives an example of how such models may be applied in a weight-of-evidence approach.[149]

5.5.3.8 Historical Data on Poisoning Incidents and Field Studies

Information on historical incidents and field studies of a previously authorized pesticide can provide further lines of evidence for the risk assessment. The study by Lopez-Antia *et al.*[103] is an example of how multiple lines of evidence obtained in the field can be combined to perform a risk assessment.

5.5.4 Future Directions

The increasing number of lines of evidence showing that registered pesticide seed treatments can exert negative effects on birds[103,122,150] indicates that it is necessary to improve current regulatory risk assessment schemes. This is also true for pesticide risk assessment in general, as it has, overall, failed to prevent registered pesticides to have detrimental effects on biodiversity and the environment.[151–153]

Improvements at the levels of ecological, landscape, and management considerations are necessary to improve the accuracy of the predictions obtained through risk assessment.[152] At the ecological level, the potential indirect effects of pesticides need to be considered, as current risk assessment schemes typically address the effects of a pesticide on each group of organisms separately. Because trophic interactions are fundamental features of ecosystems, it will be impossible to prevent indirect effects until these are considered in pesticide risk assessment.[151,154]

As mentioned earlier, another weakness of current procedures is that risk assessment is performed for one product at a time, whereas birds are, in fact, exposed to a cocktail of pesticides in real life.[153,155] The current approach also ignores the scale of use and site-specific aspects such as the connectivity to other sites and the presence of additional stressors.[152,153] Although some developments have been proposed to tackle issues such as spatial considerations[155] and population modeling,[156] there is still a long way ahead in order to ensure the accuracy of risk assessment predictions and the safety of pesticide products to birds and other wildlife species.

References

1. D. Tilman, K. G. Cassman, P. A. Matson, R. Naylor and S. Polasky, Agricultural sustainability and intensive production practices, *Nature*, 2002, **418**, 671–677.
2. J. Pretty, Intensification for redesigned and sustainable agricultural systems, *Science*, 2018, **362**, 1.
3. J. Mateo-Sagasta, S. M. Zadeh and H. Turral, *More people, more food, worse water? A global review of water pollution from agriculture*, Food and Agriculture Organization of the United Nations (FAO) and the International Water Management Institute, Rome, Italy, 2018.
4. D. Tilman, C. Balzer, J. Hill and B. L. Befort, Global food demand and the sustainable intensification of agriculture, *Proc. Natl. Acad. Sci. U. S. A.*, 2011, **108**, 20260.

5. T. Newbold, L. N. Hudson, S. L. L. Hill, S. Contu, I. Lysenko, R. A. Senior, L. Börger, D. J. Bennett, A. Choimes, B. Collen, J. Day, A. De Palma, S. Díaz, S. Echeverria-Londoño, M. J. Edgar, A. Feldman, M. Garon, M. L. K. Harrison, T. Alhusseini, D. J. Ingram, Y. Itescu, J. Kattge, V. Kemp, L. Kirkpatrick, M. Kleyer, D. L. P. Correia, C. D. Martin, S. Meiri, M. Novosolov, Y. Pan, H. R. P. Phillips, D. W. Purves, A. Robinson, J. Simpson, S. L. Tuck, E. Weiher, H. J. White, R. M. Ewers, G. M. MacE, J. P. W. Scharlemann and A. Purvis, Global effects of land use on local terrestrial biodiversity, *Nature*, 2015, **520**, 45–50.
6. F. Zabel, R. Delzeit, J. M. Schneider, R. Seppelt, W. Mauser and T. Václavík, Global impacts of future cropland expansion and intensification on agricultural markets and biodiversity, *Nat. Commun.*, 2019, **10**, 1–10.
7. D. Goulson, E. Nicholls, C. Botías and E. L. Rotheray, *Science*, 2015, **80**, 347.
8. P. H. Raven and D. L. Wagner, Agricultural intensification and climate change are rapidly decreasing insect biodiversity, *Proc. Natl. Acad. Sci. U. S. A.*, 2021, **118**, 1–6.
9. J. Rockström, J. Williams, G. Daily, A. Noble, N. Matthews, L. Gordon, H. Wetterstrand, F. DeClerck, M. Shah, P. Steduto, C. de Fraiture, N. Hatibu, O. Unver, J. Bird, L. Sibanda and J. Smith, Sustainable intensification of agriculture for human prosperity and global sustainability, *Ambio*, 2017, **46**, 4–17.
10. T. Benton, C. Bieg, H. Harwatt, R. Pudassaini and L. Wellesley, Food system impacts on biodiversity loss, *Energy, Environ. Resour. Program.*, 2021, 1–71.
11. FAO, Land use in agriculture by the numbers, http://www.fao.org/sustainability/news/detail/en/c/1274219/ (accessed 17 November 2020).
12. W. Steffen, W. Broadgate, L. Deutsch, O. Gaffney and C. Ludwig, The trajectory of the anthropocene: The great acceleration, *Anthr. Rev.*, 2015, **2**, 81–98.
13. D. Pimentel, Pesticides and Pest Control, *Integr. Pest Manag.*, 2009, **1**, 83–87.
14. F. Sanchez-Bayo, in *Ecological Impacts of Toxic Chemicals*, 2011, pp. 63–87.
15. J. Pretty and Z. P. Bharucha, Integrated pest management for sustainable intensification of agriculture in Asia and Africa, *Insects*, 2015, **6**, 152–182.
16. C. Shunthirasingham, C. E. Oyiliagu, X. Cao, T. Gouin, F. Wania, S. C. Lee, K. Pozo, T. Harner and D. C. G. Muir, Spatial and temporal pattern of pesticides in the global atmosphere, *J. Environ. Monit.*, 2010, **12**, 1650–1657.
17. S. Stehle and R. Schulz, Agricultural insecticides threaten surface waters at the global scale, *Proc. Natl. Acad. Sci. U. S. A.*, 2015, **112**, 5750–5755.

18. A. Ippolito, M. Kattwinkel, J. J. Rasmussen, R. B. Schäfer, R. Fornaroli and M. Liess, Modeling global distribution of agricultural insecticides in surface waters, *Environ. Pollut.*, 2015, **198**, 54–60.
19. V. Silva, H. G. J. Mol, P. Zomer, M. Tienstra, C. J. Ritsema and V. Geissen, Pesticide residues in European agricultural soils – A hidden reality unfolded, *Sci. Total Environ.*, 2019, **653**, 1532–1545.
20. J. F. Borsuah, T. L. Messer, D. D. Snow, S. D. Comfort and A. R. Mittelstet, Literature Review: Global Neonicotinoid Insecticide Occurrence in Aquatic Environments, *Water*, 2020, **12**, 3388.
21. F. H. M. Tang, M. Lenzen, A. McBratney and F. Maggi, Risk of pesticide pollution at the global scale, *Nat. Geosci.*, 2021, **14**, 206–210.
22. C. A. Brittain, M. Vighi, R. Bommarco, J. Settele and S. G. Potts, Impacts of a pesticide on pollinator species richness at different spatial scales, *Basic Appl. Ecol.*, 2010, **11**, 106–115.
23. M. Hahn, A. Schotthöfer, J. Schmitz, L. A. Franke and C. A. Brühl, The effects of agrochemicals on Lepidoptera, with a focus on moths, and their pollination service in field margin habitats, *Agric. Ecosyst. Environ.*, 2015, **207**, 153–162.
24. C. A. Hallmann, R. P. B. Foppen, C. A. M. Van Turnhout, H. De Kroon and E. Jongejans, Declines in insectivorous birds are associated with high neonicotinoid concentrations, *Nature*, 2014, **511**, 341–343.
25. Y. Li, R. Miao and M. Khanna, Neonicotinoids and decline in bird biodiversity in the United States, *Nat. Sustain.*, 2020, 1–9.
26. M. A. Beketov, B. J. Kefford, R. B. Schäfer and M. Liess, Pesticides reduce regional biodiversity of stream invertebrates, *Proc. Natl. Acad. Sci. U. S. A.*, 2013, **110**, 11039–11043.
27. R. Schulz, S. Bub, L. L. Petschick, S. Stehle and J. Wolfram, Applied pesticide toxicity shifts toward plants and invertebrates, even in GM crops, *Science*, 2021, **372**, 81–84.
28. M. Liess, L. Liebmann, P. Vormeier, O. Weisner, R. Altenburger, D. Borchardt, W. Brack, A. Chatzinotas, B. Escher, K. Foit, R. Gunold, S. Henz, K. L. Hitzfeld, M. Schmitt-Jansen, N. Kamjunke, O. Kaske, S. Knillmann, M. Krauss, E. Küster, M. Link, M. Lück, M. Möder, A. Müller, A. Paschke, R. B. Schäfer, A. Schneeweiss, V. C. Schreiner, T. Schulze, G. Schüürmann, W. von Tümpling, M. Weitere, J. Wogram and T. Reemtsma, Pesticides are the dominant stressors for vulnerable insects in lowland streams, *Water Res.*, 2021, **201**, 117262.
29. BirdLife International, *State of the world's birds: taking the pulse of the planet*, Cambridge, UK, 2018.
30. P. F. Donald, R. E. Green and M. F. Heath, Agricultural intensification and the collapse of Europe's farmland bird populations, *Proc. R. Soc. B Biol. Sci.*, 2001, **268**, 25–29.
31. P. F. Donald, F. J. Sanderson, I. J. Burfield and F. P. J. van Bommel, Further evidence of continent-wide impacts of agricultural intensification on European farmland birds, 1990–2000, *Agric. Ecosyst. Environ.*, 2006, **116**, 189–196.

32. I. Newton, The recent declines of farmland bird population, *Ibis*, 2004, **146**, 579–600.
33. T. Tscharntke, A. M. Klein, A. Kruess, I. Steffan-Dewenter and C. Thies, Landscape perspectives on agricultural intensification and biodiversity – ecosystem service management, *Ecol. Lett*, 2005, **8**, 857–874.
34. J. Reif, Long-Term Trends in Bird Populations: A Review of Patterns and Potential Drivers in North America and Europe, *Acta Ornithol.*, 2013, **48**, 1–16.
35. R. L. Stanton, C. A. Morrissey and R. G. Clark, Analysis of trends and agricultural drivers of farmland bird declines in North America: A review, *Agric. Ecosyst. Environ.*, 2018, **254**, 244–254.
36. Ç. H. Şekercioglu, C. D. Mendenhall, F. Oviedo-Brenes, J. J. Horns, P. R. Ehrlich and G. C. Daily, Long-term declines in bird populations in tropical agricultural countryside, *Proc. Natl. Acad. Sci. U. S. A.*, 2019, **116**, 9903–9912.
37. M. Busch, J. Katzenberger, S. Trautmann, B. Gerlach, R. Dröschmeister and C. Sudfeldt, Drivers of population change in common farmland birds in Germany, *Bird Conserv. Int.*, 2020, **30**, 335–354.
38. M. Emmerson, M. B. Morales, J. J. Oñate, P. Batáry, F. Berendse, J. Liira, T. Aavik, I. Guerrero, R. Bommarco, S. Eggers, T. Pärt, T. Tscharntke, W. Weisser, L. Clement and J. Bengtsson, How Agricultural Intensification Affects Biodiversity and Ecosystem Services, *Adv. Ecol. Res.*, 2016, **55**, 43–97.
39. K. V. Rosenberg, A. M. Dokter, P. J. Blancher, J. R. Sauer, A. C. Smith, P. A. Smith, J. C. Stanton, A. Panjabi, L. Helft, M. Parr and P. P. Marra, Decline of the North American avifauna, *Science*, 2019, **366**, 120–124.
40. T. G. Benton, J. A. Vickery and J. D. Wilson, Farmland biodiversity: Is habitat heterogeneity the key?, *Trends Ecol. Evol.*, 2003, **18**, 182–188.
41. F. Geiger, G. R. de Snoo, F. Berendse, I. Guerrero, M. B. Morales, J. J. Oñate, S. Eggers, T. Pärt, R. Bommarco, J. Bengtsson, L. W. Clement, W. W. Weisser, A. Olszewski, P. Ceryngier, V. Hawro, P. Inchausti, C. Fischer, A. Flohre, C. Thies and T. Tscharntke, Landscape composition influences farm management effects on farmland birds in winter: A pan-European approach, *Agric. Ecosyst. Environ.*, 2010, **139**, 571–577.
42. J. Traba and M. B. Morales, The decline of farmland birds in Spain is strongly associated to the loss of fallowland, *Sci. Reports*, 2019, **9**, 1–6.
43. L. Macchi, J. Decarre, A. P. Goijman, M. Mastrangelo, P. G. Blendinger, G. I. Gavier-Pizarro, F. Murray, M. Piquer-Rodriguez, A. Semper-Pascual and T. Kuemmerle, Trade-offs between biodiversity and agriculture are moving targets in dynamic landscapes, *J. Appl. Ecol.*, 2020, **57**, 2054–2063.
44. D. Pimentel and M. Burgess, Environmental and Economic Costs of the Application of Pesticides Primarily in the United States, *Integr. Pest Manag.*, 2014, **3**, 47–71.
45. F. Geiger, J. Bengtsson, F. Berendse, W. W. Weisser, M. Emmerson, M. B. Morales, P. Ceryngier, J. Liira, T. Tscharntke, C. Winqvist, S. Eggers, R. Bommarco, T. Pärt, V. Bretagnolle, M. Plantegenest,

L. W. Clement, C. Dennis, C. Palmer, J. J. Oñate, I. Guerrero, V. Hawro, T. Aavik, C. Thies, A. Flohre, S. Hänke, C. Fischer, P. W. Goedhart and P. Inchausti, Persistent negative effects of pesticides on biodiversity and biological control potential on European farmland, *Basic Appl. Ecol.*, 2010, **11**, 97–105.

46. P. Mineau and M. Whiteside, Pesticide Acute Toxicity Is a Better Correlate of U.S. Grassland Bird Declines than Agricultural Intensification, *PLoS One*, 2013, **8**, e57457.
47. A. M. Calvert, C. A. Bishop, R. D. Elliot, E. A. Krebs, T. M. Kydd, C. S. Machtans and G. J. Robertson, Synthèse des sources de mortalité aviaire d'origine anthropique au Canada, *Avian Conserv. Ecol.*, 2013, **8**, 214.
48. R. F. Shore and M. A. Taggart, Population-level impacts of chemical contaminants on apex avian species, *Curr. Opin. Environ. Sci. Heal.*, 2019, **11**, 65–70.
49. P. I. Plaza, E. Martínez-López and S. A. Lambertucci, The perfect threat: Pesticides and vultures, *Sci. Total Environ.*, 2019, **687**, 1207–1218.
50. M. I. Goldstein, T. E. Lacher, B. Woodbridge, M. J. Bechard, S. B. Canavelli, M. E. Zaccagnini, G. P. Cobb, E. J. Scollon, R. Tribolet and M. J. Hooper, Monocrotophos-induced mass mortality of Swainson's hawks in Argentina, 1995–96, *Ecotoxicology*, 1999, **8**, 201–214.
51. M. I. Goldstein, T. E. Lacher, M. E. Zaccagnini, M. L. Parker and M. J. Hooper, Monitoring and assessment of Swainson's hawks in Argentina following restrictions on monocrotophos use, 1996–97, *Ecotoxicology*, 1999, **8**, 215–224.
52. P. Mineau, *The hazard of carbofuran to birds and other vertebrate wildlife*, Ottawa, Canada, 1993, vol. 177.
53. P. Mineau, A review and analysis of study endpoints relevant to the assessment of 'long term' pesticide toxicity in avian and mammalian wildlife, *Ecotoxicology*, 2005, **14**, 775–799.
54. P. Berny, F. Buronfosse, B. Videmann and T. Buronfosse, Evaluation of the toxicity of imidacloprid in wild birds. A new high performance thin layer chromatography (HPTLC) method for the analysis of liver and crop samples in suspected poisoning cases, *J. Liq. Chromatogr. Relat. Technol.*, 1999, **22**, 1547–1559.
55. F. Millot, A. Decors, O. Mastain, T. Quintaine, P. Berny, D. Vey, R. Lasseur and E. Bro, Field evidence of bird poisonings by imidacloprid-treated seeds: a review of incidents reported by the French SAGIR network from 1995 to 2014, *Environ. Sci. Pollut. Res.*, 2017, **24**, 5469–5485.
56. C. J. Botha, E. C. Du Plessis, H. Coetser and M. Rosemann, Analytical confirmation of imidacloprid poisoning in granivorous Cape spurfowl (Pternistis capensis), *J. S. Afr. Vet. Assoc.*, 2018, **89**, 1–5.
57. K. H. Rogers, S. McMillin, K. J. Olstad and R. H. Poppenga, Imidacloprid Poisoning of Songbirds Following a Drench Application of Trees in a Residential Neighborhood in California, USA, *Environ. Toxicol. Chem.*, 2019, **38**, 1724–1727.

58. D. Goulson, Ecology: Pesticides linked to bird declines, *Nature*, 2014, **511**, 295–296.
59. A. Badry, D. Schenke, G. Treu and O. Krone, Linking landscape composition and biological factors with exposure levels of rodenticides and agrochemicals in avian apex predators from Germany, *Environ. Res.*, 2021, **193**, 110602.
60. EFSA, Guidance Document on Risk Assessment for Birds & Mammals, *EFSA J.*, 2009, 7(12), 1–139.
61. M. Friend and J. C. Franson, in *Field Manual of Wildlife Diseases General Field Procedures and Diseases of Birds*, Washington, D.C., 2008, pp. 131–176.
62. A. Lopez-Antia, M. E. Ortiz-Santaliestra, F. Mougeot and R. Mateo, Imidacloprid-treated seed ingestion has lethal effect on adult partridges and reduces both breeding investment and offspring immunity, *Environ. Res.*, 2015, **136**, 97–107.
63. E. Fernández-Vizcaíno, I. G. Fernández de Mera, F. Mougeot, R. Mateo and M. E. Ortiz-Santaliestra, Multi-level analysis of exposure to triazole fungicides through treated seed ingestion in the red-legged partridge, *Environ. Res.*, 2020, **189**, 109928.
64. M. B. Poliserpi, D. Cristos, J. M. Perez Iglesias and J. C. Brodeur, Tissue distribution and sublethal effects of imidacloprid in the South American grayish baywing (Agelaioides badius), *Chemosphere*, 2021, **284**, 131327.
65. A. Khan, L. Ahmad and M. Z. Khan, Hemato-biochemical changes induced by pyrethroid insecticides in avian, fish and mammalian species, *Int. J. Agric. Biol.*, 2012, **14**, 834–842.
66. G. J. Mayne, C. A. Bishop, P. A. Martin, H. J. Boermans and B. Hunter, Thyroid function in nestling tree swallows and eastern bluebirds exposed to non-persistent pesticides and p,p′-DDE in apple orchards of southern Ontario, Canada, *Ecotoxicology*, 2005, **14**, 381–396.
67. S. P. Pandey and B. Mohanty, The neonicotinoid pesticide imidacloprid and the dithiocarbamate fungicide mancozeb disrupt the pituitary-thyroid axis of a wildlife bird, *Chemosphere*, 2015, **122**, 227–234.
68. A. Lopez-Antia, M. E. Ortiz-Santaliestra, P. R. Camarero, F. Mougeot and R. Mateo, Assessing the Risk of Fipronil-Treated Seed Ingestion and Associated Adverse Effects in the Red-Legged Partridge, *Environ. Sci. Technol.*, 2015, **49**, 13649–13657.
69. M. L. Eng, B. J. M. Stutchbury and C. A. Morrissey, A neonicotinoid insecticide reduces fueling and delays migration in songbirds, *Science*, 2019, **365**, 1177–1180.
70. C. Narváez, J. M. Ríos, G. Píriz, J. C. Sanchez-Hernandez and P. Sabat, Subchronic exposure to chlorpyrifos affects energy expenditure and detoxification capacity in juvenile Japanese quails, *Chemosphere*, 2016, **144**, 775–784.
71. L. M. Addy-Orduna, J. C. Brodeur and R. Mateo, Oral acute toxicity of imidacloprid, thiamethoxam and clothianidin in eared doves: A

contribution for the risk assessment of neonicotinoids in birds, *Sci. Total Environ.*, 2019, **650**, 1216–1223.
72. M. B. Poliserpi, Ph.D. Thesis, Universidad de Buenos Aires, 2021.
73. N. B. Vyas, E. F. Hill, J. R. Sauer and W. J. Kuenzel, Acephate affects migratory orientation of the white-throated sparrow (Zonotrichia albicollis), *Environ. Toxicol. Chem.*, 1995, **14**, 1961–1965.
74. M. L. Eng, B. J. M. Stutchbury and C. A. Morrissey, Imidacloprid and chlorpyrifos insecticides impair migratory ability in a seed-eating songbird, *Sci. Rep.*, 2017, **7**, 1–9.
75. C. A. Bishop, P. Ng, P. Mineau, J. S. Quinn and J. Struger, Effects of Pesticide Spraying on Chick Growth, Behavior, and Parental Care in Tree Swallows (Tachycineta Bicolor) Nesting in an Apple Orchard in Ontario, Canada, *Environ. Toxicol. Chem.*, 2000, **19**, 2286.
76. A. Lopez-Antia, M. E. Ortiz-Santaliestra, F. Mougeot and R. Mateo, Experimental exposure of red-legged partridges (Alectoris rufa) to seeds coated with imidacloprid, thiram and difenoconazole, *Ecotoxicology*, 2013, **22**, 125–138.
77. A. Lopez-Antia, M. E. Ortiz-Santaliestra, F. Mougeot, P. R. Camarero and R. Mateo, Brood size is reduced by half in birds feeding on flutriafol-treated seeds below the recommended application rate, *Environ. Pollut.*, 2018, **243**, 418–426.
78. S. Ruuskanen, M. J. Rainio, M. Uusitalo, K. Saikkonen and M. Helander, Effects of parental exposure to glyphosate-based herbicides on embryonic development and oxidative status: a long-term experiment in a bird model, *Sci. Rep.*, 2020, **10**, 1–7.
79. M. Lydy, J. Belden, C. Wheelock, B. Hammock and D. Denton, Challenges in regulating pesticide mixtures, *Ecol. Soc.*, 2004, **9**, 1–15.
80. R. Verro, A. Finizio, S. Otto and M. Vighi, Predicting pesticide environmental risk in intensive agricultural areas. II: Screening level risk assessment of complex mixtures in surface waters, *Environ. Sci. Technol.*, 2009, **43**, 530–537.
81. N. D. Boatman, N. W. Brickle, J. D. Hart, T. P. Milsom, A. J. Morris, A. W. A. Murray, K. A. Murray and P. A. Robertson, Evidence for the indirect effects of pesticides on farmland birds, *Ibis*, 2004, **146**, 131–143.
82. S. J. Moreby and S. E. Southway, Influence of autumn applied herbicides on summer and autumn food available to birds in winter wheat fields in southern England, *Agric. Ecosyst. Environ.*, 1999, **72**, 285–297.
83. D. W. Gibbons, D. A. Bohan, P. Rothery, R. C. Stuart, A. J. Haughton, R. J. Scott, J. D. Wilson, J. N. Perry, S. J. Clark, R. J. G. Dawson and L. G. Firbank, Weed seed resources for birds in fields with contrasting conventional and genetically modified herbicide-tolerant crops, *Proc. R. Soc. B Biol. Sci.*, 2006, **273**, 1921–1928.
84. R. L. Taylor, B. D. Maxwell and R. J. Boik, Indirect effects of herbicides on bird food resources and beneficial arthropods, *Agric. Ecosyst. Environ.*, 2006, **116**, 157–164.

85. A. D. M. Hart, W. Rolelofs, J. Crocker and P. Mineau, Addressing Uncertainty and Variability in Pesticide Risk Assessments for Birds and Mammals, *Research Project Final Report- PS2303*, Department for Environment Food & Rural Affairs (DEFRA), UK, 2006, **5**, pp. 1–28.
86. A. J. Morris, J. D. Wilson, M. J. Whittingham and R. B. Bradbury, Indirect effects of pesticides on breeding yellowhammer (Emberiza citrinella), *Agric. Ecosyst. Environ.*, 2005, **106**, 1–16.
87. S. Breinlinger, T. J. Phillips, B. N. Haram, J. Mareš, J. A. M. Yerena, P. Hrouzek, R. Sobotka, W. M. Henderson, P. Schmieder, S. M. Williams, J. D. Lauderdale, H. D. Wilde, W. Gerrin, A. Kust, J. W. Washington, C. Wagner, B. Geier, M. Liebeke, H. Enke, T. H. J. Niedermeyer and S. B. Wilde, Hunting the eagle killer: A cyanobacterial neurotoxin causes vacuolar myelinopathy, *Science*, 2021, **371**, 1–7.
88. A. I. M. Nourou, A. K. Saidou, W. Arifa, A. O. Abdoulaye and J. B. Aune, Intensification of Pearl Millet Production in Niger through Mechanized Sowing and Weeding, Seed Priming, Seed Treatment, and Microdosing, *Agronomy*, 2020, **10**, 629.
89. M. Haque, A. R. Chowdhury, S. S. Acharya, M. Ghosh, S. Kumar and N. Kumar, Effect of Planting Geometry and Seed Treatment on Growth and Yield of Wheat (Triticum aestivum L.) under System of Wheat Intensification, *Int. J. Bio-Resource Stress Manag.*, 2015, **6**, 335.
90. C. A. Hitaj, D. J. Smith, A. Code, S. Wechsler, P. D. Esker and M. R. Douglas, Sowing Uncertainty: What We Do and Don't Know about the Planting of Pesticide-Treated Seed, *Bioscience*, 2020, **70**, 390–403.
91. Croplife, *Seed Treat. Environ. Comm. Int. Seed Fed.*, 2009, 2.
92. R. Sur and A. Stork, *Bull. Insectol.*, 2003, **56**, 35–40.
93. A. Alford and C. H. Krupke, Translocation of the neonicotinoid seed treatment clothianidin in maize, *PLoS One*, 2017, **12**, e0173836.
94. K. Banerjee, S. H. Patil, S. Dasgupta, D. P. Oulkar and P. G. Adsule, Sorption of thiamethoxam in three Indian soils, *J. Environ. Sci. Heal. – Part B Pestic. Food Contam. Agric. Wastes*, 2008, **43**, 151–156.
95. M. Fossen, *Environmental Fate of Imidacloprid*, Sacramento, United States, 2006.
96. C. A. Morrissey, P. Mineau, J. H. Devries, F. Sanchez-Bayo, M. Liess, M. C. Cavallaro and K. Liber, Neonicotinoid contamination of global surface waters and associated risk to aquatic invertebrates: A review, *Environ. Int.*, 2015, **74**, 291–303.
97. J. C. Anderson, C. Dubetz and V. P. Palace, Neonicotinoids in the Canadian aquatic environment: A literature review on current use products with a focus on fate, exposure, and biological effects, *Sci. Total Environ.*, 2015, **505**, 409–422.
98. R. L. Metcalf and W. H. Luckmann, *Introduction to insect pest management*, John Wiley & Sons, 1994.
99. D. Goulson, An overview of the environmental risks posed by neonicotinoid insecticides, *J. Appl. Ecol.*, 2013, **50**, 977–987.

100. J. A. Pascual, S. L. Fryday and A. D. M. Hart, Effects of food restriction on food avoidance and risk of acute poisoning of captive feral pigeons from fonofos-treated seeds, *Arch. Environ. Contam. Toxicol.*, 1999, **37**, 115–124.
101. H. V. McKay, P. J. Prosser, A. D. M. Hart, S. D. Langton, A. Jones, C. M'Coy, S. A. Chandler-Morris and J. A. Pascual, Do wood-pigeons avoid pesticide-treated cereal seed?, *J. Appl. Ecol.*, 1999, **36**, 283–296.
102. G. R. de Snoo and R. Luttik, Availability of pesticide-treated seed on arable fields, *Pest Manag. Sci.*, 2004, **60**, 501–506.
103. A. Lopez-Antia, J. Feliu, P. R. Camarero, M. E. Ortiz-Santaliestra and R. Mateo, Risk assessment of pesticide seed treatment for farmland birds using refined field data, *J. Appl. Ecol.*, 2016, **53**, 1373–1381.
104. C. L. Roy, P. L. Coy, D. Chen, J. Ponder and M. Jankowski, Multi-scale availability of neonicotinoid-treated seed for wildlife in an agricultural landscape during spring planting, *Sci. Total Environ.*, 2019, **682**, 271–281.
105. R. J. Lennon, W. J. Peach, J. C. Dunn, R. F. Shore, M. G. Pereira, D. Sleep, S. Dodd, C. J. Wheatley, K. E. Arnold and C. D. Brown, From seeds to plasma: Confirmed exposure of multiple farmland bird species to clothianidin during sowing of winter cereals, *Sci. Total Environ.*, 2020, **723**, 138056.
106. C. L. Roy and P. L. Coy, Wildlife consumption of neonicotinoid-treated seeds at simulated seed spills, *Environ. Res.*, 2020, **190**, 109830.
107. C. T. de Montaigu and D. Goulson, Field evidence of UK wild bird exposure to fludioxonil and extrapolation to other pesticides used as seed treatments, *Environ. Sci. Pollut. Res.*, 2021, **1**, 1–12.
108. E. Fernández-Vizcaíno, M. E. Ortiz-Santaliestra, M. Fernández-Tizón, R. Mateo, P. R. Camarero and F. Mougeot, Bird exposure to fungicides through the consumption of treated seeds: A study of wild red-legged partridges in central Spain, *Environ. Pollut.*, 2022, **292**, 118335.
109. R. Mateo, N. Petkov, A. Lopez-Antia, J. Rodríguez-Estival and A. J. Green, Risk assessment of lead poisoning and pesticide exposure in the declining population of red-breasted goose (Branta ruficollis) wintering in Eastern Europe, *Environ. Res.*, 2016, **151**, 359–367.
110. U. Turaga, S. T. Peper, N. R. Dunham, N. Kumar, W. Kistler, S. Almas, S. M. Presley and R. J. Kendall, A survey of neonicotinoid use and potential exposure to northern bobwhite (Colinus virginianus) and scaled quail (Callipepla squamata) in the Rolling Plains of Texas and Oklahoma, *Environ. Toxicol. Chem.*, 2016, **35**, 1511–1515.
111. C. Hao, M. L. Eng, F. Sun and C. A. Morrissey, Part-per-trillion LC-MS/MS determination of neonicotinoids in small volumes of songbird plasma, *Sci. Total Environ.*, 2018, **644**, 1080–1087.
112. A. M. MacDonald, C. M. Jardine, P. J. Thomas and N. M. Nemeth, Neonicotinoid detection in wild turkeys (Meleagris gallopavo silvestris) in Ontario, Canada, *Environ. Sci. Pollut. Res.*, 2018, **25**, 16254–16260.

113. H. M. Ertl, M. A. Mora, D. E. Boellstorff, D. Brightsmith and K. Carson, Potential effects of neonicotinoid insecticides on northern bobwhites, *Wildl. Soc. Bull.*, 2018, **42**, 649–655.
114. S. Humann-Guilleminot, S. Clément, J. Desprat, J. Binkowski, G. Glauser and F. Helfenstein, A large-scale survey of house sparrows feathers reveals ubiquitous presence of neonicotinoids in farmlands, *Sci. Total Environ.*, 2019, **660**, 1091–1097.
115. R. J. Lennon, R. F. Shore, M. G. Pereira, W. J. Peach, J. C. Dunn, K. E. Arnold and C. D. Brown, High prevalence of the neonicotinoid clothianidin in liver and plasma samples collected from gamebirds during autumn sowing, *Sci. Total Environ.*, 2020, **742**, 140493.
116. A. Esther, D. Schenke and W. Heim, Noninvasively Collected Fecal Samples as Indicators of Multiple Pesticide Exposure in Wild Birds, *Environ. Toxicol. Chem.*, 2022, **41**, 201–207.
117. R. K. Murton and M. Vizoso, Dressed cereal seed as a hazard to wood-pigeons, *Ann. Appl. Biol.*, 1963, **52**, 503–517.
118. R. E. R. Porter, Methiocarb protects sprouting peas from small birds, *New Zeal. J. Exp. Agric.*, 1977, **5**, 335–338.
119. G. R. De Snoo, N. M. I. Scheidegger and F. M. W. De Jong, Vertebrate wildlife incidents with pesticides: A European survey, *Pestic. Sci.*, 1999, **55**, 47–54.
120. E. Bro, F. Millot, A. Decors and J. Devillers, Quantification of potential exposure of gray partridge (Perdix perdix) to pesticide active substances in farmlands, *Sci. Total Environ.*, 2015, **521–522**, 315–325.
121. P. Prosser and A. D. M. Hart, Assessing potential exposure of birds to pesticide-treated seeds, *Ecotoxicology*, 2005, **14**, 679–691.
122. M. B. Poliserpi, D. S. Cristos and J. C. Brodeur, Imidacloprid seed coating poses a risk of acute toxicity to small farmland birds: A weight-of-evidence analysis using data from the grayish baywing Agelaioides badius, *Sci. Total Environ.*, 2021, **763**, 142957.
123. C. A. Bishop, A. J. Moran, M. C. Toshack, E. Elle, F. Maisonneuve and J. E. Elliott, Hummingbirds and bumble bees exposed to neonicotinoid and organophosphate insecticides in the Fraser Valley, British Columbia, Canada, *Environ. Toxicol. Chem.*, 2018, **37**, 2143–2152.
124. C. A. Bishop, M. B. Woundneh, F. Maisonneuve, J. Common, J. E. Elliott and A. J. Moran, Determination of neonicotinoids and butenolide residues in avian and insect pollinators and their ambient environment in Western Canada (2017, 2018), *Sci. Total Environ.*, 2020, **737**, 139386.
125. C. A. Bishop, S. G. English, F. Maisonneuve, A. J. Moran and H. A. Higo, J. Common, K. G. Hick and J. E. Elliott, Temporal and spatial patterns of systemic insecticides in avian and insect pollinators and flowers in western Canada (2018, 2019), *Environ. Adv.*, 2022, **8**, 100211.
126. A. Taliansky-Chamudis, P. Gómez-Ramírez, M. León-Ortega and A. J. García-Fernández, Validation of a QuEChERS method for analysis of neonicotinoids in small volumes of blood and assessment of

exposure in Eurasian eagle owl (Bubo bubo) nestlings, *Sci. Total Environ.*, 2017, **595**, 93–100.
127. P. Byholm, S. Mäkeläinen, A. Santangeli and D. Goulson, First evidence of neonicotinoid residues in a long-distance migratory raptor, the European honey buzzard (Pernis apivorus), *Sci. Total Environ.*, 2018, **639**, 929–933.
128. S. Humann-Guilleminot, S. Laurent, P. Bize, A. Roulin, G. Glauser and F. Helfenstein, Contamination by neonicotinoid insecticides in barn owls (Tyto alba) and Alpine swifts (Tachymarptis melba), *Sci. Total Environ.*, 2021, **785**, 147403.
129. OECD, *Guideline for Testing of Chemicals. Test No. 223: Avian Acute Oral Toxicity Test*, 2016.
130. USEPA, *Ecological Effects Test Guidelines OCSPP 850.2100: Avian Acute Oral Toxicity Test*, Washington, D.C., 2012.
131. USEPA, *Ecological Effects Test Guidelines OCSPP 850.2200: Avian Dietary Toxicity Test*, Washington, D.C., 2012.
132. OECD, *Guideline for Testing of Chemicals. Test No. 206: Avian Reproduction Test*, 1984.
133. USEPA, *Ecological Effects Test Guidelines OCSPP 850.2300: Avian Reproduction Test,* Washington, D.C., 2012.
134. USEPA, *Endocrine Disruptor Screening Program Test Guidelines OCSPP 890.2100: Avian Two-generation Toxicity Test in the Japanese Quail*, Washington, D.C., 2015.
135. USEPA, *Pesticide Assessment Guidelines Subdivision E Hazard Evaluation: Wildlife and Aquatic Organisms*, Washington, D.C., 1982.
136. P. Mineau, B. T. Collins and A. Baril, On the use of scaling factors to improve interspecies extrapolation of acute toxicity in birds, *Regul. Toxicol. Pharmacol.*, 1996, **24**, 24–29.
137. K. A. Nagy, Field metabolic rate and body size, *J. Exp. Biol*, 2005, **208**, 1621–1625.
138. P. Mineau and C. Palmer, *The Impact of the Nation's Most Widely Used Insecticides on Birds*, American Bird Conservancy, 2013.
139. T. H. Furley, J. Brodeur, H. C. Silva de Assis, P. Carriquiriborde, K. R. Chagas, J. Corrales, M. Denadai, J. Fuchs, R. Mascarenhas, K. S. Miglioranza, D. M. Miguez Caramés, J. M. Navas, D. Nugegoda, E. Planes, I. A. Rodriguez-Jorquera, M. Orozco-Medina, A. B. Boxall, M. A. Rudd and B. W. Brooks, Toward sustainable environmental quality: Identifying priority research questions for Latin America, *Integr. Environ. Assess. Manag.*, 2018, **14**, 344–357.
140. USEPA, *Methods for Acute Toxicity Tests With Fish, Macroinvertebrates, and Amphibians*, Washington, D.C., 1975.
141. ASTM, *Standard Guide for Conducting Acute Toxicity Tests on Test Materials with Fishes, Macroinvertebrates, and Amphibians*, ASTM E729-96, 2002.
142. B. S. Echols, A. J. Smith, G. M. Rand and B. C. Seda, Factors Affecting Toxicity Test Endpoints in Sensitive Life Stages of Native Gulf of Mexico Species, *Arch. Environ. Contam. Toxicol.*, 2015, **68**, 655–662.

143. J. C. Brodeur and M. B. Poliserpi, South American Cowbirds as Avian Models for Environmental Toxicity Testing, *Ecotoxicol. Genotoxicol. Non-traditional Terr. Model.*, 2017, 290–307.
144. USEPA, United States Environmental Protection Agency, 2022.
145. E. Bonneris, Z. Gao, A. Prosser and R. Barfknecht, Selecting appropriate focal species for assessing the risk to birds from newly drilled pesticide-treated winter cereal fields in France, *Integr. Environ. Assess. Manag.*, 2019, **15**, 422–436.
146. P. Prosser, *Potential exposure of birds to treated seeds*, York, London, 1999.
147. A. Lopez-Antia, M. E. Ortiz-Santaliestra and R. Mateo, Experimental approaches to test pesticide-treated seed avoidance by birds under a simulated diversification of food sources, *Sci. Total Environ.*, 2014, **496**, 179–187.
148. L. Addy-Orduna, J. Cazenave and R. Mateo, Avoidance of neonicotinoid-treated seeds and cotyledons by captive eared doves (Zenaida auriculata, Columbidae), *Environ. Pollut.*, 2022, **304**, 119237.
149. EFSA, *Opinion of the Scientific Panel on Plant protection products and their residues (PPR) on a request from EFSA related to the evaluation of pirimicarb*, EFSA Journal, 2005, vol. 240.
150. L. B. Sabin and M. A. Mora, Ecological risk assessment of the effects of neonicotinoid insecticides on northern bobwhites (*Colinus virginianus*) in the South Texas Plains Ecoregion, *Integr. Environ. Assess. Manag.*, 2021, **18**, 488–489.
151. C. A. Brühl and J. G. Zaller, Biodiversity Decline as a Consequence of an Inappropriate Environmental Risk Assessment of Pesticides, *Front. Environ. Sci.*, 2019, **7**, 2013–2016.
152. R. B. Schäfer, M. Liess, R. Altenburger, J. Filser, H. Hollert, M. Roß-Nickoll, A. Schäffer and M. Scheringer, Future pesticide risk assessment: narrowing the gap between intention and reality, *Environ. Sci. Eur.*, 2019, **31**, 1–5.
153. C. J. Topping, A. Aldrich and P. Berny, Overhaul environmental risk assessment for pesticides, *Science*, 2020, **367**, 360–363.
154. D. Gibbons, C. Morrissey and P. Mineau, A review of the direct and indirect effects of neonicotinoids and fipronil on vertebrate wildlife, *Environ. Sci. Pollut. Res.*, 2015, **22**, 103–118.
155. S. Sala, M. Cavalli and M. Vighi, Spatially explicit method for ecotoxicological risk assessment of pesticides for birds, *Ecotoxicol. Environ. Saf.*, 2010, **73**, 213–221.
156. M. Etterson, K. Garber and E. Odenkirchen, Mechanistic modeling of insecticide risks to breeding birds in North American agroecosystems, *PLoS One*, 2017, **12**, 1–23.

CHAPTER 6

Teratological Effects of Pesticides in Reptiles – A Review

A. GARCÊS*[a,b,c] AND I. PIRES[d,e]

[a] Politécnico de Viseu, Escola Superior Agrátia de Viseu, Quinta da Alagoa Estrada de Nelas, 3500-606 Viseu, Portugal; [b] Instituto de Investigação e Formação Avançada em Ciências e Tecnologias da Saúde, Instituto Universitário de Ciências da Saúde, R. Central Dada Gandra, 1317, 4585-116 Gandra, Portugal; [c] CITAB, University of Trás-os-Montes and Alto Douro, Quinta de Prados 5000-801, Vila Real, Portugal; [d] Department of Veterinary Sciences University of Trás-os-Montes and Alto Douro, Quinta de Prados 5000-801, Vila Real, Portugal; [e] CECAV, University of Trás-os-Montes and Alto Douro, Quinta de Prados 5000-801, Vila Real, Portugal
*Email: andreiamvg@gmail.com

6.1 Introduction

Studies of congenital malformations have increasingly attracted the attention of the scientific community. By definition, congenital malformations result from pathological events characterized by morphological abnormalities that occur during embryonic development, which ultimately can lead to embryo death. The occurrence of congenital malformations has been reported in all types of vertebrates.[1] Their etiology is multifactorial and includes genetic factors (*e.g.*, chromosomal aberrations, such as numerical and structural chromosomal abnormalities), environmental factors (*e.g.*, xenobiotics including agrochemicals and radiation), infectious agents or a combination of these different factors.[2]

Issues in Toxicology No. 45
Bird and Reptile Species in Environmental Risk Assessment Strategies
Edited by Guillermo Eli Liwszyc and Marcelo L. Larramendy

Published by the Royal Society of Chemistry, www.rsc.org

The knowledge of congenital malformations and their etiology, with a focus on wildlife, is a powerful weapon in the holistic/overall evaluation of the "one health status". This understanding can lead to the identification of substances harmful to humans, non-humans and the environment, and the implementation of significant sanitary measures.

The class Reptilia has a wide diversity of species with different anatomical features, adapted to the hot and humid habitats in terrestrial and aquatic environments.[3,4] This class is categorized into four orders: Squamata (*e.g.*, amphisbaenians, lizards, snakes), Testudinata (*e.g.*, turtles), Crocodilia (*e.g.*, crocodilians) and Rhynchocephalia (*e.g.*, tuataras).[5] Figure 6.1 shows examples of different reptilian species.

In reptiles, congenital abnormalities are attributed to environmental changes such as temperature and humidity changes, radiation, infection[6] and direct or indirect contact with teratogenic compounds originating from anthropogenic activities.[1,7,8] Pesticides and other chemical compounds are some examples.[8–10] These substances can act as hormone disruptors, among other functions, interfering with hormonal regulation, chemical messengers and metabolic pathways.[11] In the early stages of embryo development, exposure to these substances can lead to its abnormal development.[10,12,13] In addition, chronic exposure to these compounds in adults can result in the development of malignancies, kidney and liver injury, neurotoxicity, endocrine changes and infertility.[14–16]

Reptiles are among the least studied taxa in ecotoxicology[17] and the knowledge of the harmful effects of environmental contaminants on reptiles is still limited.[7]

Many species depend on safe environmental conditions for their good development. The exposure of animals to air, water or food contaminated with xenobiotics originating from agriculture and urban activity[3,18] makes them perfect models in ecotoxicological studies.[6]

In this chapter, we present a brief compilation of the teratogenic effects of pesticides observed in the orders and suborders of the class Reptilia.

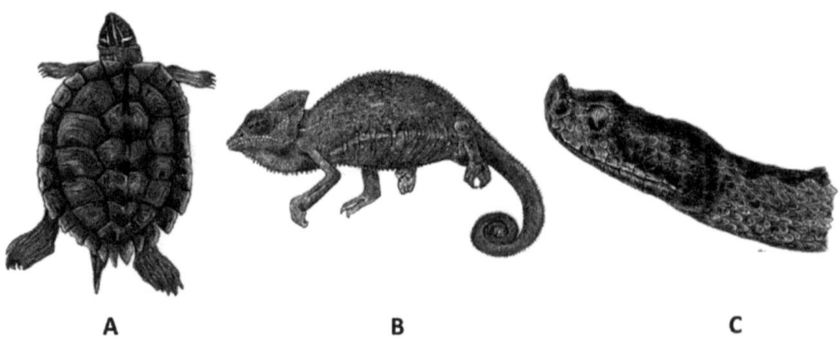

Figure 6.1 Representation of the three orders/suborders of the Reptilia class: Testudinata order (A), Sauria suborder (B) and Ophidia suborder (C). (Author illustration: Andreia Garcês).

6.2 Reptiles as Sentinel Species

Several animal species can be used as sentinels.[19] Many attributes of an animal contribute to its ability as a sentinel. The selection of a sentinel/indicator species depends on many criteria, such as its relevance to the particular geographical area, the coincidence of its territory with the area to be monitored, its abundance (preference for a species which is not listed as threatened), life expectancy, manipulation and catching abilities and consistent and measurable responsiveness to environmental change.[19–21]

Reptiles may serve as sentinels of human and ecosystem health since they meet many of these criteria.[5] Crocodilians, for instance, are excellent sentinel species due to their position in the food chain, longevity (>50 years), aquatic habitats and dependence on environmental conditions for embryos' and juveniles' normal development. Moreover, they can reflect changes in a determined area over extensive periods of time.[22] They can be used as bioindicators of contamination since they accumulate different compounds (*e.g.*, heavy metal) in their bones, muscles and organs. One example is the alterations of annual laminations of the osteoderms in the dorsal skin of the saltwater crocodile (*Crocodylus porosus*) when exposed to high levels of Pb.[22]

6.3 Teratological Effects of Pesticides on Reptiles

The teratological impacts of pesticides on reptiles are poorly known. Reported cases are rare, largely due to difficulties in handling these animals and in collecting biological samples. Furthermore, in some of these cases, a cause-and-effect relationship between pesticide exposure and adverse effects cannot be established.[23–25] In this sense, the observed results are seen, in most cases, as a mere proposal of an effect assumed to be exerted by xenobiotic agents.

6.3.1 Order Testudinata

Some studies indicate that the presence of pesticides could be associated with teratogenic effects in semi-aquatic and sea turtles, namely the common snapping turtle (*Chelydra serpentina*), the painted turtle (*Chrysemys picta*), the hawksbill sea turtle (*Eretmochelys imbricata*), the green sea turtle (*Chelonia mydas*), the olive ridley sea turtle (*Lepidochelys olivacea*) and the loggerhead sea turtle (*Caretta caretta*), among others.[26]

The common snapping turtle (*C. serpentina*) extends from south-eastern Canada, southwest to the edge of the Rocky Mountains and as far east as Nova Scotia and Florida in the USA. During the 1986–1989 period, the eggs were collected for artificial incubation from six different areas (Lake Sasajewun, Lynde Creek, Cranberry Marsh, Cootes Paradise, Hamilton Harbour, Big Creek Marsh) and in the nest areas for chemical analyses. They had significantly higher rates of unhatched eggs and malformations that include deformed carapaces, tails, limbs and the cranium/jaw/nostrils, missing claws, enlarged yolk sacs, the absence of eyes and dwarfism.[27] The analysis of the eggs revealed the presence of hexachlorobenzene (HCB), *cis*-chlordane, *t*-nonachlor, polychlorinated biphenyl (PCB), *p,p′*-dichlorodiphenyldichloroethylene (DDE), *O*-chlordane, mirex, heptachlor epoxide and dieldrin in the five areas studied. As an example, in Cootes Paradise the levels detected were 112 $ng\,g^{-1}$ for *O*-chlordane, 249 $ng\,g^{-1}$ for *t*-nonachlor, 877 $ng\,g^{-1}$ for *p,p′*-DDE, 118 $ng\,g^{-1}$ for mirex, 20 $ng\,g^{-1}$ for dieldrin, 17 $ng\,g^{-1}$ for HCB, 1618 $ng\,g^{-1}$ for PCB and 5.6 $ng\,g^{-1}$ for heptachlor epoxide.[27]

Benzo(*a*)pyrene and dibenzo(*a,h*)antracene, PCBs, As, Cu, Cd, Hg and Pb were related to the development of malformations in semi-aquatic turtles, *e.g.*, the snapping turtle (*C. serpentina*) and painted turtle (*C. picta*) from the John Heinz National Wildlife Refuge (Philadelphia, USA). During the years 2000 to 2003, eggs were collected in the nesting areas. The malformations observed and reported in the embryos included the misshapen or absence of the carapace or plastron, few or no scales, extra, missing or misshapen scouts, lack or abnormal pigmentation, a deformed tail, deformed or absent digits or even a foot or limb, a narrow body shape, the spine/notochord bent in two or more places, kyphosis or scoliosis, microcephalia, exencephaly, craniorachischisis, hydrocephaly, anophthalmia and cyclops, anomalies in the development and fusion of maxillary processes and nasal processes, the absence of the maxilla or mandible, malocclusions, gastrotrichsis, herniation of the yolk sac, ectocardia, developmental asynchrony and dwarfism.[26]

There are only a few reports of malformations in sea turtles after *in ovo* exposure to environmental stressors. However, sea turtles are valuable models for studying changes induced by toxic compounds. Their embryonic development, like other freshwater and terrestrial specimens, depends on the external environment. Extreme changes in the environment can lead to the appearance of abnormalities. Additionally, maternal transfer of xenobiotics to fertilized eggs may occur due to their lipophilic properties.[2]

Bárcenas-Ibarra *et al.*[28] studied three sea turtle species, namely *E. imbricata* ($n = 23\,559$ eggs), *C. mydas* ($n = 17\,690$ eggs) and *L. olivacea* ($n = 20\,257$ eggs), from the Mexican Pacific and Mexican Caribbean. External anatomical malformations were found in 0.2 to 2% of the newborn turtles. A total of 63 types of congenital malformations, with 38 being novel, were identified. Thirty-one types were found in *E. imbricata*, 23 in *C. mydas* and 59 in *L. olivacea*. The head region showed a higher number of malformation types. Out of the three species, the *L. olivacea* presented the highest incidence of congenital malformations. The high number of severe abnormalities was found in the craniofacial region (49%), in the carapace (45%) and in the limbs (33%). Pigmentation disorders occurred in 20% of the animals (see Figure 6.2). The malformations are described in Table 6.1 according to their anatomical location.

Another study was performed by Craven *et al.*[29] during a three-year period. A total of 2731 of *C. caretta* eggs from Wassaw, Sapelo and Jekyll islands along the Georgia coast (USA) were analyzed. One hundred and fifty embryos (7% of all embryos) had abnormalities in the head, shell and limbs. The conditions included: microcephaly, underhung jaw, anophthalmia, cyclops,

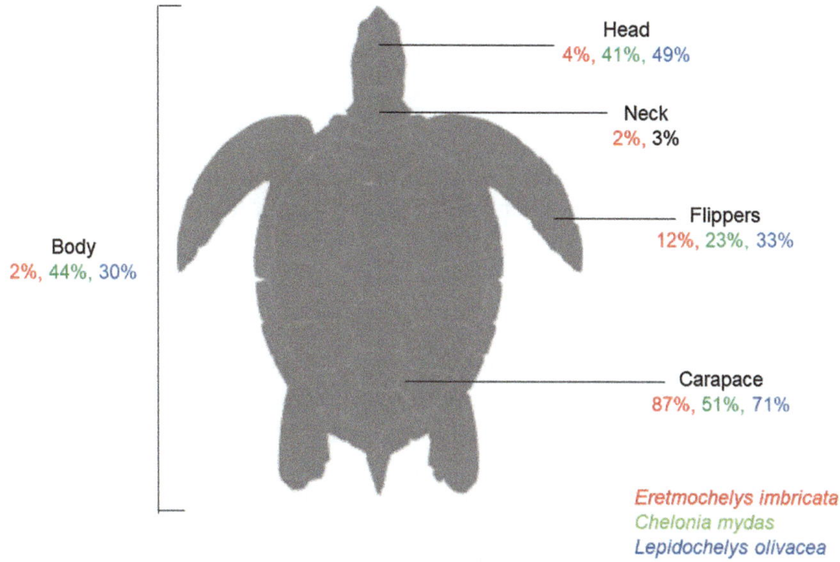

Figure 6.2 Distribution and abundance of congenital malformations by anatomical location observed in embryos of *Eretmochelys imbricata*, *Chelonia mydas* and *Lepidochelys olivacea*.

Table 6.1 Summary of congenital malformations observed in *Eretmochelys imbricata*, *Chelonia mydas* and *Lepidochelys olivacea*.

Anatomical location	Congenital malformations
Body	Leucism, conjoined twins, anasarca, dwarfism, gigantism
Head	Bicephaly, acephaly, hydrocephaly, encephalocele, skull dysostosis, anencephaly, exencephaly, eyelid hyperplasia, microphthalmia, macrophthalmia, synophthalmia, cyclopia, paroxistic cyclops, anophthalmia, rhinoschisis, arhinia, rhinocephaly, laterognathia, misshapen jaws, brachygnathia, gnathoschisis, agnathia
Neck	Cervical cyst, short neck, neck pseudofistula
Carapace	Compressed carapace, misshapen scutes, supernumerary scutes, subnumerary scutes, kyphosis, scoliosis, misshapen bones, white femoral scutes, plastron metaplasia, misshapen tail, brachyury, anury, cloacal atresia
Flippers	Lack of pigmentation, hyperflexion, micromelia, aplastic claws, macrodactyly, brachydactyly, polydactyly, adactyly, syndactyly, diplopodia, hypoplasia, malpositioned, phocomelia, hemimelia, amelia

scissor jaw, megacephaly, two-headed, no head, hypomelanism, undersized limbs and bent, twisted and inverted limbs. The most common anatomical location was the limbs.[29]

There is no information on pollution levels and the occurrence of deformities in sea turtle embryos in both studies, but this possibility should be considered. Since teratogenic deformities are multifactorial, and levels of pesticides such as dichlorodiphenyldichloroethane (DDT) and DDE have been detected in the tissues of adult sea turtles,[30–32] maternal transfer could have occurred and justified the observed congenital anomalies.[28]

6.3.2 Order Rhynchocephalia

To the best of our knowledge, studies on the teratological effects of pesticides in this group are not yet available for Tuatara (*Sphenodon punctatus*) and Guenther's tuatara (*S. guntheri*).

6.3.3 Order Crocodilia

Only a few studies report the teratogenic effects of pesticides on reptiles of the order Crocodilia, namely in the Florida alligator (*Alligator mississippiensis*), Morelet's crocodile (*Crocodylus moreletii*), the spectacled caiman (*Caiman crocodilus*), the freshwater crocodile (*Crocodylus johnstoni*), the saltwater crocodile (*C. porosus*), the broad-snouted caiman (*C. latirostris*) and the Nile crocodile (*C. niloticus*).[33,34]

Embryos of *A. mississippiensis* from Lake Apopka (USA) presented a large number malformations, associated with a chemical spill of sulfuric acid and dicofol, as well as other contaminants such as organochlorine pesticides and PCBs.[33,35] These contaminants seem to be transferred *via* the maternal route, leading to a change in lipid content and reduced viability of the embryos.[33,34]

In the same region, a study by Crain *et al.*[36] on juveniles of *A. mississippiensis* showed a significant decrease of the plasma testosterone concentrations (29.2 $pg\,mi^{-1}$ compared to 51.3 $pg\,mi^{-1}$) and gonadal–adrenal mesonephros (GAM) aromatase (7.6 $pmol\,g^{-1}\,h^{-1}$ compared to 11.4 $pmol\,g^{-1}\,h^{-1}$). An *in ovo* study of 17β-estradiol and tamoxifen caused an increase in aromatase activity, with sex reversal from male to female.[36]

In ovo exposure of *A. mississippiensis* to 0.1 $mg\,kg^{-1}$ 17α-ethynylestradiol (EE2) during the egg incubation temperature-sensitive period (TSP) resulted in a significant increase in the number of females.[37]

Eggs of *C. latirostris* topically exposed to 90 µg 17β-estradiol (E2) per egg or 9 mg bisphenol A (BPA) per egg resulted in a total sex reversal of gonads and external genitalia in all cases.[38] The application of lower doses (0.9 µg E2 per egg or 90 mg BPA per egg) resulted in disorganized seminiferous tubules.[39]

Non-viable eggs of Morelet's crocodile (*C. moreletii*) from Belize also had high concentrations of lindane, aldrin, methoxychlor, heptachlor epoxide and *p,p*′-DDT, among others. Similar chemical residues were also present in the eggs or body tissue of four other species of crocodilians (*C. crocodilus*, *C. johntsoni*, *C. porosus* and *C. niloticus*). These compounds not only cause teratogenic anomalies that reduce embryo viability but also interfere with the reproductive cycle of adults, increasing their infertility. Both phenomena contributed to the decline of the crocodilian's populations of these species during the last few years.[39]

A study performed by Rauschenberger *et al.*[40] evaluated the maternal transfer of organochlorine pesticides in *A. mississippiensis* from exposure sites. The values of organochlorine pesticides in blood were less than

0.8 pmol g^{-1} and in abdominal adipose tissue more than 44 000 pmol g^{-1}. The lipid-adjusted ratio of maternal adipose burdens to yolk burdens was nearly 1, suggesting that organochlorine pesticides in eggs were derived from abdominal adipose lipids.[40]

The exposure of reptiles to pesticides from banana plantations in Costa Rica was studied using blood samples of the spectacled caiman (*C. crocodilus*). Eldrin, permethrin, mirex, 4,4′-DDE, α-endosulfan, heptachlor epoxide, oxychlordane, heptachlor and cypermethrin levels were evaluated. Caimans living in high-intensity banana plantations had higher levels of pesticides than animals from non-altered areas. As expected, animals with higher concentrations of pesticides in their blood showed poor body condition.[41]

Genotoxic techniques could also be promising tools for investigating the impacts of pesticides on reptiles. Polleta *et al.* (2008)[42] used comet assay and micronucleus tests to determine the baseline values of DNA damage in 41 juvenile *C. latirostris* exposed to endocrine-disrupting contaminants (organochlorines). The results showed that the baseline values were quite stable among the tested animals, confirming the potential values of both tests as accurate screening tools for the evaluation of genotoxic agents in this species.[42] In 2020, a study conducted in *C. latirostris* confirmed the genotoxicity and oxidative stress effects of pesticides during the incubation period. *Caiman latirostris* embryos were exposed to different herbicide and insecticide formulations (glyphosate, chlorpyrifos and cypermethrin). After hatching, the results showed a deleterious genetic effect of these pesticides, *e.g.*, DNA damage, increased occurrence of micronuclei and oxidation of pyrimidines and purines.[43]

6.3.4 Order Squamata

Although there are reports of malformations in specimens belonging to the order Squamata, there is still no tangible evidence of their association with pollutants. However, in view of the capacity of these substances to cause congenital changes in a number of other species, consideration must be given to the fact that exposure to contaminants in the environment may be an important etiology of teratogenic changes in these animals.

6.3.4.1 Suborder Ophidia

From 2007 to 2012, at the Butantan Institute (São Paulo, Brazil), malformations in South American pit vipers (*Bothrops jararaca*) and South American rattlesnakes (*Crotalus durissus*) were assessed. Congenital abnormalities were detected in a total of 95 (2.3%) newborn *B. jararaca* and 36 (11.1%) newborn *C. durissus* from wild captured pregnant females (litters of 240 and 35, respectively).[25] The teratogenic abnormalities observed were: kyphosis, anophthalmia, scoliosis, lordosis, microphthalmia, buphthalmia, brachygnathia, scale fusion, hydrocephalus, heterotopia of scale, bicephaly, micrognathia, kinked tails, schistosomia, twisted tails, incomplete tails and dwarfism. There is a possibility that the increase in malformations in both species is related to the increased use of pesticides and insecticides in adjacent areas.[25]

Isolated case reports of congenital disorders in ophidians are rare. Brachygnathia, with the shortening of the jaw, microphthalmia of the right eye and abnormality of the pupil in the left eye, has been reported in a dark-spotted anaconda (*Eunectes deschauenseei*) from the urban area of Macapá, Brazil. Although this is a single case and no toxicological tests were performed, it has been suggested that exposure to exogenous agents such as pesticides in urban areas could be a trigger for this malformation.[44]

6.3.4.2 Suborder Sauria

Some abnormalities have been reported in captive and wild lizards in the form of kyphosis in the skink (*Marisora brachypoda*) and in the Ibiza wall lizard (*Podarcis pityusensis*) or kyphosis and scoliosis in Southern sagebrush lizard (*Sceloporus graciosus vandenburgianus*). The etiology of these cases has not been determined, but external factors such as pesticides could have played a role in the development of these abnormalities.[45]

In La Palma (Canary Islands, Spain), a study was conducted to assess the effect of exposure to organophosphate and carbamate pesticides on lacertids (*Gallotia galloti palmae*). The butyrylcholinesterase (BChE) activity was measured in the plasma from 420 lizards collected from agricultural and reference areas. The BChE activity was significantly lower in lizard samples of agricultural areas.[7] However, the teratogenic effects were not evaluated.

6.4 Final Remarks

The misuse and overuse of pesticides have negative consequences on environmental health. Contaminants affect numerous non-target species of aquatic and terrestrial habitats, including reptiles, leading to a well-known population decline.[8] Due to their biological features and habitat, reptiles have an increased risk of being exposed to pesticides used by anthropogenic activities.[46] However, toxicity data regarding the effects of pesticides on reptiles are still very limited, especially when compared to other classes of vertebrates such as amphibians and birds.[18] Furthermore, reptiles are currently not taken into account in risk assessment protocols during pesticide application procedures in certain countries.[46]

The impact of pesticides on reptiles will depend on numerous factors such as species, habitat type, food chain position and xenobiotic exposure time and concentration. Pesticides accumulate in reptile tissue or, in extreme cases, result in the animal's death.[8,9,47] Thus, they can have devastating consequences on populations since, according to the International Union for Conservation of Nature (IUCN) Red List of Threatened Species in 2015, 18% of all reptile species are considered threatened by the European Union (EU).[46] Consequently, some already vulnerable species may decline even faster due to the presence of these compounds in their habitat.

According to the literature, aquatic species and species living in close-by areas seem to be most affected, which is expected as pesticides accumulate in water after being washed by rain from agricultural fields.

Additionally, pesticide levels may increase as multiple pesticide applications may be required to successfully eradicate pests that have become resistant to pesticides.[8,47] This will increase the concentration and duration of exposure, thereby increasing the risk of health problems, infertility and mortality of reptiles.[46]

6.5 Conclusion

Congenital malformations have multifactorial causes and could involve complex interactions between genetic and environmental factors.[48,49] In reptiles, most of the reports of congenital malformations linked to the exposure to anthropogenic compounds have been documented in wild animals. However, determining the percentages of teratogenic phenomena in reptile populations and their relationship to the presence of pesticides is very challenging. In some studies, ova or embryos have been tested, but, in others, the association is merely speculative.[2] Furthermore, the large majority of animals are not observed by humans or documented in census data. In the case of congenital anomalies, it is even more difficult because the embryo may die in the early stages of development. Thus, it is possible that the causes of congenital abnormalities in reptiles, particularly those that live in aquatic environments, are higher than those reported.

Pesticides play a role in the appearance of congenital malformations in reptiles, but further studies are needed to understand their real impact on populations. Reptiles must be included in ecotoxicity research and treated as promising environmental sentinels of environmental contamination in a holistic/overall health concept.

Acknowledgements

The participation of I. Pires was supported by the projects UIDB/CVT/00772/2020 and LA/P/0059/2020 funded by the Portuguese Foundation for Science and Technology (FCT) (Project UIDB/CVT/0772/2020). The participation of A. Garcês was supported by the FCT (Project UIDB/04033/2020).

References

1. M. P. N. de Carvalho, S. S. Sant'Anna, K. F. Grego, A. C. B. D. C. Fonseca-Pinto, C. A. B. Lorigados, N. G. T. Queiroz-Hazarbassanov and J. L. Catão-Dias, Microcomputed tomographic, morphometric, and histopathologic assessment of congenital bone malformations in two Neotropical viperids, *J. Wildl. Dis.*, 2017, **53**, 804–815.
2. R. Martín-del-Campo, M. F. Calderón-Campuzano, I. Rojas-Lleonart, R. Briseño-Dueñasand and A. García-Gasca, Congenital malformations

in sea turtles: Puzzling interplay between genes and environment, *Animals*, 2021, **11**, 1–22.
3. H. Heatwole, J. A. Taylor and H. Heatwole, *Ecology of reptiles*, Surrey Beatty & Sons, Chipping Norton, N.S.W, 1987.
4. *BSAVA Manual of Reptiles*, ed. S. Girling and P. Raiti, British Small Animal Veterinary Association, Quedgeley, Gloucs, United Kingdom, 3rd edn, 2019.
5. G. V. Zychowski and C. A. J. Godard-Codding, Reptilian exposure to polycyclic aromatic hydrocarbons and associated effects: Reptiles and PAHs, *Environ. Toxicol. Chem.*, 2017, **36**, 25–35.
6. R. Martín-del-Campo, I. Sifuentes-Romero and A. García-Gasca, Hox Genes in Reptile Development, Epigenetic Regulation, and Teratogenesis, *Cytogenet. Genome Res.*, 2019, **157**, 34–45.
7. C. Sánchez-Hernández, R. Carbonell, A. Henríquez Pérez, M. Montealegre and L. Gómez, Inhibition of plasma butyrylcholinesterase activity in the lizard *Gallotia galloti palmae* by pesticides: a field study, *Environ. Pollut.*, 2004, **132**, 479–488.
8. R. Jayaraj, P. Megha and P. Sreedev, Organochlorine pesticides, their toxic effects on living organisms and their fate in the environment, *Interdiscip. Toxicol.*, 2016, **9**, 90–100.
9. S. M. Weir, M. Dobrovolny, C. Torres, C. Torres, M. Goode, T. R. Rainwater, C. J. Salice and T. A. Anderson, Organochlorine Pesticides in Squamate Reptiles from Southern Arizona, USA, *Bull. Environ. Contam. Toxicol.*, 2013, **90**, 654–659.
10. K. E. Gibbs, R. L. Mackey and D. J. Currie, Human land use, agriculture, pesticides and losses of imperiled species, *Divers. Distrib.*, 2009, **15**, 242–253.
11. D. A. Crain and L. J. Guillette, Reptiles as models of contaminant-induced endocrine disruption, *Anim. Reprod. Sci.*, 1998, **53**, 77–86.
12. C. B. Pepper, T. R. Rainwater, S. G. Platt, J. A. Dever, T. A. Anderson and S. T. McMurry, Organochlorine pesticides in chlorioallantoic membranes of Morelet's crocodile eggs from Belize, *J. Wildl. Dis.*, 2004, **40**, 493–500.
13. J. S. P. de Oliveira, L. G. Vieira, W. F. Carvalho, M. B. de Souza, A. S. de Lima Rodrigues, K. Simões, D. de Melo De Silva, J. dos Santos Mendonça, L. Q. L. Hirano, A. L. Q. Santos and G. Malafaia, Mutagenic, genotoxic and morphotoxic potential of different pesticides in the erythrocytes of Podocnemis expansa neonates, *Sci. Total Environ.*, 2020, **737**, 140304.
14. U. Tiemann, In vivo and in vitro effects of the organochlorine pesticides DDT, TCPM, methoxychlor, and lindane on the female reproductive tract of mammals: a review, *Reprod. Toxicol.*, 2008, **25**, 316–326.
15. M. Chaturvedi, C. Sharma and M. Chaturvedi, Effects of Pesticides on Human Beings and Farm Animals: A Case Study, *Res. J. Chem. Environ.*, 2013, **1**, 6.
16. A. Garcês, I. Pires and P. Rodrigues, Teratological effects of pesticides in vertebrates: a review, *J. Environ. Sci. Health, Part B*, 2020, **55**, 75–89.

17. S. M. Weir, L. G. Talent, T. A. Anderson and C. J. Salice, Insights into reptile dermal contaminant exposure: Reptile skin permeability to pesticides, *Chemosphere*, 2016, **154**, 17–22.
18. R. J. Hall, *Effects of environmental contaminants on reptiles: A review*, U.S. Fish and Wildlife Service, 1980, vol. 228.
19. Board on Environmental Studies and Toxicology, National Research Council, Division on Earth and Life Studies, Commission on Life Sciences, and Committee on Animals as Monitors of Environmental Hazard, Animals As Sentinels of Environmental Health Hazards: Committee on Animals As Monitors of Environmental Hazards, Board on Environmental Studies and Toxicology, Commission on Life Sciences, National Re, National Academies Press, 1991.
20. G. A. LeBlanc and L. J. Bain, Chronic Toxicity of Environmental Contaminants: Sentinels and Biomarkers, *Environ. Health Perspect.*, 1997, **105**(Suppl. 1), 65–80.
21. A. A. Aguirre and P. L. Lutz, Marine Turtles as Sentinels of Ecosystem Health: Is Fibropapillomatosis an Indicator?, *EcoHealth*, 2004, **1**, 275–283.
22. C. Manolis, G. J. Webb and A. R. Britton, *Crocodilians and other reptiles: bioindicators of pollution*, Australian Nuclear Science and Technology Organisation (ANSTO), Lucas Heights, NSW (Australia), Australia, 2002.
23. D. Koleska and D. Jablonski, Two cases of unclear hindlimb malformation in *Bombina variegata*, *Ecol. Montenegrina*, 2016, **9**, 56–58.
24. A. Christopoulos and P. Pafilis, Hindlimb malformation in a widely distributed skink, Chalcides ocellatus, *Herpetol. Notes*, 2020, **13**, 15–17.
25. S. S. Sant'Anna, K. F. Grego, C. A. B. Lorigados, A. C. B. C. Fonseca-Pinto, W. Fernandes, L. C. Sá-Rocha and J. L. Catão-Dias, Malformations in Neotropical Viperids: Qualitative and Quantitative Analysis, *J. Comp. Pathol.*, 2013, **149**, 503–508.
26. B. Bell, J. R. Spotila and J. Congdon, High incidence of deformity in aquatic turtles in the John Heinz National Wildlife Refuge, *Environ. Pollut.*, 2006, **142**, 457–465.
27. C. A. Bishop, R. J. Brooks, J. H. Carey, P. Ng, R. J. Norstrom and D. R. S. Lean, The case for a cause-effect linkage between environmental contamination and development in eggs of the common snapping turtle (*Chelydra serpentina*) from Ontario, Canada, *J. Toxicol. Environ. Health*, 1991, **33**, 521–547.
28. A. Bárcenas-Ibarra, H. de la Cueva, I. Rojas-Lleonart, F. A. Abreu-Grobois, R. I. Lozano-Guzmán, E. Cuevas and A. García-Gasca, First approximation to congenital malformation rates in embryos and hatchlings of sea turtles, *Birth Defects Res., Part A*, 2015, **103**, 203–224.
29. K. S. Craven, S. Sheppard, L. B. Stallard, M. Richardson and C. N. Belcher, Investigating a link between head malformations and lack of pigmentation in loggerhead sea turtle embryos (*Caretta caretta*, Linnaeus, 1758) in the southeastern United States, 7.
30. K. E. Clukey, C. A. Lepczyk, G. H. Balazs, T. M. Work, Q. X. Li, M. J. Bachman and J. M. Lynch, Persistent organic pollutants in fat of

three species of Pacific pelagic longline caught sea turtles: Accumulation in relation to ingested plastic marine debris, *Sci. Total Environ.*, 2018, **610–611**, 402–411.
31. K. E. Clukey, C. A. Lepczyk, G. H. Balazs, T. M. Work, Q. X. Li, M. J. Bachman and J. M. Lynch, Persistent organic pollutants in fat of three species of Pacific pelagic longline caught sea turtles: Accumulation in relation to ingested plastic marine debris, *Sci. Total Environ.*, 2018, **610–611**, 402–411.
32. C. R. Rodenbusch, C. Baptistotte, M. R. Werneck, T. T. Pires, M. T. D. Melo, M. W. de Ataíde, P. Testa, M. M. Alieve and C. W. Canal, Fibropapillomatosis in green turtles *Chelonia mydas* in Brazil: characteristics of tumors and virus, *Dis. Aquat. Org.*, 2014, **111**, 207–217.
33. R. H. Rauschenberger, *Developmental mortality in American alligators (Alligator mississippiensis) exposed to organochlorine pesticides*, PhD thesis, University of Florida, 2004.
34. M. S. Sepúlveda, J. J. Wiebe, D. C. Honeyfield, H. R. Rauschenberger, J. P. Hinterkopf, W. E. Johnson and T. S. Gross, Organochlorine Pesticides and Thiamine in Eggs of Largemouth Bass and American Alligators and Their Relationship with Early Life-stage Mortality, *J. Wildl. Dis.*, 2004, **40**, 782–786.
35. Z. Khan and F. Law, Adverse effects of pesticides and related chemicals on enzyme and hormone systems of fish, amphibians and reptiles: A review, *Proc. Pak. Acad. Sci.*, 2005, **42**, 315–323.
36. D. A. Crain, L. J. Guillette, A. A. Rooney and D. B. Pickford, Alterations in Steroidogenesis in Alligators (*Alligator mississippiensis*) Exposed Naturally and Experimentally to Environmental Contaminants, *Environ. Health Perspect.*, 1997, **105**, 6.
37. R. K. Bhandari, S. L. Deem, D. K. Holliday, C. M. Jandegian, C. D. Kassotis, S. C. Nagel, D. E. Tillitt, F. S. V. Saal and C. S. Rosenfeld, Effects of the environmental estrogenic contaminants bisphenol A and 17α-ethinyl estradiol on sexual development and adult behaviors in aquatic wildlife species, *Gen. Comp. Endocrinol.*, 2015, **214**, 195–219.
38. C. Stoker, F. Rey, H. Rodriguez, J. G. Ramos, P. Sirosky, A. Larriera, E. H. Luque and M. Muñoz-de-Toro, Sex reversal effects on *Caiman latirostris* exposed to environmentally relevant doses of the xenoestrogen bisphenol A, *Gen. Comp. Endocrinol.*, 2003, **133**, 287–296.
39. T. H. Wu, T. R. Rainwater, S. G. Platt, S. T. Mcmurry and T. A. Anderson, Organochlorine contaminants in Morelet crocodile (*Crocodylus moreletii*) eggs from Belize, *Chemosphere*, 2000, **40**, 671–678.
40. R. H. Rauschenberger, M. S. Sepúlveda, J. J. Wiebe, N. J. Szabo and T. S. Gross, Predicting maternal body burdens of organochlorine pesticides from eggs and evidence of maternal transfer in *Alligator mississippiensis*, *Environ. Toxicol. Chem.*, 2004, **23**, 2906–2915.
41. P. B. C. Grant, M. B. Woudneh and P. S. Ross, Pesticides in blood from spectacled caiman (*Caiman crocodilus*) downstream of banana plantations in Costa Rica, *Environ. Toxicol. Chem.*, 2013, **32**, 2576–2583.

42. G. L. Poletta, A. Larriera, E. Kleinsorge and M. D. Mudry, *Caiman latirostris* (broad-snouted caiman) as a sentinel organism for genotoxic monitoring: Basal values determination of micronucleus and comet assay, *Mutat. Res., Genet. Toxicol. Environ. Mutagen.*, 2008, **650**, 202–209.
43. L. M. Odetti, E. C. López González, M. L. Romito, M. F. Simoniello and G. L. Poletta, Genotoxicity and oxidative stress in *Caiman latirostris* hatchlings exposed to pesticide formulations and their mixtures during incubation period, *Ecotoxicol. Environ. Saf.*, 2020, **193**, 110312.
44. R. Tavares-Pinheiro and C. E. Costa-Campos, Snout and abnormal eyes: a rare case of malformations reported in Dark Spotted *Anaconda Eunectes deschauenseei* Dunn & Conant, 1936 (Serpentes: Boidae) from the Amapá State, Northern Brazil, *Herpetol. Notes*, 2021, **14**, 873–875.
45. C. Arrivillaga, Kyphosis in a free-living *Marisora brachypoda* (Squamata: Scincidae) from Utila Island, Honduras, *Herpetol. Bull.*, 2019, 43–44.
46. V. Mingo, S. Lötters and N. Wagner, Risk of pesticide exposure for reptile species in the European Union, *Environ. Pollut.*, 2016, **215**, 164–169.
47. L. Freitas, J. Paranaíba, A. Peréz, M. Machado and F. Lima, Toxicity of pesticides in lizards, *Hum. Exp. Toxicol.*, 2020, **39**, 596–604.
48. R. Seifer, in *Encyclopedia of Infant and Early Childhood Development*, ed. M. M. Haith and J. B. Benson, Academic Press, San Diego, 2008, pp. 333–343.
49. E. Ujházy, M. Mach, J. Navarová, I. Brucknerová and M. Dubovický, Teratology – past, present and future, *Interdiscip. Toxicol.*, 2012, **5**, 163–168.

CHAPTER 7

Combined Impact of Pesticides and Other Environmental Stressors on Reptile Diversity in Irrigation Ponds Compared to Other Animal Taxa

HIROSHI C. ITO*[a,b] AND NORIKO TAKAMURA[a]

[a] National Institute for Environmental Studies, Tsukuba, Ibaraki 305-8506, Japan; [b] Department of Evolutionary Studies of Biosystems, The Graduate University for Advanced Studies, SOKENDAI, Hayama, Kanagawa 240-0193, Japan

*Email: hiroshibeetle@gmail.com

7.1 Introduction

Freshwater ecosystems provide a wide variety of ecosystem services,[1] some of which are irreplaceable.[2] Freshwater systems occupy only 0.01% of the world's water and 0.8% of the Earth's surface,[3] but contain 6% of all described species and 1/3 of all vertebrate species.[4,5] However, freshwater ecosystems are among the most threatened types of ecosystems.[6,7] The current rapid biodiversity loss in freshwater ecosystems implies that they are degrading at a critical rate, because the loss of biodiversity tends to exponentially reduce the efficiencies and temporal stabilities of ecosystem functions.[8] Major environmental stressors on freshwater ecosystems include

Issues in Toxicology No. 45
Bird and Reptile Species in Environmental Risk Assessment Strategies
Edited by Guillermo Eli Liwszyc and Marcelo L. Larramendy

Published by the Royal Society of Chemistry, www.rsc.org

overexploitation, water pollution, flow modification, destruction or degradation of habitats and invasion by alien species.[4,9–11] These stressors have serious impacts on many organisms, but their effects vary greatly among taxa,[12,13] and hence different measures may be required for each taxon. It is therefore necessary to clarify the stressors' separate and combined impacts on each taxon. However, we lack information about these impacts on taxa that are not major components of freshwater ecosystems such as rivers, lakes and wetlands. This is particularly the case for reptiles, a taxon that includes many endangered species,[13] yet stressors' impacts on this taxon are unknown compared to other taxa.[14] Especially, regarding chemical pollution, reptiles are not included in the mandatory toxicity tests for environmental risk assessments of pesticides.[15] Hence standards for chemical pollution may not be properly set for reptiles. Reptiles as well as amphibians tend to be more sensitive than birds or mammals to the acute toxicity caused by pyrethroids or pyrethrins, and less sensitive to rodenticides, organophosphate insecticides and explosives.[15] In addition, reptiles could be more severely affected than other taxa by chemical contamination that affects physiological properties specific to reptiles (*e.g.*, sex determination).[16,17]

In the field, when chemical pollution causes a decline or local extinction of a particular taxon, this could be due to direct effects (effects on the physiology, behaviour, reproduction and survival of the organism) as well as indirect effects (reduction in food availability, disadvantage over competitors for the same food, increase in the number of natural enemies, increased susceptibility to predation by natural enemies, *etc.*).[18] In order to elucidate such direct and indirect effects, we need to combine different methods and targets, including field studies and laboratory experiments, in such a way that they compensate for each other's disadvantages.

A preferable system for field studies in freshwater is the large number of irrigation ponds (or farm ponds) that are maintained for agricultural and irrigation purposes[19] (see Figure 7.1). Not only do irrigation ponds play an

Figure 7.1 Irrigation ponds in Japan. (a) A pond of traditional and typical landscape. (b) A pond with concrete bank protection as part of agricultural modernization.

important role in providing a stable supply of water for agriculture,[20] but they also maintain considerably higher biodiversity per unit area than lakes and rivers.[21] Therefore, irrigation ponds enable studies on impacts of stressors at the community level comprising high diversities and complex interactions. In addition, the high independence of each pond from others allows us to find sufficient numbers and variations of ponds in intensities of stressors, so that comparative field studies are statistically effective.[22–25]

In this chapter, after a short overview of biological communities in irrigation ponds, we describe the current knowledge of how multiple stressors (*i.e.*, pesticide pollution, eutrophication, habitat destruction, habitat modification, vegetation change and invasion by alien species) affect reptile diversities in comparison with other animal taxa. Then we discuss the usefulness of reptiles as indicator organisms for the stressors' combined impacts on animal diversities, as well as test organisms in long-term toxicity studies and biomarkers of environmental pollution.

7.2 Biological Communities in Irrigation Ponds

7.2.1 Biological Communities

There currently exist about 160 000 irrigation ponds in Japan,[26] 69% of which were built before the 1860s or of unknown date.[27] As irrigation ponds are normally smaller in size and shallower in depth than natural lakes,[23,28] they are subject to greater magnitudes of environmental fluctuations and disturbances in water temperature, water quality, water depth, *etc.* These characteristics are shared with floodplain wetlands, and hence biological communities in irrigation ponds are characterized by the richness of species adapted to floodplain wetlands, such as aquatic insects.[19] Since floodplain wetlands have largely been lost over the past century,[19] irrigation ponds function as the secondary nature that harbour many endangered species originally inhabiting floodplain wetlands.[19,29] Despite their small size, irrigation ponds have diverse habitats so that considerably more species per unit area are maintained than lakes or rivers.[21] Consequently, the contribution of irrigation ponds to regional biodiversity is significant.[30–33] However, like other ecosystems, they are subject to various stressors.[19,22–24]

Animals inhabiting or utilizing irrigation ponds include birds, reptiles, fishes, amphibians, molluscs, crustaceans, insects, annelids and zooplankton (see Figure 7.2), while plants include algae and macrophytes (*i.e.*, submerged plants, floating-leaved plants and emergent plants).[19,22–24] Birds, reptiles, amphibians and insects include species that also utilize the surrounding environment of irrigation ponds, such as forests, woodlands and farmlands.[34]

7.2.2 Reptiles

The representative reptile species in irrigation ponds in the world would be turtles, with snakes and crocodilians also present in some regions.[28,35,36] Many

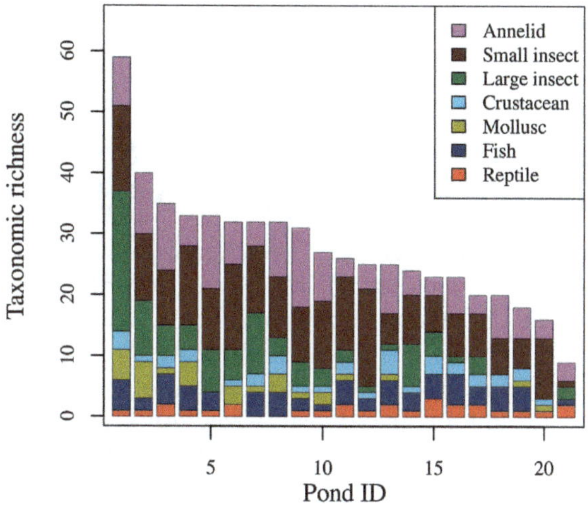

Figure 7.2 Animals in Japanese irrigation ponds, plotted from the data of Ito *et al.*[23] Animals were sampled with stationary nets, baited traps, D-frame dip-nets and bottom sediment samplers, and grouped into reptiles, fishes, molluscs, crustaceans, large insects (other than Diptera), small insects (Diptera) and annelids (consisting mainly of annelids and also bryozoans and sponges).

turtle species are endangered all over the world due to the destruction or modification of habitats, overfishing and water pollution.[37] In Japan, common reptiles inhabiting irrigation ponds are three turtle species: *Mauremys japonica* (Japanese pond turtle), *Mauremys reevesii* (Reeves' pond turtle) and *Trachemys scripta elegans* (red-eared slider turtle). A survey in 518 ponds (western Japan, 2010–2014) found 4347 individuals, of which 15.4% were *M. japonica*, 45.6% were *M. reevesii* and 36.9% were *T. scripta elegans.*[35] *Mauremys japonica* is a native species of Japan, mainly found in foothills and mountainous areas (rivers, lakes, ponds, swamps and rice paddies),[38,39] preferring flowing waters and being cold tolerant.[39,40] This species is omnivorous feeding on fishes, amphibians, insects, crustaceans, molluscs, annelids, animal carcasses, algae, macrophytes and terrestrial plants.[38,40] *Mauremys reevesii* is thought to have been introduced to Japan from the Asian continent hundreds of years ago.[39,41] This species is mainly found in lowland areas, preferring slow water current, and is omnivorous.[38,39] *Trachemys scripta elegans* is thought to have been introduced to Japan from the Americas (distributed from Illinois (USA) to Mexico) in the 1950s.[42] This species is mainly found in lowland areas,[38,39] preferring slow water current,[38,39] and is omnivorous,[38,43] with a strong phytophagic tendency.[44,45] This species may be tolerant of water pollution associated with eutrophication,[38,43] while basking is essential for its maintaining preferable body temperature (28–29 °C).[46] *Trachemys scripta elegans* has been invading and expanding its distribution in many regions all over the world.[47] In Japan, the expansion of its distribution may be causing the decline of *M. japonica* and *M. reevesii.*[39,48] These three turtle species can be preyed upon by mammals, birds, bullfrogs and large predatory insects (*e.g.*, the hemiptera *Kirkaldyia deyrollei*).[49–52]

7.3 Combined Impact of Multiple Stressors on Reptile Diversity Compared to Other Animal Taxa

It is important to comprehensively investigate multiple stressors on biological communities to evaluate the relative severity of each and their combined impacts.[53,54] Major stressors for biological communities in irrigation ponds are: (1) destruction and modification of habitats (reclamation or disappearance of ponds, reduction of pond water depth, concrete bank protection and decline of macrophyte vegetation), (2) eutrophication and pesticide pollution and (3) invasion by alien species.[19] The concrete bank protection denotes the bank protection with concrete dikes (see Figure 7.1b). The water depth reduction is caused by the frequency decline of bottom mud removal (due to the decline in rice farming), which eventually results in disappearance of the ponds.[55,56]

Most previous studies of multiple stressors in irrigation ponds focused on invertebrates,[24,25] but only Ito *et al.*[23] surveyed for a wide range of animal taxa comprising reptiles. Nakanishi *et al.*[25] surveyed for aquatic insects (including Odonata [dragonflies], Coleoptera [beetles] and Hemiptera [true bugs]) and environmental factors in 21 irrigation ponds in Shiga Prefecture (Japan). The statistical analysis with canonical correspondence analysis (CCA) and permutation tests showed that the number of emergent plant species, the rate (percentage) of concrete bank protection, the presence/absence of litter sediments and the perimeter length of ponds were significantly correlated with the species composition of the aquatic insect community. In addition, the number of aquatic insect species was negatively correlated with the rate of concrete bank protection and positively correlated with the number of emergent plant species. The research group also showed in the same region that rice fields with pesticide application had significantly lower biomasses of a mud-snail species *Cipangopaludina chinensis laeta* than those without pesticide application.[57]

Kadoya *et al.*[24] surveyed for macroinvertebrates, aquatic plants and environmental factors (not including pesticides) in 64 irrigation ponds in Hyōgo Prefecture (Japan) and conducted a hierarchical Bayesian analysis to construct an integrated biodiversity indicator incorporating the combined effects of multiple environmental stressors. The constructed index was a linear combination of negative effects of eutrophication (chlorophyll-a concentration), habitat destruction (rate of concrete bank protection) and invasive alien fish (abundance of *Lepomis macrochirus* [bluegill]). Actually, the index could integratively explain the behaviours of several different biodiversity indicators (including total richness, endangered species richness and functional diversity), respectively, for aquatic macrophytes, dragonflies (Odonata) and benthic macroinvertebrates.

Ito *et al.*[23] surveyed in 21 irrigation ponds in Hyōgo Prefecture (Japan) for animals including reptiles, amphibians, fishes, molluscs, crustaceans, insects and annelids (sampled with stationary nets, baited traps, D-frame dipnets and bottom sediment samplers) and for 48 environmental factors including pond water depth, pond area, the rate of concrete bank protection, macrophytes' coverages, water temperature, pH, concentrations of pesticides, indicator substances for eutrophication (*i.e.*, N, P, chlorophyll-a), frequencies of pond drainage (removal of bottom mud) and the presence/absence of four predatory invasive alien species: *Micropterus salmoides* (black bass), *L. macrochirus* (bluegill), *Lithobates catesbeianus* (bullfrog) and *Procambarus clarkii* (red swamp crayfish). As shown in Figure 7.2, the sampled animals were classified into seven groups: reptiles, fishes, molluscs, crustaceans, large insects (other than Diptera), small insects (Diptera) and annelids (consisting mainly of annelids and also bryozoans and sponges), where amphibians were absent because the only amphibian species sampled was an alien species, the bullfrog, which was treated as one of the environmental factors. The statistical analysis involved model selection and permutation tests. Each group's taxonomic richness (number of identified taxa) was explained by a subset of 48 environmental factors in the best model. The analyses showed that taxonomic groups other than reptiles had a significantly negative correlation with one or more of six environmental factors: insecticide (2-*sec*-butylphenyl methylcarbamate (BPMC) concentration), fungicide (probenazole concentration), depth reduction (*i.e.*, pond shallowness), concrete bank protection, decline of macrophytes (floating-leaved plants) and invasion by alien fish (presence/absence of bluegill). These six factors were treated as the major stressors. BPMC, a carbamate insecticide also known as fenobucarb, had a significantly negative correlation with the taxonomic richness of large insects and that of the whole invertebrates (molluscs, crustaceans, large insects, small insects and annelids), whereas probenazole, a benzothiazole fungicide, had a significantly negative correlation uniquely with the taxonomic richness of fishes. The taxonomic richness of large insects was estimated to have declined by an average of 77% (from that under the hypothetical unstressed condition) due to the combined impact of BPMC, macrophyte decline and alien fish. In addition, the taxonomic richness of invertebrates was also estimated to have declined by an average of 15% due to the combined impact of BPMC, macrophyte decline and concrete bank protection. These results on invertebrates accord with those of Nakanishi *et al.*[25] and Kadoya *et al.*[24] described previously. With regard to reptiles, their taxonomic richness was not significantly correlated with any environmental factor, possibly because their maximum taxonomic richness was low (three species at maximum). However, in the biplot of the redundancy analysis, only reptiles were placed in such a way that their taxonomic richness had positive correlations with most stressors.[23]

In the survey data of Ito *et al.*,[23] the regression of each group's taxonomic richness on the major six stressors shows that only reptiles have positive correlations with all these stressors, whereas all sampled animals have negative correlations with all the stressors (see Table 7.1). Excluding the alien species *T. scripta elegans* from reptiles keeps the signs of correlations with all stressors other than macrophyte decline and alien fish. Furthermore, the variance among taxonomic groups in their responses (regression coefficients) to each stressor is greatest in the BPMC (BPMC: 0.93, concrete bank protection: 0.70, macrophyte decline: 0.40, probenazole: 0.28, depth reduction: 0.25 and alien fish: 0.13).

We can roughly estimate the stressors' combined impact on each group in each pond as the loss or gain of taxonomic richness due to the stressors (see the caption of Table 7.1). Each cell in Table 7.2 shows the correlation coefficient between two groups in the estimated combined impacts on them among ponds, where −1 corresponds to the largest limit for the difference in combined impacts on them. The combined impacts on reptiles differ significantly from those on the other taxa except for fishes and crustaceans, as marked with "*" in Table 7.2. These differences are mainly due to their different responses to BPMC, concrete bank protection, depth reduction and alien fish ("B", "C", "D" and "A" in Table 7.2). Note that Tables 7.1 and 7.2 do show possibilities of some causal relationships but do not clarify them. For clarification, manipulative experiments would be needed.

7.4 Impact of Each Stressor on Reptile Diversity Compared to Other Animal Taxa

As shown in the previous section, in irrigation ponds, the taxonomic richness of turtles differs considerably from those of other taxa in their response to multiple environmental stressors. Here we discuss the impact of each environmental stressor in irrigation ponds, comparing turtles with other animal taxa.

7.4.1 Concrete Bank Protection, Water Depth Reduction and Macrophyte Decline

The number of irrigation ponds in Japan has been decreasing for several decades due to the progress of urbanization, the decline of rice farming and the structural modification of irrigation ponds with steel and concrete.[19] In fact, the use of steel and concrete has facilitated the construction of large irrigation ponds and dams by destroying small ones. The present trend on the remaining ponds is to have a concrete bank protection, which coupled with agricultural modernization practices enables effective maintenance.[19] Such a habitat modification significantly reduced habitat diversity, with a negative impact particularly on benthic invertebrates.[23,24] In contrast, the response of turtles' taxonomic richness to concrete bank protection has a positive and high correlation with the concrete bank protection, differently from other taxa (see Tables 7.1 and 7.2). This may be attributed to basking, which is the unique behaviour of turtles. It has been reported that turtles

Table 7.1 Responses (regression coefficients) of animal groups' taxonomic richness to major six stressors. The Poisson regression mixed model was fitted to explain each group's taxonomic richness with the stressors' intensities (and with a pond-specific Gaussian random effect). Italic, upright and boldface texts indicate "smaller than −0.1", "larger than −0.1 and smaller than 0.1" and "larger than 0.1", respectively. Each stressor's intensities among ponds are standardized so that they range from 0 to 1. The analysis is specifically described as follows. For the *i*th group (1: reptiles, 2: fishes, 3: molluscs, 4: crustaceans, 5: large insects, 6: small insects, 7: annelids), its taxonomic richness y_{ij} in the *j*th pond was estimated by $Y_i(\boldsymbol{x}_j) = \exp\left(\acute{a}_i + \sum_{k=1}^{6} \hat{a}_{ik} x_{jk}\right)$ with $\boldsymbol{X}_j = (x_{j1}, \cdots, x_{j6})$, where x_{jk} describes the *k*th stressor's intensity (1: BPMC, 2: probenazole, 3: concrete bank protection, 4: depth reduction, 5: macrophyte decline, 6: alien fish) in the *j*th pond. $\acute{a}_i$ is the intercept for the *i*th group, and $\hat{a}_{ik}$ is its regression coefficient for the *k*th stressor. The stressors' combined impact on the *i*th group in the *j*th pond was calculated as the difference between its estimated taxonomic richness under the observed condition, $Y_i(\boldsymbol{x}_j)$, and that under the hypothetical stress-free condition $Y_i(0)$, denoted by $I_{ij} = Y_i(\boldsymbol{x}_j) - Y_i(0) = \exp(\acute{a}_i)\left[\exp\left(\sum_{k=1}^{6} \hat{a}_{ik} x_{jk}\right) - 1\right]$. The analysis was conducted with statistical software R (version 3.4.4) and its packages glmmML-1.0 and pforeach-1.3.

Taxonomic group	BPMC (insecticide)	Probenazole (fungicide)	Concrete bank protection	Depth reduction	Macrophyte decline	Alien fish
Reptiles	**0.71**	**0.17**	**0.85**	**0.39**	**0.28**	**0.18**
Fishes	0.03	*−0.92*	**0.20**	*−0.26*	*−0.11*	−0.02
Molluscs	*−2.21*	0.05	*−1.87*	**0.13**	*−1.10*	*−0.83*
Crustaceans	0.05	*−0.89*	0.05	*−0.40*	**0.13**	**0.18**
Large insects	*−1.35*	*−0.19*	*−0.83*	*−0.96*	*−1.60*	*−0.52*
Small insects	*−0.27*	**0.41**	*−0.72*	*−0.27*	*−0.55*	0.02
Annelids	*−0.55*	*−0.24*	*−0.58*	**0.36**	*−0.43*	*−0.19*
All sampled	*−0.45*	−0.08	*−0.48*	*−0.33*	*−0.55*	*−0.12*

Table 7.2 Differences among taxonomic groups in stressors' combined impacts on them, indicated with correlation coefficients. "Self-excluded" means all sampled animals excluding the focal group. The statistical significance for the difference (*i.e.*, smallness of correlation) between the *i*th and *i'*th groups is denoted by "*", indicating that the *p*-value < 0.05 (see "test 1", explained later). The statistical significance for the difference between the *i*th and *i'*th groups in their responses to each stressor was also tested, as denoted by B: BPMC, P: probenazole, C: concrete bank protection, D: depth reduction and A: alien fish, indicating that the *p*-value < 0.05 (see "test 2", explained later). Test 1: the significance was tested by generating a probability distribution for the correlation coefficient in combined impacts by random 10 000 resampling of species (or identified taxa) for the *i*th and *i'*th groups, following the frequency distribution $[y_{ij} + y_{i'j}] \Big/ \sum_{j=1}^{21} [y_{ij} + y_{i'j}]$ over the ponds, and keeping the total taxonomic richnesses for the *i*th and *i'*th groups equal to $\sum_{j=1}^{21} y_{ij}$ and $\sum_{j=1}^{21} y_{i'j}$, respectively (see also the caption of Table 7.1). Test 2: the significance was tested analogously to test 1, where a probability distribution for the correlation coefficient was generated by random 10 000 permutations of linear regressor residuals of the focal stressor's intensities on the other stressors.[89] The analysis was conducted with statistical software R (version 3.4.4) and its packages glmmML-1.0 and pforeach-1.3.

	Reptiles	Fishes	Molluscs	Crustaceans	Large insects	Small insects	Annelids	Self-excluded
Reptiles	1.00	0.20	-0.83^{*D}	-0.03	-0.57^{*BA}	-0.70^{*C}	-0.69^{*BC}	-0.78^{*BCA}
Fishes		1.00	0.05	0.88	0.39*	-0.50^{*P}	-0.07	0.06*
Molluscs			1.00	0.12	0.87	0.65	0.66	0.90
Crustaceans				1.00	0.33*	-0.38^{*P}	-0.07	0.04
Large insects					1.00	0.48^{*P}	0.50*	0.67
Small insects						1.00	0.55	0.50^{*P}
Annelids							1.00	0.59

prefer concrete banks for their basking.[58] The monotonous physical structures of concrete banks may allow turtles to quickly escape into the water, against natural enemies such as terrestrial mammals.[58] Therefore, concrete bank protection possibly has a positive effect on turtles.

The depth reduction of irrigation ponds has been caused by a decline in frequency of bottom mud removal, which eventually results in disappearance of the ponds.[55,56] In the survey data of Ito *et al.*,[23] the minimum, average and maximum depths of the studied ponds are 0.3 m, 1.37 m and 4.83 m, respectively. The depth reduction (*i.e.*, pond shallowness) has a significantly negative correlation with the taxonomic richness of "large animals" consisting of reptiles, fishes, molluscs, crustaceans and large insects.[23] A possible reason for this relationship is that the shallower ponds may be subject to greater environmental fluctuations and higher predation pressures by birds and mammals. However, although turtles also belong in large animals, Table 7.1 implies that turtles might benefit from the depth reduction in comparison with other taxa. This might be explained by their adaption to terrestrial life.

The decline of macrophyte vegetation is thought to have been caused by urbanization, concrete bank protection and herbicides.[55,59] In freshwater systems, macrophytes provide food resources and habitats for various animals.[60] Herbicides and their residues can reduce macrophytes, which in turn negatively affects arthropods.[61] In irrigation ponds, the decline of macrophytes may also seriously reduce the taxonomic diversity of animals including aquatic insects.[23–25] In contrast, the taxonomic richness of turtles is positively correlated with macrophyte decline (see Table 7.1), although the effect of macrophyte decline on reptiles in irrigation ponds is unknown. It is even possible that *T. scripta elegans* (more phytophagous species than *M. japonica* and *M. reevesii*) is reducing macrophytes.[62]

7.4.2 Eutrophication and Pesticide Pollution

Eutrophication is caused by fertilizers on agricultural land and by domestic and industrial wastewater.[63] In freshwater systems, eutrophication causes an increase in algal blooms, which leads to decreases of macrophytes and dissolved oxygen levels.[63] In addition, wastewater contains a variety of toxic chemicals.[64] In irrigation ponds, eutrophication is thought to reduce the diversity of macrophytes and benthic invertebrates.[24] Although the impact of eutrophication on turtles in irrigation ponds is unknown, turtles may be less dependent on water qualities as they are also adapted to terrestrial life. Therefore, turtles may be more tolerant of water pollution than other taxonomic groups in irrigation ponds. For example, large populations of *Mauremys leprosa* (Spanish pond turtle) were found even in sewers where fishes were not found.[65] Therefore, for a certain degree of eutrophication and associated water pollution, turtles in irrigation ponds may also have a relative advantage over other taxa.

With regard to pesticides, many irrigation ponds are adjacent to agricultural land and can therefore be subject to serious pesticide pollution. Pesticide concentrations are expected to be higher in the ponds built by digging through flat land than the ponds built by damming valleys, because the water source for the former may be paddy fields (or another irrigation pond that uses paddy fields as its water source).[66] In these ponds, extremely high concentrations of pesticides can occur in pulses at their each application.[23]

According to toxicity tests, herbicides have negative effects mainly on algae and macrophytes, insecticides on invertebrates and fungicides on amphibians.[61,64] Furthermore, pesticide pollution has been shown to reduce the regional diversities of invertebrates in freshwater systems worldwide.[67,68] In irrigation ponds and rivers in Japan, BPMC may have seriously reduced the taxonomic richness of invertebrates including insects.[23] Probenazole may also have reduced the taxonomic richness of fishes.[23]

The response of the taxonomic richness of turtles to BPMC is significantly different from those of other taxa (see Table 7.2) and positively correlated (see Table 7.1). Thus, at some degree of BPMC pollution, turtles may have a relative advantage over other taxonomic groups, which may imply that turtles have higher tolerance than the other taxa. In general, however, there seems to exist a trade-off between the capacity for detoxification and the efficiency for growth and reproduction. For example, water snakes (*Nerodia fasciata*) in polluted habitats have high metabolic rates, presumably due to detoxification, which reduces their efficiency of growth and reproduction.[69] In addition, the expression of detoxification enzymes is higher in individuals inhabiting degraded water quality conditions in Spanish pond turtles (*M. leprosa*).[70] Therefore, a trade-off similar to that for snakes is expected for turtles. Although the detoxification capacity for BPMC can reduce the growth and reproductive efficiency of turtles, they may have a relative advantage in irrigation ponds with a certain degree of BPMC pollution, if the negative impact of BPMC on them is less than those on the other taxa.

7.4.3 Invasive Alien Species

In Japanese irrigation ponds, predatory alien species such as the black bass, bluegill, bullfrog and red swamp crayfish are thought to have particularly strong negative impacts on native organisms.[19,24] These invasive alien species may have been reducing native species through predation and competition.[23,24,71–73] In particular, the bluegill is thought to have serious impacts on native invertebrates.[23,24] These invasive species could also compete with turtles, although their relationship in irrigation ponds is unknown.

Among turtle species in Japanese irrigation ponds, the invasion and expansion of the distribution of the non-native *T. scripta elegans* may be reducing *M. japonica* and *M. reevesii*.[39,48] In many aquatic systems all over the world, *T. scripta elegans* has invaded and expanded its distribution, whereas it has been reported that this species is not likely to cause complete extinction of native species; *T. scripta elegans* is unlikely to establish dominant populations in hill or forest streams nor in large rivers and reservoirs, but is more likely to establish in lowland, vegetated, slow-moving or static water bodies, *e.g.*, in canals, ponds and lakes.[47] According to the fossil record, invasions of species from one geographical region to another cause local extinctions of native species but rarely lead to their complete extinction.[74] Therefore, if an invasive alien species is rapidly expanding its distribution to the point of complete extinction of the native species,[75] it is possibly stemming from that habitats favouring the native species over the invasive species are rapidly decreasing due to habitat modification and/or pollution.[76] Thus, in areas where *T. scripta elegans* continues to expand its distribution rapidly, undetected habitat modification or pollution might be progressing rapidly, causing the decline of habitats where the native species have an advantage over *T. scripta elegans*.

In molluscs, the snail species *Melanoides tuberculatus*, which was artificially introduced into freshwater systems in the Caribbean and West Indies, has lower rates for growth and reproduction than the native gastropod species *Biomphalaria glabrata*, but is more tolerant of heavy metals (Cd), pesticides (malathion), temperature extremes and desiccation. This trade-off relationship is thought to have caused the displacement of the native gastropod (*B. glabrata*) by the introduced snail (*M. tuberculatus*).[76] Analogously, in North American freshwater invertebrates, over 35% of native species are intolerant of organic pollution, whereas all invasive species are at least moderately tolerant.[77]

7.5 Perspective: Usefulness of Turtles as Bioindicators

Aquatic organisms are exposed to complex mixtures of pesticides with high concentrations occurring in pulses associated with spray and rain events.[78] However, due to technical and financial constraints, the number of measurable pesticides and their residues is limited. Furthermore, pulses of pesticide concentrations often occur at intervals shorter than the sampling time interval, and therefore the measurement of the highest concentrations of those pesticides is not easy.[23,79] Such an uncertainty of information on pesticide concentrations in the environment causes difficulties in statistical analyses on the combined impacts of multiple stressors including pesticides. A potential solution for this problem is the use of bioindicators. A bioindicator is defined as a living organism or its part or group of organisms, the product of an organism (*e.g.*, enzyme) and the process of biological means which can be used to generate information on environmental quality,[80,81] enabling efficient estimation for the intensities of environmental stressors and their impacts on biological communities. Although uncertainty is a problem still in this case, the uncertainty could be controlled by appropriately combining different bioindicators.

In freshwater systems, the abundance and diversity of aquatic insects of the orders Ephemeroptera, Plecoptera and Trichoptera have been used as indicators for the extent of water pollution and community richness for decades.[82,83] SPEAR, an index based on pesticide sensitivity data of aquatic insects, has also been a useful indicator for the extent of pesticide pollution load in freshwater systems and the associated changes in biota.[84] In the survey data of irrigation ponds by Ito *et al.*,[23] the taxonomic richness of large insects (containing Ephemeroptera, Plecoptera and Trichoptera) is a good estimator for the stressors' combined impact on the total taxonomic richness of all the sampled animals (see Table 7.3). Its estimation performance is significantly improved when reptiles (*i.e.*, turtles) as well as large insects are used (see Table 7.3).

Compared to aquatic insects, turtles can be easily collected and identified, and they have long lifespans and large body sizes. Hence their body tissues are also suitable as indicators (*i.e.*, biomarkers) of concentrations of toxic substances in the environment.[85] Moreover, turtles are also suitable for long-term toxicity studies,[86,87] since breeding techniques for turtles are well established.[88] Thus, integrated knowledge could be obtained by using turtles, firstly as biomarkers for the pollution in the environment, secondly as test organisms for long-term toxicity studies and thirdly as indicator organisms for the impacts of environmental stressors on biodiversity. Such knowledge would contribute to the understanding of the current situations and impacts of chemical pollution ongoing in many regions all over the world, and its combined effects with other stressors.

Table 7.3 Indicators for stressors' combined impacts on the taxonomic richnesses of all sampled animals in Ito *et al.*,[23] constructed with taxonomic richnesses of large insects and reptiles. In each row, the middle column shows the indicator formula for the combined impacts on the taxonomic richness of the focal group (the left column), where $\Delta y_{\mathrm{Large\_insect}}$ and $\Delta y_{\mathrm{Reptile}}$ describe the changes (or differences among ponds) of taxonomic richnesses for large insects and reptiles, respectively. The definition of the combined impact is described in the caption of Table 7.1. Each indicator formula was constructed by fitting the combined impact as a linear function of the taxonomic richnesses of large insects ($y_{\mathrm{Large\_insect}}$) and reptiles ($y_{\mathrm{Reptile}}$). The indicator formula in the second row (both large insects and reptiles used) is significantly better than that in the first row (only large insects used) with the p-value = 0.009, tested by the permutation-of-regressor-residual test.[89] The analysis was conducted with statistical software R (version 3.4.4) and its packages glmmML-1.0 and pforeach-1.3.

Focal group	Indicator formula for combined impact (I)	Coefficient of determination (r^2)
All sampled	$1.26\Delta y_{\mathrm{Large\_insect}}$	0.69
All sampled	$1.15\Delta y_{\mathrm{Large\_insect}} - 3.22\Delta y_{\mathrm{Reptile}}$	0.77
All sampled (excluding large insects)	$0.52\Delta y_{\mathrm{Large\_insect}}$	0.47
All sampled (excluding large insects and reptiles)	$0.49\Delta y_{\mathrm{Large\_insect}} - 2.39\Delta y_{\mathrm{Reptile}}$	0.62

7.6 Conclusion

- Irrigation ponds are the secondary nature, harbouring the constituent species of floodplain ecosystems, and making a significant contribution to biodiversity, although they are subject to multiple environmental stressors (pesticides, eutrophication, concrete bank protection, water depth reduction, macrophyte decline, invasive alien species).
- In irrigation ponds, reptiles (turtles) are significantly different from other animal taxa in the estimated combined impact of major stressors on their taxonomic richness.
- Turtles may be useful not only as biomarkers of environmental pollution and test organisms in long-term toxicity studies, but also as bioindicators for stressors' combined impacts on animal diversity in irrigation ponds.

Acknowledgements

The authors are grateful to the editors of this book for their careful reading of the manuscript and their valuable comments and suggestions.

References

1. R. Costanza, R. d'Arge, R. de Groot, S. Farber, M. Grasso, B. Hannon, K. Limburg, S. Naeem, R. V. O'Neill, J. Paruelo, R. G. Raskin, P. Sutton and M. van den Belt, The value of the world's ecosystem services and natural capital, *Nature*, 1997, **387**, 253–260.
2. A. Covich, K. Ewel, P. Giller, W. Goedkoop, R. Hall and D. Merritt, in *Sustaining Biodiversity and Ecosystem Services in Soil and Sediments*, ed. D. H. Wall, Island Press, Washington D.C., 2004, pp. 45–73.
3. J. G. Lundberg, M. Kottelat, G. R. Smith, M. L. J. Stiassny and A. C. Gill, So many fishes, so little time: an overview of recent ichthyological discovery in continental waters, *Ann. Mo. Bot. Gard.*, 2000, **87**, 26–62.
4. D. Dudgeon, A. H. Arthington, M. O. Gessner, Z.-I. Kawabata, D. J. Knowler, C. Lévêque, R. J. Naiman, A.-H. Prieur-Richard, D. Soto, M. L. J. Stiassny and C. A. Sullivan, Freshwater biodiversity: importance, threats, status and conservation challenges, *Biol. Rev.*, 2006, **81**, 163–182.
5. A. J. Reid, A. K. Carlson, I. F. Creed, E. J. Eliason, P. A. Gell, P. T. J. Johnson, K. A. Kidd, T. J. MacCormack, J. D. Olden, S. J. Ormerod, J. P. Smol, W. W. Taylor, K. Tockner, J. C. Vermaire, D. Dudgeon and S. J. Cooke, Emerging threats and persistent conservation challenges for freshwater biodiversity, *Biol. Rev.*, 2019, **94**, 849–873.
6. *Living planet report 2018: aiming higher*, ed. M. Grooten and R. E. A. Almond, WWF–World Wide Fund for Nature, Gland, Switzerland, 2018.
7. A. Ricciardi and J. B. Rasmussen, Extinction rates of North American freshwater fauna, *Conserv. Biol.*, 1999, **13**, 1220–1222.
8. B. J. Cardinale, J. E. Duffy, A. Gonzalez, D. U. Hooper, C. Perrings, P. Venail, A. Narwani, G. M. Mace, D. Tilman, D. A. Wardle, A. P. Kinzig,

G. C. Daily, M. Loreau, J. B. Grace, A. Larigauderie, D. S. Srivastava and S. Naeem, Biodiversity loss and its impact on humanity, *Nature*, 2012, **486**, 59–67.
9. E. Malaj, P. C. von der Ohe, M. Grote, R. Kühne, C. P. Mondy, P. Usseglio-Polatera, W. Brack and R. B. Schäfer, Organic chemicals jeopardize the health of freshwater ecosystems on the continental scale, *Proc. Natl. Acad. Sci. U. S. A.*, 2014, **111**, 9549–9554.
10. P. Nõges, C. Argillier, Á. Borja, J. M. Garmendia, J. Hanganu, V. Kodeš, F. Pletterbauer, A. Sagouis and S. Birk, Quantified biotic and abiotic responses to multiple stress in freshwater, marine and ground waters, *Sci. Total Environ.*, 2016, **540**, 43–52.
11. R. A. Relyea, The impact of insecticides and herbicides on the biodiversity and productivity of aquatic communities, *Ecol. Appl.*, 2005, **15**, 618–627.
12. K. Barrett and C. Guyer, Differential responses of amphibians and reptiles in riparian and stream habitats to land use disturbances in western Georgia, USA, *Biol. Conserv.*, 2008, **141**, 2290–2300.
13. B. Collen, F. Whitton, E. E. Dyer, J. E. M. Baillie, N. Cumberlidge, W. R. T. Darwall, C. Pollock, N. I. Richman, A.-M. Soulsby and M. Böhm, Global patterns of freshwater species diversity, threat and endemism, *Glob. Ecol. Biogeogr.*, 2014, **23**, 40–51.
14. L. Fitzgerald, D. Walkup, K. Chyn, E. Buchholtz, N. Angeli and M. Parker, in *Reference Module in Earth Systems and Environmental Sciences*, Elsevier, Amsterdam, 2017.
15. M. E. Ortiz-Santaliestra, J. P. Maia, A. Egea-Serrano and I. Lopes, Validity of fish, birds and mammals as surrogates for amphibians and reptiles in pesticide toxicity assessment, *Ecotoxicology*, 2018, **27**, 819–833.
16. D. W. Sparling, G. Linder, C. A. Bishop and S. Krest, *Ecotoxicology of Amphibians and Reptiles*, CRC Press, Boca Raton, 2nd edn, 2010.
17. E. Willingham, Embryonic exposure to low-dose pesticides: effects on growth rate in the hatchling red-eared slider turtle, *J. Toxicol. Environ. Health, Part A*, 2001, **64**, 257–272.
18. J. W. Fleeger, How do indirect effects of contaminants inform ecotoxicology? A review, *Processes*, 2020, **8**, 1659.
19. N. Takamura, in *Biodiversity Observation Network in Asia-Pacific region: Towards Further Development of Monitoring Activities*, Springer, Berlin, 2012, pp. 133–148.
20. B. López Felices, J. A. Aznar-Sánchez, J. Velasco Muñoz and M. Piquer-Rodríguez, Contribution of irrigation ponds to the sustainability of agriculture: a review of worldwide research, *Sustainability*, 2020, **12**, 5425.
21. P. Williams, M. Whitfield, J. Biggs, S. Bray, G. Fox, P. Nicolet and D. Sear, Comparative biodiversity of rivers, streams, ditches and ponds in an agricultural landscape in Southern England, *Biol. Conserv.*, 2004, **115**, 329–341.
22. M. Akasaka and N. Takamura, Hydrologic connection between ponds positively affects macrophyte α and γ diversity but negatively affects β diversity, *Ecology*, 2012, **93**, 967–973.

23. H. C. Ito, H. Shiraishi, M. Nakagawa and N. Takamura, Combined impact of pesticides and other environmental stressors on animal diversity in irrigation ponds, *PLoS One*, 2020, **15**, e0229052.
24. T. Kadoya, M. Akasaka, T. Aoki and N. Takamura, A proposal of framework to obtain an integrated biodiversity indicator for agricultural ponds incorporating the simultaneous effects of multiple pressures, *Ecol. Indic.*, 2011, **11**, 1396–1402.
25. K. Nakanishi, T. Nishida, M. Kon and H. Sawada, Effects of environmental factors on the species composition of aquatic insects in irrigation ponds, *Entomol. Sci.*, 2014, **17**, 251–261.
26. T. Kizuka, S. Ishida, T. Kadoya, M. Akasaka and N. Takamura, Distribution and abundance of small lentic water bodies for wildlife habitats throughout Japan inferred from geospatial data (in Japanese), *Jpn. J. Conserv. Ecol.*, 2016, **21**, 181–192.
27. E. Amano, Towards urgent disaster prevention measures for irrigation ponds: Passage of the draft law on the management and conservation of irrigation ponds (in Japanese), *Legis. Res.*, 2019, **414**, 65–75.
28. S. Wolinsky, Farm Pond Ecosystems, in *Fish and Wildlife Habitat Management Leaflet*, Wildlife Habitat Management Institute (U.S.), 2005, vol. 29, pp. 1–12.
29. P. Nicolet, J. Biggs, G. Fox, M. J. Hodson, C. Reynolds, M. Whitfield and P. Williams, The wetland plant and macroinvertebrate assemblages of temporary ponds in England and Wales, *Biol. Conserv.*, 2004, **120**, 261–278.
30. E. T. Chester and B. J. Robson, Anthropogenic refuges for freshwater biodiversity: Their ecological characteristics and management, *Biol. Conserv.*, 2013, **166**, 64–75.
31. S. Declerck, T. De Bie, D. Ercken, H. Hampel, S. Schrijvers, J. Van Wichelen, V. Gillard, R. Mandiki, B. Losson, D. Bauwens, S. Keijers, W. Vyverman, B. Goddeeris, L. De Meester, L. Brendonck and K. Martens, Ecological characteristics of small farmland ponds: Associations with land use practices at multiple spatial scales, *Biol. Conserv.*, 2006, **131**, 523–532.
32. K. Kawano, Aquatic insects in the ponds of Hirata area (Izumo city) in Shimane Prefecture, Japan (in Japanese), *Bull. Hoshizaki Green Found.*, 2006, **9**, 13–37.
33. J. H. Kim, H.-Y. Chung, S. Kim and J. G. Kim, The influence of water characteristics on the aquatic insect and plant assemblage in small irrigation ponds in Civilian Control Zone, Korea, *J. Wetlands Res.*, 2016, **18**, 331–341.
34. K. Katoh, S. Sakai and T. Takahashi, Factors maintaining species diversity in satoyama, a traditional agricultural landscape of Japan, *Biol. Conserv.*, 2009, **142**, 1930–1936.
35. M. Taniguchi, S. Ueno, K. Mine and N. Kamezaki, Distribution and density of freshwater turtles in western Japan (in Japanese), *Bull. Herpetol. Soc. Jpn.*, 2015, 144–157.
36. Z. Topalcengiz, L. K. Strawn and M. D. Danyluk, Microbial quality of agricultural water in Central Florida, *PLoS One*, 2017, **12**, e0174889.

37. C. B. Stanford, J. B. Iverson, A. G. J. Rhodin, P. Paul van Dijk, R. A. Mittermeier, G. Kuchling, K. H. Berry, A. Bertolero, K. A. Bjorndal, T. E. G. Blanck, K. A. Buhlmann, R. L. Burke, J. D. Congdon, T. Diagne, T. Edwards, C. C. Eisemberg, J. R. Ennen, G. Forero-Medina, M. Frankel, U. Fritz, N. Gallego-García, A. Georges, J. W. Gibbons, S. Gong, E. V. Goode, H. T. Shi, H. Hoang, M. D. Hofmeyr, B. D. Horne, R. Hudson, J. O. Juvik, R. A. Kiester, P. Koval, M. Le, P. V. Lindeman, J. E. Lovich, L. Luiselli, T. E. M. McCormack, G. A. Meyer, V. P. Páez, K. Platt, S. G. Platt, P. C. H. Pritchard, H. R. Quinn, W. M. Roosenburg, J. A. Seminoff, H. B. Shaffer, R. Spencer, J. U. Van Dyke, R. C. Vogt and A. D. Walde, Turtles and tortoises are in trouble, *Curr. Biol.*, 2020, **30**, R721–R735.
38. E. Takada and T. Ohya, *Primary Colour Reptile and Amphibian Search Book (in Japanese)*, Hokuryukan, Tokyo, 2011.
39. S. Kagayama, D. Ogano, M. Taniguchi, K. Mine, S. Ueno, H. Takahashi, N. Kamezaki and M. Hasegawa, Species distribution modeling provides new insights into different spatial distribution patterns among native and alien freshwater turtles in Japan, *Curr. Herpetol.*, 2020, **39**, 147–159.
40. Y. Yasukawa, Taxonomy and natural history of the genus *Mauremys* and its relatives: Part 2 (in Japanese), *Creeper*, 2007, **40**, 30–67.
41. D. Suzuki, H. Ota, H.-S. Oh and T. Hikida, Origin of Japanese populations of Reeves' Pond Turtle, *Mauremys reevesii* (Reptilia: Geoemydidae), as inferred by a molecular approach, *Chelonian Conserv. Biol.*, 2012, **10**, 237–249.
42. N. Kamezaki, Overview of the problems concerning freshwater turtles in Japan, especially *Trachemys scripta elegans* (in Japanese), *Bull. Herpetol. Soc. Jpn.*, 2015, 123–133.
43. B. Ferronato, T. Marques, I. Guardia, A. B. Longo, C. Piña, J. Bertoluci and L. Verdade, The turtle *Trachemys scripta elegans* (Testudines, Emydidae) as an invasive species in a polluted stream of southeastern Brazil, *Herpetol. Bull.*, 2009, **109**, 29–34.
44. T. Nagasaka, *Reptiles and Amphibians 800 Species Picture Book (in Japanese)*, Pisces, Yokohama, 3rd edn, 2002.
45. Y. Yasukawa, Taxonomy and natural history of the genus *Trachemys* (sliders) 1 (in Japanese), *Creeper*, 2007, **36**, 30–50.
46. C. H. Ernst and J. E. Lovich, *Turtles of the United States and Canada*, Johns Hopkins Univ Pr, Baltimore, 2nd edn, 2009.
47. N. Ramsay, P. Ng, R. O'Riordan and L. Chou, in *Biological Invaders in Inland Waters: Profiles, Distribution, and Threat*, Springer, Berlin, 2007, pp. 161–174.
48. M. Taniguchi, J. Lovich, K. Mine, S. Ueno and N. Kamezaki, Unusual population attributes of invasive red-eared slider turtles (*Trachemys scripta elegans*) in Japan: Do they have a performance advantage?, *Aquat. Invasions*, 2017, **12**, 97–108.
49. W. M. V. Haegen, S. L. Clark, K. M. Perillo, D. P. Anderson and H. L. Allen, Survival and causes of mortality of head-started western

pond turtles on Pierce National Wildlife Refuge, Washington, *J. Wildl. Manage.*, 2009, **73**, 1402–1406.
50. A. Haskell, T. Graham, C. Griffin and J. Hestbeck, Size related survival of headstarted redbelly turtles (*Pseudemys rubriventris*) in Massachusetts, *J. Herpetol.*, 1996, **30**, 524.
51. R. Jones, Reproduction and nesting of the endangered ringed map turtle, *Graptemys oculifera*, in Mississippi, *Chelonian Conserv. Biol.*, 2006, **5**, 195–209.
52. S. Ohba, Field observation of predation on a turtle by a giant water bug, *Entomol. Sci.*, 2011, **14**, 364–365.
53. M. C. Jackson, C. J. G. Loewen, R. D. Vinebrooke and C. T. Chimimba, Net effects of multiple stressors in freshwater ecosystems: a meta-analysis, *Global Change Biol.*, 2016, **22**, 180–189.
54. B. Spears, D. Chapman, L. Carvalho, C. Feld, M. Gessner, J. Piggott, L. Banin, T. Gutiérrez-Cánovas, A. Solheim, J. Richardson, R. Schinegger, P. Segurado, S. Thackeray and S. Birk, Making Waves. Bridging theory and practice towards multiple stressor management in freshwater ecosystems, *Water Res.*, 2021, **196**, 116981.
55. N. Takamura, in *Biodiversity monitoring: collaboration to build capacity for ecosystem management (in Japanese)*, ed. I. Washitani and S. Kito, University of Tokyo Press, Tokyo, 2007, pp. 49–69.
56. H. Tsunoda, Ecological impacts of pond losses and abandonments on regional aquatic biodiversity: What will happen in the depopulating Japan? (in Japanese), *Wildl. Human Soc.*, 2017, **5**, 5–15.
57. K. Nakanishi, K.-I. Takakura, R. Kanai, K. Tawa, D. Murakami and H. Sawada, Impacts of environmental factors in rice paddy fields on abundance of the mud snail (*Cipangopaludina chinensis laeta*), *J. Molluscan Stud.*, 2014, **80**, 460–463.
58. M. Lambert, S. Nielsen, A. Wright, R. Thomson and H. Shaffer, Habitat features determine the basking distribution of introduced red-eared sliders and native western pond turtles, *Chelonian Conserv. Biol.*, 2013, **12**, 192–199.
59. M. Akasaka, N. Takamura, H. Mitsuhashi and Y. Kadono, Effects of land use on aquatic macrophyte diversity and water quality of ponds, *Freshwater Biol.*, 2010, **55**, 909–922.
60. S. Thomaz and E. Cunha, The role of macrophytes in habitat structuring in aquatic ecosystems: Methods of measurement, causes and consequences on animal assemblages' composition and biodiversity, *Acta Limnol. Bras.*, 2010, **22**, 218–236.
61. F. Sánchez-Bayo, Indirect Effect of Pesticides on Insects and Other Arthropods, *Toxics*, 2021, **9**, 177.
62. S. Arima, A. Suzuki, S.-H. Zheng, A. Tanaka, M. Okuzono and I. Nishimura, Survey and control of feeding damage by red-eared slider turtle (in Japanese), *Coastal Bioenviron. (Coastal Bioenviron. Center, Saga Univ.)*, 2008, **12**, 53–57.
63. M. F. Chislock, E. Doster, R. Zitomer and A. E. Wilson, Eutrophication: causes, consequences, and controls in aquatic ecosystems, *Nat. Educ. Knowl.*, 2013, **4**, 10.

64. S. M. Harmon and F. E. Wiley, Effects of pollution on freshwater organisms, *Water Environ. Res.*, 2010, **82**, 1945–2000.
65. M. S. El Hassani, E. M. El Hassan, T. Slimani and X. Bonnet, Morphological and physiological assessments reveal that freshwater turtle (Mauremys leprosa) can flourish under extremely degraded-polluted conditions, *Chemosphere*, 2019, **220**, 432–441.
66. N. Takamura, *Aquatic organisms and their environments in irrigation ponds (in Japanese)*, National Institute of Environmental Studies (NIES), Tsukuba, 2011.
67. M. A. Beketov, B. J. Kefford, R. B. Schäfer and M. Liess, Pesticides reduce regional biodiversity of stream invertebrates, *Proc. Natl. Acad. Sci. U. S. A.*, 2013, **110**, 11039–11043.
68. S. Stehle and R. Schulz, Agricultural insecticides threaten surface waters at the global scale, *Proc. Natl. Acad. Sci. U. S. A.*, 2015, **112**, 5750–5755.
69. W. A. Hopkins, C. L. Rowe and J. D. Congdon, Elevated trace element concentrations and standard metabolic rate in banded water snakes (*Nerodia fasciata*) exposed to coal combustion wastes, *Environ. Toxicol. Chem.*, 1999, **18**, 1258–1263.
70. L. Héritier, A.-L. Meistertzheim and O. Verneau, Oxidative stress biomarkers in the Mediterranean pond turtle (*Mauremys leprosa*) reveal contrasted aquatic environments in Southern France, *Chemosphere*, 2017, **183**, 332–338.
71. T. Atobe, Y. Osada, H. Takeda, M. Kuroe and T. Miyashita, Habitat connectivity and resident shared predators determine the impact of invasive bullfrogs on native frogs in farm ponds, *Proc. R. Soc. London, Ser. B*, 2014, **281**, 20132621.
72. T. Kizuka, M. Akasaka, T. Kadoya and N. Takamura, Visibility from roads predict the distribution of invasive fishes in agricultural ponds, *PLoS One*, 2014, **9**, e99709.
73. N. Usio, M. Imada, M. Akasaka and N. Takamura, Effects of pond management on the distributions of aquatic invaders in Japanese farm ponds (in Japanese), *Jpn. J. Limnol.*, 2009, **70**, 261–266.
74. D. Tilman, Diversification, biotic interchange, and the universal trade-off hypothesis, *Am. Nat.*, 2011, **178**, 355–371.
75. C. Bellard, P. Cassey and T. M. Blackburn, Alien species as a driver of recent extinctions, *Biol. Lett.*, 2016, **12**, 20150623.
76. S. M. Weir and C. J. Salice, High tolerance to abiotic stressors and invasion success of the slow growing freshwater snail, *Melanoides tuberculatus*, *Biol. Invasions*, 2012, **14**, 385–394.
77. A. Y. Karatayev, L. E. Burlakova, D. K. Padilla, S. E. Mastitsky and S. Olenin, Invaders are not a random selection of species, *Biol. Invasions*, 2009, **9**, 2009–2019.
78. N. Cedergreen and J. J. Rasmussen, in *Pesticide Dose: Effects on the Environment and Target and Non-Target Organisms*, American Chemical Society, Washington D.C., 2017, vol. 1249, pp. 167–187.

79. S. Lorenz, J. J. Rasmussen, A. Süß, T. Kalettka, B. Golla, P. Horney, M. Stähler, B. Hommel and R. B. Schäfer, Specifics and challenges of assessing exposure and effects of pesticides in small water bodies, *Hydrobiologia*, 2017, **793**, 213–224.
80. A. Saha, R. K. Ghosh, P.-K. Jesna and P. P. Choudhury, in *Sustainable Agriculture Reviews 48: Pesticide Occurrence, Analysis and Remediation Vol. 2 Analysis*, ed. Inamuddin, M. I. Ahamed and E. Lichtfouse, Springer, Berlin, 2021, pp. 185–231.
81. A. A. H. Siddig, A. M. Ellison, A. Ochs, C. Villar-Leeman and M. K. Lau, How do ecologists select and use indicator species to monitor ecological change? Insights from 14 years of publication in Ecological Indicators, *Ecol. Indic.*, 2016, **60**, 223–230.
82. D. R. Lenat, Water quality assessment of streams using a qualitative collection method for benthic macroinvertebrates, *J. North Am. Benthol. Soc.*, 1988, **7**, 222–233.
83. P. O. Raburu, F. O. Masese and K. S. Tonderski, Use of macroinvertebrate assemblages for assessing performance of stabilization ponds treating effluents from sugarcane and molasses processing, *Environ. Monit. Assess.*, 2017, **189**, 79.
84. A. A. Ganatra, F. J. Kandie, U. Fillinger, F. McOdimba, B. Torto, W. Brack, M. Liess, H. Hollert and J. M. Becker, Calibration of the SPEARpesticides bioindicator for cost-effective pesticide monitoring in East African streams, *Environ. Sci. Eur.*, 2021, **33**, 58.
85. S. Yu, R. S. Halbrook, D. W. Sparling and R. Colombo, Metal accumulation and evaluation of effects in a freshwater turtle, *Ecotoxicology*, 2011, **20**, 1801–1812.
86. R. M. Goodman, E. D. Carter and D. L. Miller, Influence of herbicide exposure and ranavirus infection on growth and survival of juvenile red-eared slider turtles (*Trachemys scripta elegans*), *Viruses*, 2021, **13**, 1440.
87. S. Soltanian, Effect of atrazine on immunocompetence of red-eared slider turtle (*Trachemys scripta*), *J. Immunotoxicol.*, 2016, **13**, 804–809.
88. R. Gurley, *Keeping and Breeding Freshwater Turtles*, Living Art Pub, Ada, Okla, 2002.
89. W. Werft and A. Benner, Glmperm: a permutation of regressor residuals test for inference in generalized linear models, *R. J.*, 2010, **2**, 39.

CHAPTER 8

Current Progress in Developing Standardized Methods for Reptilian Toxicity Testing to Inform Ecological Risk Assessment

SCOTT M. WEIR,*[a] MONICA R. YOUSSIF,[a] TAYLOR ANDERSON[b] AND CHRISTOPHER J. SALICE[b]

[a] Queens University of Charlotte, 1900 Selwyn Ave, Charlotte, NC 28274, USA;
[b] Towson University 8000 York Rd, Towson, MD 21252, USA
*Email: weir.scott.m@gmail.com

8.1 Background on Reptile Toxicity Testing

8.1.1 Lack of Reptile Toxicity Data in the Literature

While it has now been over a decade since it was assessed in 2010, reptiles have remained the least studied vertebrate taxon in ecotoxicology.[1] This lack of experimental focus translates to a notable lack of data regarding the effects of chemicals on reptiles, even for endpoints as standard as LD50s.[2] A great deal of the research on reptiles is the reporting of contaminant body burdens without connection to potential effects.[3] Body residue data also tend to be more focused on turtles, in part because they are long-lived and generally stay in the same area, so they are good bioindicators of local

Issues in Toxicology No. 45
Bird and Reptile Species in Environmental Risk Assessment Strategies
Edited by Guillermo Eli Liwszyc and Marcelo L. Larramendy

Published by the Royal Society of Chemistry, www.rsc.org

contamination.[4] Many of the toxicity data available tend to be more focused on lizards, with some occasional data available for snakes and turtles.[2] The LD50 data that do exist are focused on particular contaminant groups such as explosives and organophosphate insecticides. The available data, therefore, lack the breadth of potential toxicants that could be negatively impacting reptilian species in areas of high contamination. A lack of reptile toxicity testing hinders the process of ecological risk assessment for any sites that have reptilian receptors of interest. In turn, the significant data gap in reptile ecotoxicology stems from a lack of standardized tests and laboratory model organisms to use for toxicity testing.

8.1.2 Lack of Standardized Methods

A primary reason for the lack of standardized reptile toxicity testing is the lack of regulatory requirements for testing of reptiles. Reptile toxicity testing is not required for (1) the registration of pesticides under the Federal Insecticide, Fungicide, and Rodenticide Act (FIFRA) in the United States of America (USA); (2) control of manufacture or distribution of chemicals under the Toxic Substance Control Act; or (3) the pesticide regulatory system of the European Food Safety Authority (EFSA), though there is movement to begin considering amphibians and reptiles in Europe.[5] The general consideration is that risk assessments performed for birds and mammals are likely protective of terrestrial amphibians and reptiles. At least some of this belief is likely based on the high metabolic demands of birds and mammals necessitated by endothermic physiology. Ecological risk assessors tend to focus on the dietary route of exposure (except when other routes are considered to be important), and therefore, the high metabolic demand of birds and mammals will likely lead to increased oral exposure rates.[2]

8.2 Progress in Standardized Methods for Reptile Ecotoxicology/Ecological Risk Assessment

8.2.1 Early Efforts in Reptile Toxicology

Early efforts to understand the effects of chemicals on reptiles were often either haphazard, or of little environmental or management relevance and were instead focused on mechanisms of toxicity. Munro reported the effect of dichlorodiphenyltrichloroethane (DDT) powder on cottonmouth snakes after an attempt at mite eradication.[6] Herald followed with a similar reporting of preliminary studies on the use of DDT oil and its effects on reptiles and amphibians.[7] These studies utilized formulations of DDT, which made them relevant but of poor experimental design. For example, Munro used only two individuals and the dose was not quantified.[6] In the case of Herald, more individuals ($n = 21$) were exposed but they were exposed collectively (*i.e.*, pseudoreplication) and were a mix of species and the dose was once again not quantified and is unknown.[7] On the other hand, Coulson and

Hernandez used intraperitoneal (IP) injection to investigate the sublethal effects of cycloheximide on the caiman (*Caiman latirostris*).[8] This work had experimental rigor but lacked environmental relevance because IP injection clearly does not occur in wild animals inhabiting contaminated sites. One of the first papers to investigate the toxicity of reptiles in a controlled manner was that of Hall and Clark Jr. using green anoles (*Anolis carolinensis*); the authors[9] dissolved their organophosphate insecticides in corn oil and delivered doses *via* stomach intubation but did not provide any additional clarity in the dosing methods.[9] Without further details, it appears that these methods are generally similar to the standard methods for dosing birds for pesticide registration under the FIFRA.[10]

Brooks *et al.* were one of the first to investigate the dermal toxicity of contaminants to a reptile using brown tree snakes (*Boiga irregularis*) as their focal species.[11] In addition to dermal exposure, the authors also used oral exposure routes for their toxicity testing. Initial oral and dermal dosing used ethanol as a carrier. For oral dosing, contaminants were dissolved in ethanol and applied using a syringe (with a ball-tipped feeding needle) that was inserted into the entrance of the esophagus.[11] After some lethargy was noted due to the use of ethanol, propylene glycol was then tested as a solvent for some of the contaminants tested, but this was only done *via* oral exposure. Similarly, the authors tested the use of pyrethrin and pyrethroid fogging products on brown tree snakes. Snakes were placed into either cloth bags or wire cages and the fogging products were used as directed. Brooks *et al.* found significant toxicity only when dermal exposures could occur to snakes in the wire cages, but no toxicity occurred due to inhalation exposures only (*i.e.*, snakes in cloth bags).[12] Alexander *et al.* also investigated dermal exposure for two lizard species to deltamethrin (a pyrethroid) in a controlled setting as well as a more natural field environment.[13] In the controlled setting, deltamethrin was applied using a backpack sprayer to simulate a realistic spraying scenario at two application rates. The lizards were exposed either directly or indirectly. Direct exposure involved cooling the lizards until immobile and then spraying the experimental area. Indirect exposure involved spraying the soil and then releasing the lizards into the enclosure. Pyrethroids are more toxic at lower metabolic rates and the cooling to induce mobility in the direct exposures may have contributed to mortality. However, even the indirect exposure resulted in a higher mortality rate two months after the spraying event.[13] Finally, Talent performed slightly different dermal exposures by dipping the body (except for the head) of anole lizards into a formulated product for 2 s.[14]

Attempts at standardization of reptile toxicity methods began in the late 2000s with reports of toxic effects of explosives on lizards.[15–18] These papers were the first to report a relatively quick dosing method that did not require intubation and standardized the dosing regime (see Section 8.2.3.4.1 for a description of the "pseudo-gavage" method). Importantly, these studies tested chemicals that were soluble in water and the concerns about the use of solvents *vs.* gelatin capsules were not a necessary consideration. Further, these studies and the later studies by Weir *et al.* were possible due to the development and consistent availability of a captive reared test species of the western fence lizard (*Sceloporus occidentalis*).[19–21,22] A colony of western fence lizards was maintained for research purposes at Oklahoma State University. Unfortunately, the colony is no longer active.

Several papers have focused on experimental protocols involving reptile eggs. A common mode of exposure in controlled laboratory settings with eggs involves topical application of the contaminant directly to the egg. For example, Sparling *et al.* exposed red-eared slider eggs to a pesticide formulation containing glyphosate by submerging eggs into dilutions of the formulation.[23] Schaumburg *et al.* provided a topical application of Roundup® to tegu (*Salvator merinae*) eggs to investigate genotoxicity.[24] More commonly, eggs are incubated in contaminated substrates such as soil or vermiculite and effects on hatching, growth, *etc.*, are observed. Incubation in contaminated substrates has been tested with fertilizers, As, Cd, and mixtures of pesticides and fertilizers.[25–28] Although there are few published studies in which eggs were exposed to toxicants in substrates, it should be noted that this can be a highly relevant exposure route for reptiles inhabiting areas with contaminants in soil.

Weir *et al.* attempted to improve the standardization of dosing techniques for lizards by investigating both dermal and oral dosing.[20] Under both exposure routes, toxicity was tested using a system with a solvent carrier and without a solvent carrier. For oral exposure there was the use of a gelatin capsule (non-solvent) *vs.* the use of a carrier solvent (acetone). Similarly, for dermal exposure, there was the use of pure powder applied to the dorsal skin surface (non-solvent) *vs.* dosing with a solvent (acetone) on the ventral surface, respectively. While some compounds did not elicit toxicity, in some cases the use of a solvent increased toxicity (*e.g.*, lambda-cyhalothrin *via* oral exposure) but there were also instances in which the solvent did not increase toxicity (*e.g.*, endosulfan *via* oral exposure).[20] Further, Weir *et al.* investigated multiple methodological effects on dermal toxicity.[20] Methodological changes that were investigated were: (a) the use of propylene glycol as a more inert solvent than acetone, (b) dosing on the dorsal surface with acetone rather than the ventral surface, and (c) providing a heat lamp following dosing. The use of propylene glycol did not affect LD50 estimates, but dosing on the dorsal surface and providing a heat lamp both decreased the toxicity (*i.e.*, increased the LD50).[20] The oral dosing methods using gelatin capsules were later applied to a preliminary risk assessment of pesticides and their potential toxicity to lizards in New Zealand.[21]

8.2.2 Oral Dosing Methods Using Gelatin Capsules to Accommodate Small Reptiles

As New Zealand continues its efforts to eradicate all invasive species by 2050, pesticides remain an important tool for contributing to this goal. As reptiles remain understudied, the pesticides used to control invasive species need to be tested to ensure that native reptiles will not be negatively impacted in

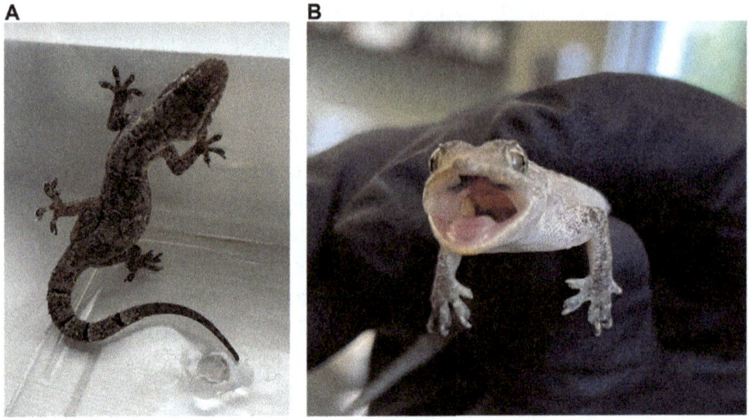

Figure 8.1 Brook's house gecko (*Hemidactylus brookii*).

New Zealand.[1] While there has been progress in dosing methods in reptiles, a standardized, rigorously tested, and approved method has not yet been fully developed.[18,20] Since research surrounding reptiles is lacking, steps to standardize these methods will allow for progress in the field of ecotoxicology, specifically regarding reptiles.

Previous studies have reported that some of the herbicides and rodenticides employed by the New Zealand government to control invasive species are not harmful to reptiles at environmentally relevant levels.[21] However, Weir *et al.* used the western fence lizard, which is not closely related to the geckos and skinks that are native to New Zealand.[21] Fence lizards are a *Sceloporus* species in the iguanid branch of the squamate phylogenetic tree; they are on a completely separate branch from the groups containing skinks and geckos.[29] The Brook's house gecko (*Hemidactylus brookii*, see Figure 8.1A and B) represents a commercially available surrogate species for the native geckos found in New Zealand. We are using Brook's house geckos in toxicity testing to determine if the results for western fence lizards will be confirmed in a species closely related to those in New Zealand.

However, the Brook's house gecko is significantly smaller in size compared to previously studied western fence lizards with an average weight between 3 and 4 g. Standard #9 gelatin capsules (Torpac Inc., New York, USA) are the ideal size to dose the common house gecko (see Figure 8.2). The #9 capsules are small with a diameter of 2.70 mm and a length, when locked, of 8.4 mm. However, when orally dosing such small animals, issues arise when weighing out the amount of the pure toxicant on basic commercial microbalances to put into the capsule. These microbalances may not be sensitive enough to accurately weigh out small enough masses needed to create the desired dose.

Our research using Brook's house geckos uses the up-and-down method.[30] When following the up-and-down method, the standard doses include 5.5, 17.5, 55, 175, 550, and 1750 $\mu g\, g^{-1}$.[30] As an example of the difficulty in

Figure 8.2 The Torpac® size #5 and #9 gel capsules (Torpac Inc., New York, USA) in relation to the size of a Brook's house gecko's head and mouth. The larger #5 capsules will not work for this smaller lizard and the #9 capsules are needed.

dosing small lizards, when weighing out one of the lower standard doses such as 17.5 $\mu g\,g^{-1}$ for a small reptile weighing 3.5 g, the mass of contaminant needed is 61.25 μg; a value that is too small for the sensitivity of a basic microbalance, which can measure accurately to only hundredths of a milligram. Due to this issue, it may be beneficial to dissolve a larger amount of the toxicant in a solvent to deliver the appropriate mass of the toxicant into the gel capsules and then evaporate the solvent. By creating a concentrated solution in solvent, we were able to weigh out a mass appropriate for our balance. For example, we created a 350 $mg\,mL^{-1}$ solution of a pesticide in acetone. We then pipetted 24 μL of the stock solution into a capsule to achieve a small dose of 8 mg and then allow the solvent to evaporate. However, while following this method, capillarity becomes an issue.

The small size and narrow diameter of Torpac® #9 capsules make it difficult to avoid capillarity when using a pipette to fill. Capillarity is the tendency of a liquid to flow along the side of a surface with or in opposition to gravity as a result of the liquid's surface tension. Capillarity within the capsule results in the solution escaping the body of the capsule and crystals of the toxicant appearing on its exterior (see Figure 8.3). When capillary action is combined with a volatile solvent (*e.g.*, acetone), solute crystallization may occur on the brim and exterior of the capsule's body due to the solvent evaporating and the pesticide remaining.

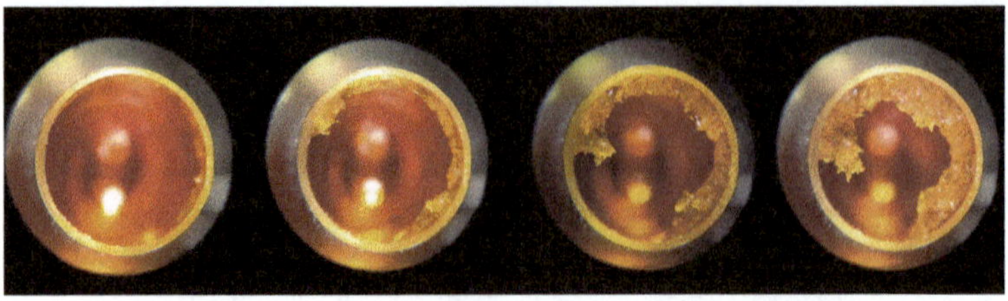

Figure 8.3 Examples of crystal formation when pipetting into a Torpac® #9 capsule with acetone as the solvent. The images show a time series that progresses from left to right as follows: immediately after dosing, 1 min after dosing, 2 min after dosing and 3 min after dosing. In some cases, crystals form on the edge/rim of the capsule and are easily knocked off, making it difficult to ensure precise doses.

Crystallization of the solution is problematic for three reasons. First, crystallized chemicals can form on the brim and exterior of the gel capsule. This makes it difficult to place the crystals into the capsule or to ensure that the exact amount is in the capsule without loss outside of the capsule. Second, crystals can easily break off and fall, which affects the mass of solute that ends up in the capsule and therefore affects accurate dosing. Third, it is uncertain if the final mass weighed is purely the remaining compound or if it includes remnants of unevaporated solvent.

We have created a system to attempt to reduce problems associated with capillary action using a hole in the top of the Torpac® #9 capsule cap. Using a 31G needle syringe, we punctured a hole in the top center of the cap then removed the needle. While still gently but firmly securing the capsule in its base stand with forceps, we then widened the hole with a dissection probe by inserting it into the existing hole. To deliver the pesticide into the capsule, we used the same syringe that was used to create the hole. We first removed the plunger from the syringe barrel and set it aside. Then, we placed the needle back into the then-widened hole of the cap and pipetted the appropriate volume of solvent/pesticide solution into the barrel of the syringe. We replaced the plunger and expelled the solution slowly from the syringe into the receiving capsule. It is important to maintain slow and steady pressure while plunging the contents of the needle into the capsule to prevent the solution from escaping through the opening at the top of the capsule. Once filled, we kept the capsule upright, ideally in its base, to prevent spilling and to promote evaporation of the solvent.

The rate of evaporation was recorded for two solvents: acetone and acetonitrile (see Table 8.1). We used the herbicide pendimethalin dissolved in each solvent to test the evaporation rates and the remaining pesticide material. The goal was to get 8.6 mg of pendimethalin in the capsules. For both solvents, we created a solution that was 350 mg mL^{-1} and we pipetted 24.6 μL, which should leave behind 8.6 mg of pesticide if 100% of the solvent evaporates. The final weights of the capsules that contained acetone as the solvent were relatively consistent (8.5 mg, 8.9 mg, 9 mg), whereas the final weights of the capsules

Table 8.1 Weights (in mg) of concentrated pesticide solutions in two different solvents pipetted into a Torpac® #9 capsule. The initial weight represents the solvent plus the dissolved pesticide and the changes in weight are due to the evaporation of the solvent over time. If all of the solvent evaporated, the final weight of the pesticide should be 8.6 mg. The data were more consistent with the acetone solvent, and full evaporation occurred more quickly.

	Time (h)					
Solvent	Initial	1	2	21	24	45
Acetone	20.4	13	8.8	8.5	8.9	8.5
	20.1	10.6	9.6	9.1	8.9	9
	19.2	10.7	9	8.7	8.5	8.9
Acetonitrile	20.2	16.9	13.1	6.8	6.1	7
	20.5	19.4	17.6	7.3	6.8	7.3
	20.3	17.5	14	6.4	5.7	6.3

containing acetonitrile as the solvent were not (6.3 mg, 7 mg, 7.3 mg). In addition to the higher variability among capsules, the acetonitrile treatment resulted in a lower final weight of pesticide residues, and it is unknown why this would be the case. This does not appear to be related to increased volatility in acetonitrile, as the boiling point of acetonitrile is higher (82 °C) than that of acetone (56 °C). This disparity in the data suggests that acetone is a better solvent than acetonitrile for our purposes.

8.2.3 Standardization of Methods and Developing Breeding Assay Using *Anolis* Species in Reptile Ecotoxicology

8.2.3.1 Background/Introduction

As mentioned earlier, an ongoing issue in reptile ecotoxicology has been the availability and use of species suitable for testing. In the absence of readily available organisms from a captive-bred colony, field-collected organisms have often been used in the lab. Indeed, field-collected organisms have been used in many laboratory studies on reptile ecology and physiology. As mentioned, there are a few examples of lizards used in ecotoxicology. Although relatively common, lizards from *Anolis* spp. have not been commonly used in ecotoxicology. Three notable exceptions include those studied by Hall and Clark Jr.[9] and Talent[14] as mentioned earlier, as well as McMurry *et al.*[31] We have focused on developing and improving standardization and determining the toxicity levels of per- and polyfluoroalkyl substances (PFASs) using *Anolis sagrei* and *A. carolinensis* species. These methods are based on those developed for the western fence lizard with some modifications. Importantly, we have also been successful in developing a breeding colony and have included some reproductive endpoints in several experiments. Using what is now a standard method in reptile ecotoxicology techniques termed the "pseudo-gavage method", we have been successful in generating robust, repeatable toxicity data for subadult and mature male and female lizards.

We have also successfully included some reproductive endpoints although more work is required to better refine reproductive toxicity testing in anoles.

8.2.3.2 *Per- and Polyfluoroalkyl Substances: Background*

PFASs represent a large class of manufactured, persistent chemicals that are and have been used commonly in both industrial and commercial applications. PFASs have a multitude of uses, with properties including water, oil, and fire resistance, making them valuable for a wide range of applications.[32,33] Importantly, PFASs have been used in aqueous film-forming foams on military installations for decades.[34] Due to their resistance to degradation and ability to bioaccumulate, local soil, ground, and surface waters have become contaminated with a variety of PFASs, including predominantly perfluorooctane sulfonate (PFOS), perfluorohexane sulfonate (PFHxS), and perfluorooctanoic acid (PFOA) as well as a variety of other PFASs.[33,35–37]

A multitude of studies have demonstrated that wildlife has been exposed to PFASs with measurable concentrations and potential adverse effects in birds, mammals, and even reptiles.[38–42] Furthermore, laboratory toxicity studies in which study organisms were exposed to different PFASs have been conducted on aquatic, avian, and terrestrial vertebrates for common PFASs such as PFOS.[36,43,44] Collectively, toxicity data point to concern for some PFASs, especially PFOS. As an example, Larson *et al.* conducted an ecological risk assessment with avian receptors and concluded that some avian species are likely at risk due to the consumption of PFASs from contaminated environments.[38] When we consider reptiles, however, there are few data that can be used to assess risk to species exposed to PFASs. One notable study was published in which the toxicity of PFOA was investigated in the Mongolian racerunner (*Eremias argus*).[45] Interestingly, their results showed that PFOA exposure induced a trade-off between reproduction and self-maintenance. Another relevant study, although it was conducted on amphibians, is that of Flynn *et al.* in which salamanders were exposed to four relevant PFASs *via* contaminated crickets.[46] Their study design yielded compelling results; the authors found that PFAS exposure impacted growth and reduced body condition, and was negatively correlated with PFOS bioaccumulation, specifically. Generating robust, defensible toxicity data for reptiles is necessary to conduct ecological risk assessments that can ideally lead to protection of species from this taxonomic group.

8.2.3.3 Anolis *Species: Background*

Lizards of *Anolis* species are fairly well-studied reptilian species and have been used commonly in ecology and evolutionary studies.[47,48] With over 400 known species, they are quite diverse and provide an excellent example of the concept of adaptive radiation.[49] Anoles occupy a wide range of habitats from the Southern USA to the Caribbean tropics, through Central America and even Asia; with a variety of niches including tree and bush dwellers to ground, rock, and stream dwellers.[48] Due to their abundance in nature and ability to adapt to their environment, *Anolis* spp. are easily obtained and they tolerate lab conditions well.[50,51] In North America, the green anole, *A. carolinensis*, is the only native species. Several introductions of other *Anolis* species, however, have occurred in North America, and several are considered to be invasive. In particular, the brown anole, *A. sagrei*, has become well established in the Southeastern USA and poses a threat to other anole species, such as the green anole. Not only do brown anoles displace green anoles from their native habitats, but they also compete for territory and food, and may be a transmitter for parasites.[52] Given that the brown anole is an invasive species in the USA, and that the species has been used in a number of studies in ecology and evolution, a key research goal was to develop the brown anole as a research model in ecotoxicology.

8.2.3.4 Methods

8.2.3.4.1 Oral Dosing Experiments. Oral dosing experiments are carried out *via* the now relatively standardized dosing technique called pseudo-gavage.[18,53] This method consists of orally exposing *Anolis* spp. using a micropipette to directly administer a fluid solvent amended with contaminant into their mouths by gently pulling down their dewlap (see Figure 8.4). This method is akin to actual gavage which is used in rodent and avian toxicity models. Lizards, however, cannot tolerate actual gavage in which the contaminant slurry would be administered directly into their stomachs. For successful pseudo-gavage, one person is needed to handle the specimen and gently pull down its dewlap, while a second person doses the lizard *via* a micropipette. A similar handling method can be used that appears to be even less invasive, in which the handler gently taps the snout of the animal, and then positions their finger in between the lizard's mouth in order to hold it open (see Figure 8.5). The micropipette should be positioned at the very corner of the anole's mouth, where it can slide easily in and out without

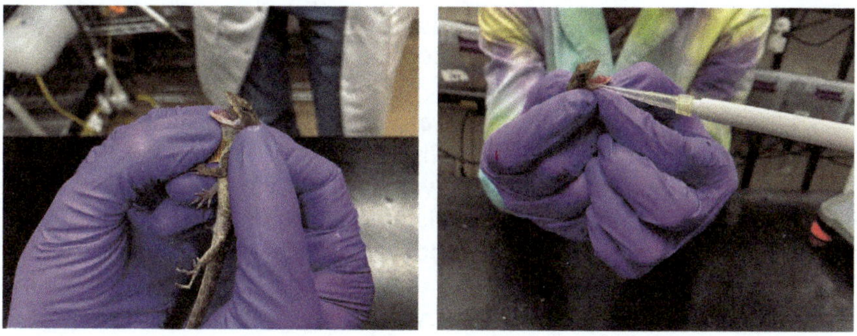

Figure 8.4 The pseudo-gavage method in action with *Anolis sagrei* specimens. The handler gently pulls down the dewlap of the animal to open its mouth. Once successfully opened, the doser gently slides the micropipette in the very corner of the animal's mouth, avoiding any teeth.

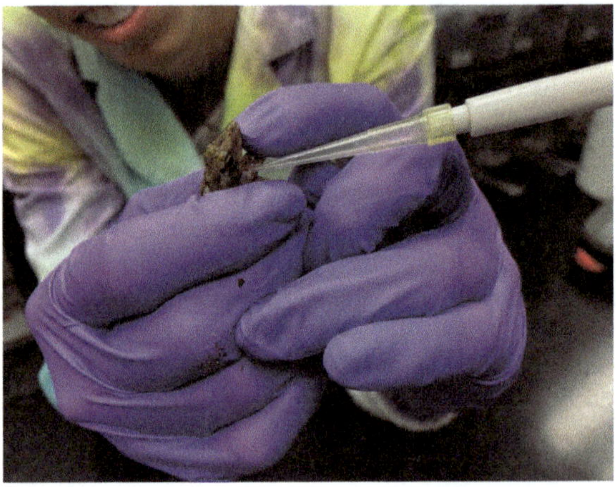

Figure 8.5 The handler gently taps the snout of the animal, initiating its mouth to open. Once successfully opened, the handler positions their finger in between the top and bottom of the animal's mouth in order to hold it open.

interference from the animal's teeth. To be as accurate as possible, the doser will also determine the amount of contaminated solution that was successfully dosed, based on a simple, qualitative percentage scale of 0–100% success rate. Additionally, lizards are dosed only three times per week to minimize handling and stress to animals.

As of this effort, our laboratory has successfully conducted more than ten studies in which male and/or female *A. sagrei* species were dosed with PFAS-containing solutions for thirty-five days or longer. One study was also successfully completed on males and females from both *A. sagrei* and *A. carolinensis*. Relevant exposure durations for PFAS are commonly considered chronic because the chemicals are very persistent and generally at relatively low concentrations. Hence, our laboratory studies have focused on relevant high-end exposure concentrations for relatively long exposure durations. For several definitive chronic studies, exposures lasted 90 days. In all cases, we have had excellent survival and growth of controls and low dose treatments, suggesting that the methods are successful.

8.2.3.4.2 Developing a Reproductive Assay with *Anolis*. Common endpoints for toxicity studies involving lizards include survival and growth as apical endpoints and a suite of behavioral (*e.g.*, sprint speed, prey capture) and sub-organismal endpoints. To date, we have been aware of only one study in which reproductive endpoints were included in a toxicity study.[31] Due to their easy maintenance, with the correct environmental conditions, *Anolis* spp. are ideal for in-laboratory breeding colonies. We successfully developed and have maintained a breeding colony for *A. sagrei*, and have begun to expand it to include *A. carolinensis* as well.

The typical breeding season for anoles in North America occurs from April to August, with males establishing territory that females may enter to breed.

Mating occurs during the entire breeding season, with females laying one to two eggs every seven to 14 days. Eggs are laid in substrates where they hatch on their own with no additional parental care. Hatchlings hatch 4–6 weeks later. In-lab housing of *Anolis* species requires mimicking that of their wild conditions and breeding season. The room temperature should be within 27–29 °C and heat tape should be used under one end of each tank to provide a gradient for thermoregulation. Humidity should be within 55–75%, which can be obtained by supplementing ambient humidity with room humidifiers. To mimic diurnal cycles, light sources are on a 14 : 10 light : dark cycle, with supplemental UVB light sources on from 8 am to 11 am during what would be peak basking times. Individual anoles are housed in 5–6 gallon enclosures containing soil, plants/vines, a PVC hide as well as a mesh "hammock" and two dowel sticks to simulate tree branches (see Figure 8.6). Hides are placed at the back of the enclosures near the heat tape, while the hammocks are placed at the front to allow the lizard's choice of comfortable location based on the temperature gradient. Enclosures are wrapped with paper covering one side to decrease male–male aggression between tanks. Females are given a 1 L plastic "egg cup", which contains vermiculite $\frac{2}{3}$ of the way filled and a hole in the lid to allow females privacy and ease to lay eggs.

Males and females should be acclimated to the room and enclosure for two weeks prior to mating. Once acclimated, females and males may be paired based on their weights, with the highest weight males being paired with the highest weight females. Females that are under 2 g are typically not paired, out of possible aggression from males that may result in female injury or mortality. Females will be placed in males' tanks, also to lessen the possibility of male aggression, for at least 72 h and up to one week. Females should not remain with males longer than one week, and should be carefully evaluated daily to confirm the safety of females. Ensure that both the female and male are receiving two appropriately sized crickets every Monday,

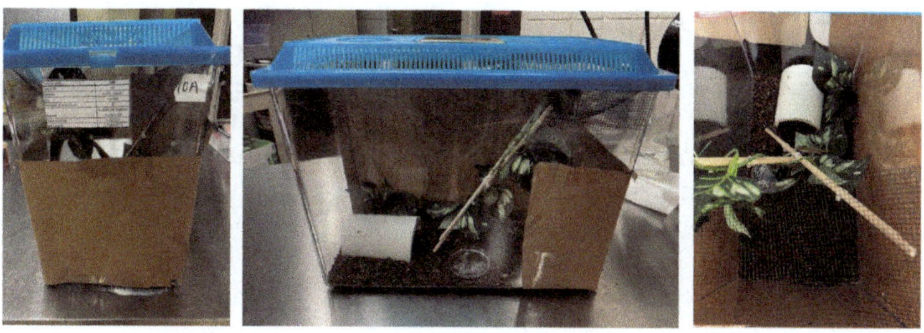

Figure 8.6 Enclosures for in-lab breeding colony of *Anolis* lizards. Each enclosure is labeled with a tank number and ID to differentiate the animals, as well as relevant information (including species, sex, date acquired, and origin). Each enclosure consists of a base layer of soil, a 60 mm$\times$15 mm petri dish for water, a PVC hide, vines/plants, a mesh hammock, and towel dowel sticks to simulate tree branches. Females will also receive a plastic egg cup as described in the text.

Wednesday, and Friday. Crickets should be 3–4 weeks of age, or about $\frac{1}{4}$ to $\frac{1}{2}$ inch in size and should be no larger than the size of the animal's head that is being fed. 60 mm×15 mm petri dishes should be provided for water consumption with daily water changes. Petri dishes should be cleaned or replaced every week, or whenever considerably soiled.

8.2.3.4.3 Reproductive Endpoints. As a way to further standardize the use of *Anolis* spp. in toxicity testing, we expanded our experiments to include reproductive endpoints. This allows us to gain a better understanding of the potential effects toxicants may have on relevant apical endpoints of the *Anolis* species. We were able to incorporate the use of the pseudo-gavage method by dosing gravid female anoles and determining the effects it has on their eggs. We also expanded our use of the breeding colony by exposing eggs in contaminated substrates to different PFAS chemicals. These two methods allowed us to view transmission between females to eggs, as well as eggs laid by females that have not been exposed.

8.2.3.4.3.1 Assessing Impacts on Eggs Directly Exposed to Contaminants. Currently, the breeding colony allows us a unique opportunity to assess the impacts on *Anolis* eggs directly exposed to PFAS contaminants. Non-exposed, mature *A. sagrei* females and males are paired and mated, and once separated, the females are able to lay their eggs in the "egg cups" as mentioned earlier. These eggs are then collected and distributed into small, round glass Pyrex® cups with holes poked through the lids to allow oxygen flow. The glass cups contain either control or experimentally treated substrate. There are one of three possible substrates, vermiculite, sand, or soil. Since vermiculite is quite porous and holds onto PFAS chemicals well, we have focused our future studies on sand and soil exposure, as these are the most environmentally relevant exposure routes.

8.2.3.4.3.2 Assessing Impacts on Egg Production. Reproductively mature females are mated with mature males, and then exposed to PFASs *via* pseudo-gavage. This allows us to assess impacts *via* egg production with endpoints including the number of viable eggs produced by a single female, the size and apparent health of the eggs, and the eventual hatching success of the eggs. Multiple studies have already been conducted assessing the number of viable eggs and the apparent health of the eggs. Preliminary studies have begun in our breeding colony involving hatching success, but more studies are needed to concisely determine PFAS impacts.

8.2.3.4.3.3 Assessing Impacts on Breeding and Egg Production. We have begun to develop a more representative life-cycle scenario of PFAS exposure for *Anolis* species, looking at both breeding success and eventual egg and hatchling production. Our goal is to expose both male and female *Anolis* spp. to PFAS solutions *via* the pseudo-gavage method or through dietary exposure (*e.g.*, PFAS-contaminated crickets) prior to breeding, and then observe mating in order to determine possible effects on breeding behavior and success. After breeding, the study would continue to assess impacts on egg production and viability, as well as eventual hatchling success as mentioned earlier.

8.2.3.5 Results

8.2.3.5.1 Dosing Studies to Assess Impacts on Survival and Growth. Our pseudo-gavage dosing studies' main goal has been to assess the impact on the growth and survival of *Anolis* spp. To date, we have been successful in yielding consistent results in over 10 studies that PFAS exposure is negatively correlated with growth and survival at high concentrations of PFASs. Not only does our dosing method yield consistent results, but it is also less invasive than typical gavage methods and, when done properly, should minimize stress for the animal.

8.2.3.5.2 Reproductive Studies. Our studies assessing female reproductive and egg success have yielded consistent results as well, though more data are needed to fully grasp reproductive endpoints for *Anolis* spp. Our preliminary results have shown that the combined use of the pseudo-gavage method coupled with breeding yields results related to relevant reproductive behavior, such as egg production success, egg size, and egg viability.

8.2.3.5.3 Egg Studies. Expanding upon our reproductive endpoints, directly exposing *Anolis* eggs in contaminated substrates has proven to be a successful and environmentally relevant method to better assess PFAS effects on the lifecycle of a species. Although preliminary, we have been successful in exposing eggs *via* contaminated substrates and evaluating their growth to the point of hatchling success. More data are needed to fully understand the direct effects of PFASs on *Anolis* eggs, but we believe that our methods are robust enough to give consistent results over time. A future endeavor will also involve successfully exposing eggs, hatching the hatchlings, and then evaluating the hatchlings' growth over time.

8.2.3.6 Conclusion

The use of *Anolis* species and more standardized dosing methods helps us to not only assess the effects on this taxonomic group, but also gives us insight into future successful reptile ecotoxicology studies by laying out clear and concise methods. Incorporating reproductive endpoints and utilizing a laboratory-based breeding colony gives us more insight into the life-cycle effects of PFASs on *Anolis*, as well as relevant exposure impacts that may occur in the wild. Although we believe our methods are robust, there are a number of other endpoints worth exploring, including behavioral, immunotoxic, and other sub-organismal responses.

8.2.4 Moving Beyond Oral Exposure Dosing: The Potential Importance of Dermal Exposure and Toxicity

Dermal exposure tends to be ignored in risk assessment or assumed to be fairly minor in comparison to dietary/oral exposure.[54] The assumption is often that dietary exposure is going to be more important for terrestrial species commonly used in risk assessment such as birds and mammals. Birds and mammals have high metabolic rates and subsequently have high field ingestion rates. However, there is growing evidence that dermal exposure is a significant exposure route in birds.[54–56] If the importance of dermal exposure in birds continues to be supported by published studies, it is likely even more important for reptiles given their lower overall oral ingestion rates. Weir *et al.* created a theoretical model of dermal exposure that (while not yet verified experimentally) suggests that dermal exposure may be significant for reptiles as is already known for amphibians.[2,57] The proof of concept of dermal exposure in reptiles was later confirmed by Weir *et al.* using a controlled dosing study.[19] The authors used phthalates as surrogates for other contaminants to investigate the role of physico-chemical properties on dermal absorption and distribution. It was reported that low log K_{ow} compounds were not well absorbed, while higher log K_{ow} compounds were found more frequently, and with higher residues, in the organs of fence lizards following exposures.[19] This work was extended by Weir *et al.* to understand the mechanisms that control the absorption of chemicals across the skin membrane in lizards.[58] There was again a relationship between the lipophilicity of the chemical and its absorption across lizard skin, which has also been shown in other organisms such as mammals and amphibians.[57,58] Finally, much of the previously cited work on reptile ecotoxicology has shown that dermal toxicity is a potentially important consideration for the effects of chemicals on reptiles;[11,13,20] although further experimentation is needed to determine under what conditions dermal exposure will be relatively important and when it may not be an important consideration (*e.g.*, very hydrophilic compounds).

8.3 Conclusions

The majority of toxicity data reported for reptiles have been focused on lizards. There are multiple reports of dosing methods for oral, dermal, and egg exposures for lizards. There are far fewer reports available for snakes and turtles. When building dosing methods for reptiles, taxonomic relationships are an important consideration. There are significant differences between turtles, lizards, and snakes in relation to anatomy and metabolic physiology. These differences have implications for routes of exposure, and subsequently, dosing methods for toxicity testing. For example, snakes are more likely to experience dermal exposure to toxicants since they have high contact with surfaces due to their absence of limbs. They also eat less frequently which can have important impacts on whether dietary exposure is an important consideration for risk assessment to these organisms. Potentially, there could be large differences between both dietary and dermal exposure (and thus, dosing methods) between a lizard and a snake. Conversely, turtles are more likely to experience exposure orally and less likely to experience exposure dermally due to the protection of their shells, although this has not been adequately tested.

While progress has been made in the field of reptile ecotoxicology, we are still a long way from developing a vetted, standardized toxicity test for reptiles. We provide successful additions to the previously published literature on dosing methods for lizards including methods for exposures to brown anole eggs *via* contaminated substrates, pseudo-gavage exposures for PFASs in brown anoles, pseudo-gavage exposures during breeding events in brown anoles, and oral exposure *via* gelatin capsules in Brook's house geckos. The development of a breeding colony of *A. sagrei* has allowed for the greater development of methods. We report here that the pseudo-gavage method has been successfully employed in PFAS chronic toxicity tests with good survival of controls and low dose groups. We further report that the breeding colony provides opportunities to assess chemicals throughout the lifecycle of the brown anole including exposures prior to, and during, breeding with subsequent effects on egg development. A breeding colony also allows isolation of the eggs through exposure to contaminated substrates. We report methods of directly exposing eggs to substrates contaminated with per- and polyfluoroalkyl substances at environmentally relevant levels. Finally, the development of oral exposure methods *via* gelatin capsules produces consistent and reliable data that are comparable to other standard toxicity tests. For example, the avian acute oral toxicity test required under the FIFRA for pesticide registration states that use of gelatin capsules is the preferred dosing method for oral exposures.[10]

While the success of these methods provides some opportunity to regulators and managers, they have not been validated and fully converted to a standard testing protocol. For example, performing an international ring test would be an important step to establishing a standardized toxicity testing system for government and industry. While commonly overlooked, dermal exposure is also supported by growing evidence to be an important exposure route for birds and lizards and is well established as important for amphibians already.[19,55,57] Much more progress in developing oral and dermal dosing methods must be made before we reach this point. However, providing an internationally recognized and standardized reptile toxicity testing method is an important endeavor to improve the ecological risk assessment process for reptiles.

Acknowledgements

The New Zealand Department of Conservation provided funding to S. M. Weir and M. R. Youssif under grant DOCDM 6525310. The Strategic Environmental Research and Development Program provided funding to C. J. Salice (ER-2627) for conducting PFAS research in reptiles.

References

1. D. W. Sparling, G. Linder, C. A. Bishop and S. K. Krest, Recent advances in amphibian and reptile ecotoxicology, in *Ecotoxicology of Amphibians and Reptiles*, ed. D. W. Sparling, G. Linder, C. A. Bishop and S. K. Krest, CRC Press, Boca Raton, 2nd edn, 2010, ch. 1, pp. 1–11.
2. S. M. Weir, J. G. Suski and C. J. Salice, Ecological risk of anthropogenic pollutants to reptiles: Evaluating assumptions of sensitivity and exposure, *Environ. Pollut.*, 2010, **158**, 3596–3606.
3. K. R. Campbell and T. S. Campbell, A logical starting point for developing priorities for lizard and snake ecotoxicology: a review of available data, *Environ. Toxicol. Chem.*, 2002, **21**, 894–898.
4. R. D. Nagle, C. L. Rowe and J. D. Congdon, Accumulation and selective maternal transfer of contaminants in the turtle *Trachemys scripta* associated with coal ash deposition, *Arch. Environ. Contam. Toxicol.*, 2001, **40**, 531–536.
5. C. Ockleford, P. Adriaanse, P. Berny, T. Brock, S. Duquesne, S. Grilli, A. F. Hernandez-Jerez, S. H. Bennekou, M. Klein, T. Kuhl, R. Laskowski, K. Machera, O. Pelkonen, S. Pieper, M. Stemmer, I. Sundh, I. Teodorovic, A. Tiktak, C. J. Topping, G. Wolterink, A. Aldrich, C. Berg, M. Ortiz-Santaliestra, S. Weir, F. Streissl and R. H. Smith, Scientific Opinion on the state of the science on pesticide risk assessment for amphibians and reptiles, *EFSA J.*, 2018, **16**, e50125.
6. D. F. Munro, Effect of DDT powder on small cottonmouths, *Herpetologica*, 1949, **5**, 71–72.
7. E. S. Herald, Effects of DDT-oil solutions upon amphibians and reptiles, *Herpetologica*, 1949, **5**, 117–120.
8. R. A. Coulson and T. Hernandez, Catabolic effects of cycloheximide in the living reptile, *Comp. Biochem. Physiol. B: Comp. Biochem.*, 1971, **40**, 741–749.
9. R. J. Hall and D. R. Clark Jr., Responses of the iguanid lizard *Anolis carolinensis* to four organophosphorus pesticides, *Environ. Pollut. Ser. A, Ecol. Biol.*, 1982, **28**, 45–52.
10. USEPA, Ecological Effects Test Guidelines OCSPP 850.2100: Avian Acute Oral Toxicity Test, 2012, EPA 712-C-025.
11. J. E. Brooks, P. J. Savarie and J. J. Johnston, The oral and dermal toxicity of selected chemicals to brown tree snakes (*Boiga irregularis*), *Wildl. Res.*, 1998, **25**, 427–435.
12. J. E. Brooks, P. J. Savarie, J. J. Johnston and R. L. Bruggers, Toxicity of pyrethrin/pyrethroid fogger products to brown tree snakes (*Boiga irregularis*) in cargo containers, *Snake*, 1998, **28**, 33–36.
13. G. J. Alexander, D. Horne and S. A. Hanrahan, An evaluation of the effects of deltamethrin on two non-target lizard species in the Karoo, South Africa, *J. Arid Environ.*, 2002, **50**, 121–133.
14. L. G. Talent, Effect of temperature on toxicity of a natural pyrethrin pesticide to green anole lizards (*Anolis carolinensis*), *Environ. Toxicol. Chem.*, 2005, **24**, 3113–3116.

15. C. A. McFarland, M. J. Quinn Jr., M. A. Bazar, A. K. Remick, L. G. Talent and M. S. Johnson, Toxicity of oral exposure to 2,4,6-trinitrotoluene in the western fence lizard (*Sceloporus occidentalis*), *Environ. Toxicol. Chem.*, 2008, **27**, 1102–1111.
16. C. A. McFarland, M. J. Quinn Jr., M. A. Bazar, L. G. Talent and M. S. Johnson, Toxic effects of oral hexahydro-1,3,5-trinitro-1,3,5-triazine in the western fence lizard (*Sceloporus occidentalis*), *Environ. Toxicol. Chem.*, 2009, **28**, 1043–1050.
17. C. A. McFarland, M. J. Quinn Jr., J. Boyce, E. M. LaFiandra, M. A. Bazar, L. G. Talent and M. S. Johnson, Toxic effects of oral 2-amino-4,6-dinitrotoluene in the Western fence lizard (*Sceloporus occidentalis*), *Environ. Pollut.*, 2011, **159**, 466–473.
18. J. G. Suski, C. J. Salice, J. T. Houpt, M. A. Bazar and L. G. Talent, Dose-related effects following oral exposure of 2,4-dinitrotoluene on the western fence lizard, *Sceloporus occidentalis*, *Environ. Toxicol. Chem.*, 2008, **27**, 352–359.
19. S. M. Weir, L. G. Talent, T. A. Anderson and C. J. Salice, Unraveling the relative importance of oral and dermal contaminant exposure in reptiles: insights from studies using the western fence lizard (*Sceloporus occidentalis*), *PLoS One*, 2014, **18**, e99666.
20. S. M. Weir, S. Yu, L. G. Talent, J. D. Maul, T. A. Anderson and C. J. Salice, Improving reptile ecological risk assessment: Oral and dermal toxicity of pesticides to a common lizard species (*Sceloporus occidentalis*), *Environ. Toxicol. Chem.*, 2015, **34**, 1778–1786.
21. S. M. Weir, S. Yu, A. Knox, L. G. Talent, J. M. Monks and C. J. Salice, Acute toxicity and risk to lizards of rodenticides and herbicides commonly used in New Zealand, *N. Z. J. Ecol*, 2016, **40**, 342–350.
22. L. G. Talent, J. N. Dumont, J. A. Bantle, D. M. Janz and S. G. Talent, Evaluation of western fence lizard (*Sceloporus occidentalis*) and eastern fence lizard (*Sceloporus undulatus*) as laboratory reptile models for toxicological investigations, *Environ. Toxicol. Chem.*, 2002, **21**, 899–905.
23. D. W. Sparling, C. Matson, J. Bickham and P. Doelling-Brown, Toxicity of glyphosate as Glypro® and LI700 to red-eared slider (*Trachemys scripta elegans*) embryos and early hatchlings, *Environ. Toxicol. Chem.*, 2006, **25**, 2768–2774.
24. L. G. Schaumburg, P. A. Siroski, G. L. Poletta and M. D. Mudry, Genotoxicity induced by Roundup® (Glyphosate) in tegu lizard (*Salvator merianae*) embryos, *Pestic. Biochem. Physiol.*, 2016, **130**, 71–78.
25. A. Marco, M. López-Vicente and V. Pérez-Mellado, Arsenic uptake by reptile flexible-shelled eggs from contaminated nest substrates and toxic effect on embryos, *Bull. Environ. Contam. Toxicol.*, 2004, **72**, 983–990.
26. A. Marco, J. Hidalgo-Vila and C. Díaz-Paniagua, Toxic effects of ammonium nitrate fertilizer on flexible-shelled lizard eggs, *Bull. Environ. Contam. Toxicol.*, 2004, **73**, 125–131.
27. S. M. Brasfield, K. Bradham, J. B. Wells, L. G. Talent, R. P. Lanno and D. M. Janz, Development of a terrestrial vertebrate model for assessing

bioavailability of cadmium in the fence lizard (*Sceloporus undulatus*) and in ovo effects on hatchling size and thyroid function, *Chemosphere*, 2004, **54**, 1643–1651.
28. S. R. de Solla, P. A. Martin and P. Mikoda, Toxicity of pesticide and fertilizer mixtures simulating corn production to eggs of snapping turtles (*Chelydra serpentina*), *Sci. Total Environ.*, 2011, **409**, 4306–4311.
29. R. Estes, K. de Queiroz and J. Gauthier, Phylogenetic Relationships within Squamata, in *Phylogenetic relationships of the lizard families: essays commemorating Charles L. Camp*, ed. R. Estes and G. Pregill, Stanford University Press, Stanford, CA, USA, 1988, pp. 119–281.
30. OECD, Acute oral toxicity – Up-and-Down procedure (UDP), OECD Guidelines for the Testing of Chemicals 425, 2008.
31. S. T. McMurry, L. E. Jones, P. N. Smith, G. P. Cobb, T. A. Anderson, M. B. Lovern, S. Cox and X. Pan, Accumulation and effects of octahydro-1,3,5,7-tetranitro-1,3,5,7-tetrazocine (HMX) exposure in the green anole (*Anolis carolinensis*), *Ecotoxicology*, 2012, **21**, 304–314.
32. C. A. Moody, J. W. Martin, W. C. Kwan, D. C. G. Muir and S. A. Mabury, Monitoring perfluorinated surfactants in biota and surface water samples following an accidental release of fire-fighting foam into Etobicoke Creek, *Environ. Sci. Technol.*, 2002, **36**, 545–551.
33. C. W. Noorlander, S. P. J. van Leeuwen, J. D. teBiesebeek, M. J. B. Mengelers and M. J. Zeilmaker, Levels of perfluorinated compounds in food and dietary intake of PFOS and PFOA in the Netherlands, *J. Agric. Food Chem.*, 2011, **59**, 7496–7505.
34. R. H. Anderson, G. C. Long, R. C. Porter and J. K. Anderson, Occurrence of select perfluoroalkyl substances at U.S. Air Force aqueous film-forming foam release sites other than fire-training areas: Field-validation of critical fate and transport properties, *Chemosphere*, 2016, **150**, 678–685.
35. M. Haukas, U. Berger, H. Hop, B. Gulliksen and G. Gabrielsen, Bioaccumulation of per- and polyfluorinated alkyl substances (PFAS) in selected species from the Barents Sea food web, *Environ. Pollut.*, 2007, **148**, 360–371.
36. Y. Wang, J. Yao, J. Dai, L. Ma, D. Liu, H. Xu, Q. Cui, J. Ma and H. Zhang, Per- and polyfluoroalkyl substances (PFASs) in blood of captive Siberian tigers in China: Occurrence and associations with biochemical parameters, *Environ. Pollut.*, 2020, **265**, 114805.
37. A. East, R. H. Anderson and C. J. Salice, Per- and polyfluoroalkyl substances (PFAS) in surface water near US Air Force bases: prioritizing individual chemicals and mixtures for toxicity testing and risk assessment, *Environ. Toxicol. Chem.*, 2021, **40**, 859–870.
38. E. S. Larson, J. M. Conder and J. A. Arblaster, Modeling avian exposures to perfluoroalkyl substances in aquatic habitats impacted by historical aqueous film forming foam releases, *Chemosphere*, 2018, **201**, 335–341.
39. P. A. Fair and M. Houde, Poly- and Perfluoroalkyl Substances in Marine Mammals, in *Marine Mammal Ecotoxicology: Impacts of Multiple Stressors*

on Population Health, ed. M. C. Fossi and C. Panti, Academic Press, London, UK, 2018, ch. 5, pp. 117–145.
40. G. Sciancalepore, G. Pietroluongo, C. Centelleghe, M. Milan, M. Bonato, G. Corazzola and S. Mazzariol, Evaluation of per- and poly-fluorinated alkyl substances (PFAS) in livers of bottlenose dolphins (*Tursiops truncatus*) found stranded along the northern Adriatic Sea, *Environ. Pollut.*, 2021, **291**, 118186.
41. T. J. Bangma, M. J. Ragland, T. R. Rainwater, J. A. Bowden, J. W. Gibbons and J. L. Reiner, Perfluoroalkyl substances in diamondback terrapins (*Malaclemys terrapin*) in coastal South Carolina, *Chemosphere*, 2019, **215**, 305–312.
42. C. Wood, G. H. Balazs, M. Rice, T. M. Work, T. T. Jones, E. Sterling, T. M. Summers, J. Brooker, L. Kurpita, C. S. King and J. M. Lynch, Sea turtles across the North Pacific are exposed to perfluoroalkyl substances, *Environ. Pollut.*, 2021, **279**, 116875.
43. I. A. Rodríguez-Jorquera, C. R. Colli-Dula, K. Kroll, B. S. Jayasinghe, M. V. Parachu Marco, C. Silva-Sanchez, G. S. Toor and N. D. Denslow, Blood transcriptomics analysis of fish exposed to perfluoro alkyls substances: Assessment of a non-lethal sampling technique for advancing aquatic toxicology research, *Environ. Sci. Technol.*, 2019, **53**, 1441–1452.
44. N. M. Dennis, S. Subbiah, A. Karnjanapiboonwong, M. L. Dennis, C. McCarthy, C. J. Salice and T. A. Anderson, Species- and tissue-specific avian chronic toxicity values for perfluorooctane sulfonate (PFOS) and a binary mixture of PFOS and perfluorohexane sulfonate, *Environ. Toxicol. Chem.*, 2021, **40**, 899–909.
45. L. Zhang, Z. Meng, L. Chen, G. Zhang, W. Zhang, Z. Tian, Z. Wang, S. Yu, Z. Zhou and J. Diao, Perfluorooctanoic acid exposure impact a trade-off between self-maintenance and reproduction in lizards (*Eremias argus*) in a gender-dependent manner, *Environ. Pollut.*, 2020, **262**, 114341.
46. W. R. Flynn, T. D. Hoskins, M. Iacchetta, C. de Perre, L. S. Lee, J. T. Hoverman and M. S. Sepulveda, Dietary exposure and accumulation of per- and polyfluoroalkyl substances alters growth and reduces body condition of post-metamorphic salamanders, *Sci. Total Environ.*, 2021, **765**, 142730.
47. J. B. Losos, J. C. Marks and T. W. Schoener, Habitat use and ecological interactions of an introduced and a native species of *Anolis* lizard on Grand Cayman, with a review of the outcomes of anole introductions, *Oecologia*, 1993, **95**, 525–532.
48. S. Poe, A. Nieto-montes de oca, O. Torres-carvajal, K. De Queiroz, J. A. Velasco, B. Truett, L. N. Gray, M. J. Ryan, G. Köhler, F. Ayala-varela and I. Latella, A phylogenetic, biogeographic, and taxonomic study of all extant species of *Anolis* (Squamata; Iguanidae), *Syst. Biol.*, 2017, **66**, 663–697.
49. J. B. Losos and C. J. Schneider, *Anolis* lizards, *Curr. Biol.*, 2009, **19**, R316.
50. M. B. Lovern, M. M. Holmes and J. Wade, The Green Anole (*Anolis carolinensis*): A reptilian model for laboratory studies of reproductive morphology and behaviour, *ILAR J.*, 2004, **45**, 54–64.

51. T. J. Sanger, P. Hime, M. Johnson, J. Diani and J. B. Losos, Laboratory protocols for husbandry and embryo collection of *Anolis* lizards, *Herpetol. Rev.*, 2008, **39**, 58–63.
52. Texas State University, *Texas Invasive Species Institute, Brown Anole: Anolis sagrei*, Retrieved September 25, 2021, from http://www.tsusinvasives.org/home/database/anolis-sagrei
53. C. J. Salice, J. G. Suski, M. A. Bazar and L. G. Talent, Effects of inorganic lead on Western fence lizards (*Sceloporus occidentalis*), *Environ. Pollut.*, 2009, **157**, 3457–3464.
54. P. Mineau, Barking up the wrong perch: Why we should stop ignoring non dietary routes of pesticide exposure in birds, *Integr. Environ. Assess. Manage.*, 2011, **7**, 297–299.
55. C. J. Driver, D. B. Drown, M. W. Ligotke, P. Van Voris, B. D. McVeety and B. J. Greenspan, Routes of uptake and their relative contribution to the toxicologic response of Northern bobwhite (*Colinus virginianus*) to an organophosphate pesticide, *Environ. Toxicol. Chem.*, 1991, **10**, 21–33.
56. N. B. Vyas, J. W. Spann, C. S. Hulse, S. Gentry and S. L. Borges, Dermal insecticide residues from birds inhabiting an orchard, *Environ. Monit. Assess.*, 2007, **133**, 209–214.
57. A. Quaranta, V. Bellantuono, G. Cassano and C. Lippe, Why amphibians are more sensitive than mammals to xenobiotics, *PLoS One*, 2009, **4**, e7699.
58. S. M. Weir, L. G. Talent, T. A. Anderson and C. J. Salice, Insights into reptile dermal contaminant exposure: reptile skin permeability to pesticides, *Chemosphere*, 2016, **154**, 17–22.

CHAPTER 9

Morphological and Molecular Evidence of Active Principle Glyphosate Toxicity on the Liver of the Field Lizard Podarcis siculus

MARIAILARIA VERDERAME, TERESA CHIANESE AND ROSARIA SCUDIERO*

Department of Biology, University Federico II – via Cintia, 80125, Napoli, Italy
*Email: rosaria.scudiero@unina.it

9.1 Introduction

Glyphosate (*N*-phosphonomethyl-glycine, GLY) is the most commonly used active ingredient present in agrochemical formulations worldwide to maximize crop yields in intensive agriculture. It is a herbicide belonging to the group of organophosphates. The agrochemical acts as a potent herbicide by inhibiting the enzyme 5-enolpyruvylshikimate-3-phosphate synthase (EPSPS), a key component of the shikimate pathway, the inhibition of which blocks the biosynthesis of aromatic amino acids in plants, resulting in their death.[1] It is a non-selective herbicide and translocates primarily to metabolic sinks, killing meristematic tissues away from the application site. Its phloem-mobile properties and slow action in killing weeds allow the herbicide to move throughout the plant to kill all meristems, making it

Issues in Toxicology No. 45
Bird and Reptile Species in Environmental Risk Assessment Strategies
Edited by Guillermo Eli Liwszyc and Marcelo L. Larramendy

Published by the Royal Society of Chemistry, www.rsc.org

effective for perennial weed control. Since its commercialization starting in 1974, its use has grown to dominate the herbicide market, due also to the introduction of novel GLY-tolerant crop varieties (*e.g.*, corn, soybean and cotton), which are genetically engineered or modified to be sprayed with GLY without dying.[2] In addition, the absence of the shikimate pathway in animals led to the assumption that GLY would not affect non-target species.

Nowadays, the safety profile of the herbicide and its commercial formulations is controversial and is a matter of great debate.[3,4] Studies performed by employees of companies producing and commercializing GLY-based herbicides (GBHs) claim that GLY is safe for animals and humans. The results collected on their effects on human health continue to be controversial;[3–6] GLY is classified by the United States Environmental Protection Agency (US EPA) as *non-toxic and not an irritant*, and by the European Food Safety Authority (EFSA) as *non-carcinogenic to humans.* The US EPA classified the herbicide as an E carcinogen, claiming that the herbicide is unlikely to be carcinogenic to humans.[7,8] In March 2015, the International Agency for Research on Cancer (IARC), an intergovernmental agency forming part of the WHO of the United Nations, classified GLY as *probably carcinogenic to humans* (Group 2A). This was based on limited evidence of cancer in humans (from real-world exposures that actually occurred) and sufficient evidence of cancer in experimental animals (from studies of pure GLY). Furthermore, the IARC also concluded that there was "strong" evidence for genotoxicity, both for pure GLY and for GLY-based formulations.[9,10] Traces of GLY have been detected in human urine samples, highlighting its persistence, bioaccumulation and potential health risk.[11] On the other hand, research studies conducted by independent academic scientists report serious health risks to wildlife.[12,13] For this reason, GBHs, such as other organophosphate pesticides, are receiving increased attention as potential human carcinogens and as toxicants to terrestrial and aquatic organisms and ecosystems.[14–16] First of all, GLY alters the soil texture and microbial diversity by reducing the microbial richness and increasing the population of phytopathogenic fungi.[17] Impacts of GLY, and more generally, GBHs, on aquatic environments include eutrophication, as demonstrated by the over-development of toxic algae.[18] GLY-exposed fish show increased levels of plasma glucose and cortisol, endocrine dysregulation and alterations in hepatocytes' antioxidant enzyme systems and transcriptome.[19,20] Among Tetrapoda, in amphibians GBHs cause teratogenic effects and developmental failures impacting both larval and adult stages;[21,22] and in reptiles DNA breakage and alterations in the immune parameters, in plasma proteins and in growth.[23–25] High rates of crossed beaks, keratin disorder and morpho-physiological alterations of the male genital system have also been observed in GBH-exposed birds.[26,27]

9.2 Lizards in Contaminated Environments

Agricultural expansion causes a constant increase of habitat fragmentation and a great degradation of ecological patches with deep consequences for biodiversity that occupy these habitats. The use of pesticides, phytopharmaceuticals and other xenobiotics has dramatically increased in the last few decades, in order to satisfy growing demands.[28] The constant application of these chemical compounds is generating a decline in diversity and size of several wild species populations, including reptiles.[29,30]

Habitat loss, fragmentation and alteration associated with agricultural expansion play a great role in reptiles' decline. Although this taxon is usually not taken into consideration in environmental risk assessments, the International Union for Conservation of Nature (IUCN) considers that about 21% of reptiles are endangered or vulnerable to extinction.[31] In recent years, biologists have been active by increasing toxicity studies using reptiles and not surrogate animal models, such as birds and mammals. However, there have been no standard protocols for reptile toxicity testing so far.

Among reptiles, lizards are particularly endangered by habitat loss and soil contamination. They are an important link between invertebrates and higher predators and represent the group of reptiles on which ecotoxicological studies have focused the most.[32] The field lizard *Podarcis siculus* (see Figure 9.1) is the most abundant lizard species present in southern Italy.[33,34]

As prey for many small mammals and birds, the field lizard plays an important role also as a soil top predator contributing, in several areas, to the control of agricultural pests. Obviously, being part of wild fauna, it can be affected by contamination from anthropogenic activities. Considered as Less Concern (LC) species for the IUCN list due to its wide distribution and tolerance to a wide variety of habitats, it is distributed in Italy south of the Alps as well as in Sicily, Sardinia and Lampedusa islands, in southern Switzerland, in Corsica and on the Adriatic coast from Slovenia up to Montenegro. Other introduced populations have been identified in France, Turkey, Spain, Tunisia, the United States of America and North Africa as well. *Podarcis siculus* can live from sea level up to 2200 m of altitude.[33]

Figure 9.1 Adult specimens of the Italian field lizard *Podarcis siculus* (Rafinesque-Schmaltz, 1810).

In Italy, its place of origin, it is abundantly distributed all over the country, except in the Po valley where many populations have become extinct due to the disappearance of the suitable habitat.[34]

The high plasticity of *P. siculus* for living in natural and anthropogenically modified habitats, including cultivated areas and urban parks, its fidelity to these environments and its omnivorous diet make *P. siculus* an appropriate species to be employed as a biological model to monitor the effects of environmental contamination on both adult and embryo specimens. The health and reproductive fitness of *P. siculus* have been studied following several physical and chemical changes in its environment.[33,35–46] The effect of environmental stressors was evaluated on embryos by studying the effects of physical stress (*i.e.*, changes in temperature incubation)[41,44,45] and chemical stress (*i.e.*, eggs incubated in cadmium-contaminated soil);[37,40,41,43] in *in ovo* embryo development and on adults experimentally exposed to Cd, pesticides and endocrine-disrupting compounds.[35,36,38,39,42,46]

9.3 Effects of Pure GLY on the Liver of *P. siculus* Specimens

The hepatotoxic effects of GBHs have been clearly demonstrated in human cell lines.[47] Furthermore, hepatic abnormalities at cellular levels have been studied by using other vertebrates as experimental models in ecotoxicological studies. In anurans, at both larval and adult stages, exposure to pure GLY or GLY-formulated products causes an increase in melanomacrophages and in catabolic pigments (*e.g.*, melanin, hemosiderin and lipofuscin), tissular lesions such as lipidosis and hepatic congestion as well as erythrocyte nuclear abnormalities.[48,49]

This chapter describes the toxic effects of active ingredient GLY on the liver of adult *P. siculus*. The adverse effects of metals, pesticides and other pollutants on the morphology and function of the liver of exposed *Podarcis* specimens are well known.[32,36,39,42,46] In this particular species, as well as in all other oviparous vertebrates, the liver, in addition to its important metabolic and detoxifying function, is a fundamental organ in the reproductive process, especially in females, being the production site of vitellogenin (VTG), the major component of eggs' yolk.[50] Consequently, any liver injury endangers reproduction and therefore the survival of the entire population.

The onset of tissular and molecular damage to the liver was evaluated after oral administration of pure GLY to mimic a realistic situation of exposure to the herbicide in the surrounding environment. Two different solutions of pure GLY (0.05 and 0.5 ng g^{-1} bw) were administered to the lizard specimens of a natural population inhabiting a pristine area from southern Italy (Campania). An aqueous solution of pure GLY was administered daily for 3 weeks by direct pipetting of the test solution into the oral cavity. Control animals, run in parallel with GLY-exposed specimens, received only the vehicle (100 μl tap water). These concentrations are much lower than those used in previous studies investigating the GLY or GBH effects in mammals. For example, mild liver damage was found in rats exposed to 56 μg GBH per g for 35 days.[51] In human cell lines, inhibition of transcriptional activities of estrogen receptors (ERs) and cytotoxicity were observed after administration of 2 ppm and 10 ppm GBH, respectively.[52] Furthermore, it is well known that the active ingredient GLY is much less toxic than any other formulation containing GLY as the active ingredient due to the absence of certain substances such as the non-ionic surfactant poly-oxyethyleneamine (POEA) and ammonium sulfate salts. Given the potential differential toxicity of the components in GBHs, when lethality is employed as the toxicity end-point, the range of acute LC50 values varies considerably. In juvenile specimens of the fish *Oncorhynchus keta*, the LC50 values reported range from 10 up to 211 $mg\,L^{-1}$ for pure GLY and from 11 to 33 $mg\,L^{-1}$ for the GBH Roundup®, depending on differences in water sources (*i.e.*, different levels of water hardness). Similarly, in premetamorphic tadpoles of the amphibian *Rana clamitans* (developmental Gosner stage 25), the LC50 value reached a concentration equivalent to 18 $mg\,L^{-1}$ and about 2.0 $mg\,L^{-1}$ for pure GLY and GLY-based formulations containing 15% POEA, respectively.[53] These observations clearly pinpoint that, in general terms, species exposed to the active ingredient GLY are less sensitive to the lethal effect exerted by the herbicide than when exposed to a GLY-based commercial formulation.

9.4 GLY-induced Changes in *P. siculus* Liver Histology

Light microscopy examination clearly shows the onset of dose-dependent alterations in liver morphology. In control animals, belonging to an unexploited pristine area, the liver structure is regular; normal intrahepatic blood vessels and bile ducts are evident, as well as some pigmented areas due to the presence of melanomacrophages (see Figure 9.2A and B); and thin connective fibers surround the sinusoids and the hepatic vessels (see Figure 9.3A and B). In lizards receiving the lower GLY dose, the liver parenchyma is characterized by the presence of more connective tissues principally around the blood vessels (see Figure 9.2C and D), which, in specimens receiving the higher GLY dose, become nodular/cystic formations consisting of abundant connective fibers (see Figure 9.2E and F).

The increase of GLY-induced liver fibrosis is clearly demonstrated by picrosirius red staining (see Figure 9.3C–F), which allowed clearly visualizing collagen fibers, which appear as well-defined, red-stained fibrillary elements, standing out from yellow-stained non-collagen components.[54] In detail, this is evident in animals treated with the high dose of GLY, which show the presence of a large quantity of collagen which forms fibrotic structures with regular edges such as red-colored cysts (see Figure 9.3E and F).

Collagen detection in histological samples is an important procedure to estimate tissue localization and the abundance of connective fibers. This is a very relevant task in pathological conditions such as fibrosis, which results from an imbalance between collagen deposition and reabsorption, generally due to chronic inflammatory processes.[55,56] Liver fibrosis is a dynamic process

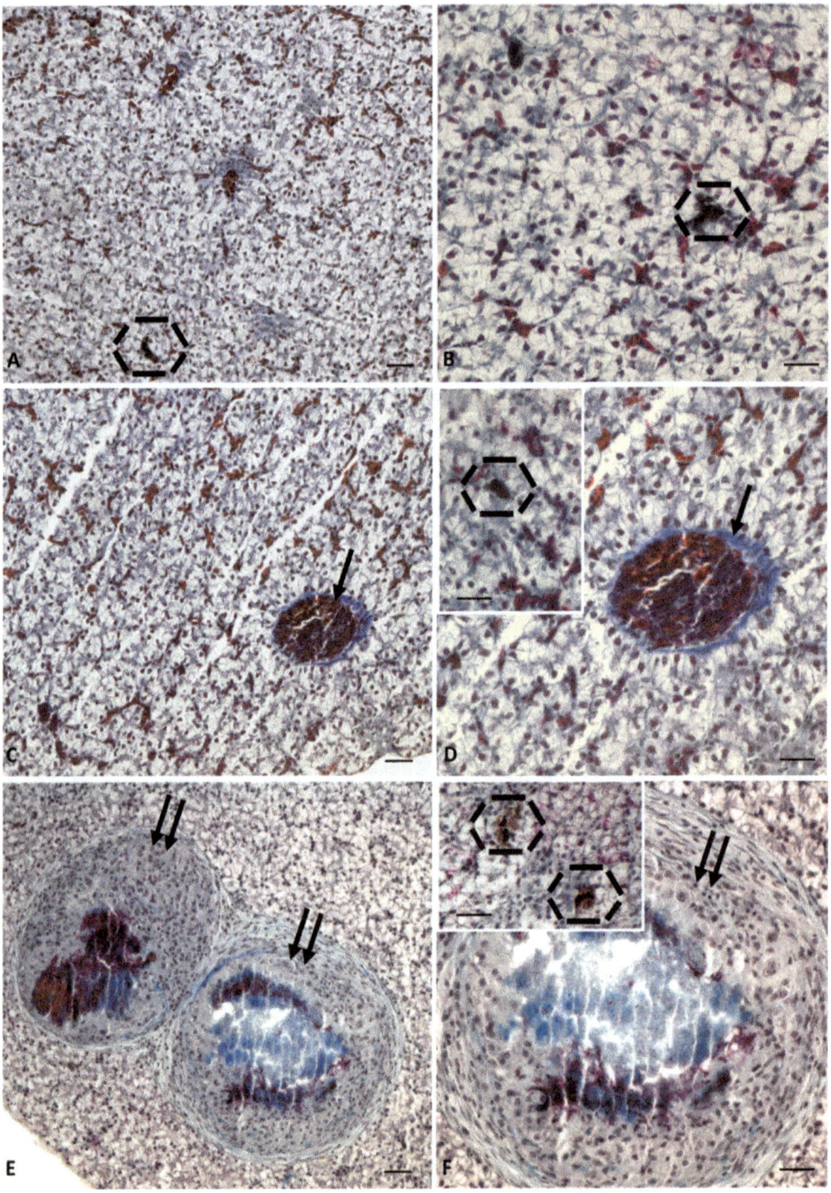

Figure 9.2 Mallory's trichrome staining of the *P. siculus* liver: (A and B) untreated lizard, (C and D) lizards exposed to 0.05 ng GLY g^{-1} bw and (E and F) lizards exposed to 0.5 ng GLY g^{-1} bw. Melanomacrophages (dashed circle); blood vessel (arrow) and fibrotic structures (double arrow). Scale bars: A, C and E – 50 μm; B, D, inset in D, F and inset in F – 20 μm.

triggered in response to an inflammatory process and is mediated by several cell types, myofibroblasts, macrophages, hepatocytes, immune cells and cellular mediators such as cytokines, lipids, hormones and reactive oxygen

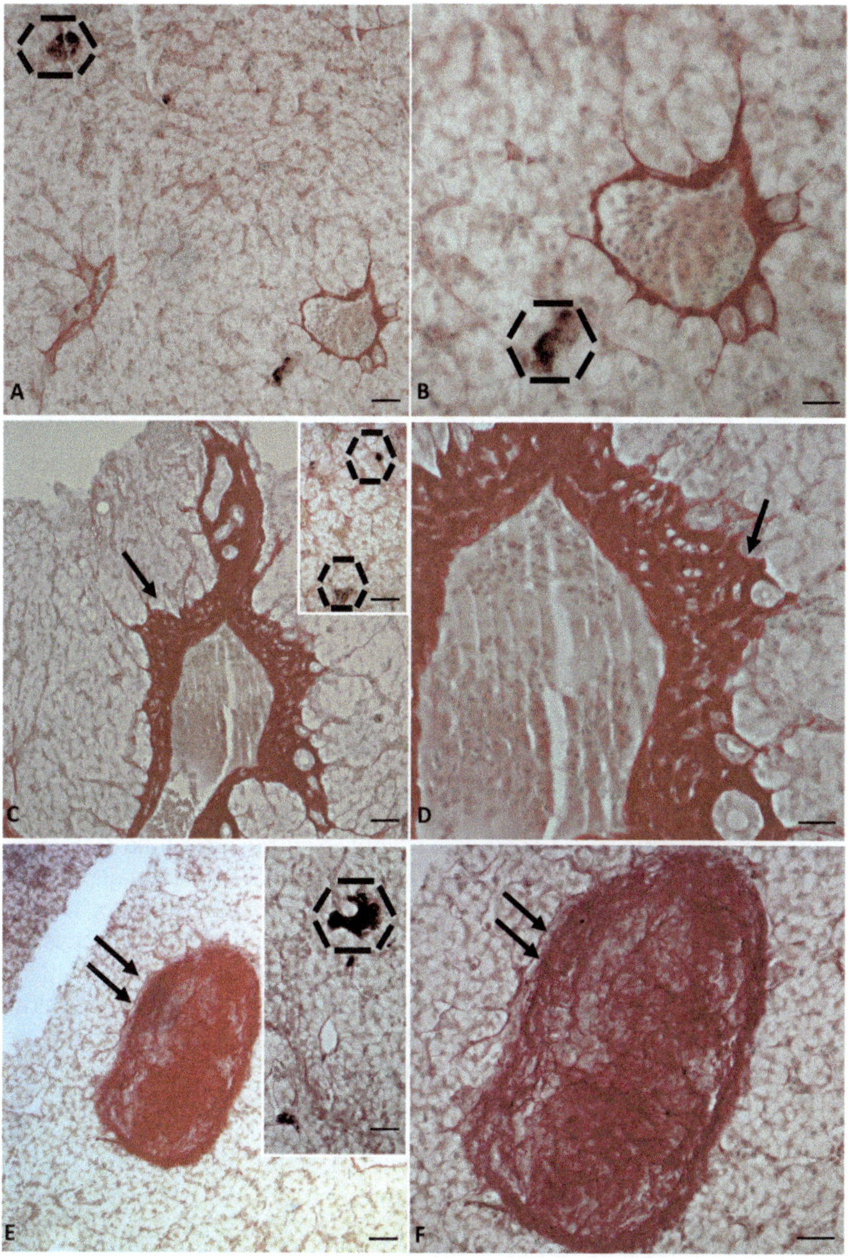

Figure 9.3 Picrosirius staining of *P. siculus* liver: (A and B) untreated lizard, (C and D) lizards exposed to 0.05 ng GLY g^{-1} bw and (E and F) lizards exposed to 0.5 ng GLY g^{-1} bw. Melanomacrophages (dashed circle); deposition of connective tissue around a blood vessel (arrow) and fibrotic structures (double arrow). Scale bars: A, C, C inset, E and E inset – 50 μm; B, D and F – 20 μm.

Table 9.1 Presence of granules in *P. siculus* hepatocytes.[a]

	Treatment		
Type of granule	Control	Glyphosate (0.05 ng g^{-1} bw)	Glyphosate (0.5 ng g^{-1} bw)
Melanin	+	++	++
Lipofuscin	n.d.	+	+
Glycogen	+	++	++

[a]n.d. = not detected.

species. This process leads to an excessive accumulation of the extracellular matrix which compromises the functional architecture of the organ. Following liver injury, hepatic stellate cells acquire a myofibroblastic phenotype, capable of proliferation, contraction and fibrogenesis.[55,56]

Parallel to the fibrogenesis, the hepatic cells of animals exposed to GLY also show an increase in melanin and glycogen deposits and the presence of lipofuscin granules as well (see Table 9.1). These cellular deposits are probably to be ascribed to GLY-induced oxidative stress in the liver, since the presence of melanin and lipofuscin is considered to be a cellular response to remove the free radicals.[57,58]

Melanin-pigmented cells found in the liver of fish, amphibians and reptiles are called melanomacrophages for their phagocytic activity. They are responsible for the phagocytosis of damaged organelles, clearance of worn-out erythrocytes and absorption of free radicals and other potentially toxic agents resulting from cellular catabolism.[59] For all these reasons, they are often referred to as indicators of environmental contamination and tissular damage.[48,60]

Likewise, the endogenous lipofuscin pigment can be used as a non-invasive biomarker of chronic liver disease. It is a highly oxidized cross-linked aggregate consisting of oxidized protein and lipid clusters and its formation depends on the rate of oxidative damage to proteins, the functionality of mitochondrial repair systems, the proteasomal system and the functionality and effectiveness of lysosomes.[61] The lipofuscin aggregates have a redox-active surface thanks to their ability to bind to metals, such as Fe; consequently, lipofuscin, resulting from tissue damage, can cause other toxic effects and facilitate its own formation by driving lipid peroxidation and protein oxidation.[61]

Finally, regarding glycogen accumulation observed in the liver of GLY-treated lizards, disruptions in glycogen metabolism have been shown to be closely related to a certain level of dysfunction in the liver and other organs, such as the heart, kidneys and brain, and muscles. Excessive hepatic glycogen is associated with impaired hepatic glycogenolysis and may be due to enzymatic disorder and fibrosis.[62]

9.5 GLY-induced Changes in the Expression and Synthesis of Proteins in the *P. siculus* Liver

The harmful effects of free radicals in cells are known to be balanced by the activity of antioxidant enzymes such as superoxide dismutases (SODs) and glutathione peroxidases (GPxs). SODs are a family of metalloenzymes that catalyze the dismutation of the superoxide radical into molecular oxygen and hydrogen peroxide; Cu/Zn-superoxide dismutase (SOD1) is the most common SOD found in vertebrates, located in the cellular cytoplasm.[63] GPxs are a family of oxidoreductases that catalyze the reduction of hydrogen peroxide or peroxide radicals to oxygen and water or alcohols, respectively; the first purified and most abundant member in vertebrates is glutathione peroxidase 1 (GPx1).[64]

Immunohistochemical investigations show a different localization of these two antioxidant enzymes in the liver of control animals and those exposed to GLY. While in controls the proteins are localized in the cytosol of the hepatocytes, in the treated animals, regardless of the concentration, the proteins localize in the nucleus (see Table 9.2).

The distribution of proteins in cellular compartments plays a key role in their function. Under normal conditions, GPx1 is localized in the cytoplasm and to a lesser extent in the mitochondria. Notwithstanding, it has been demonstrated that GPx1 translocates into the nucleus in liver cancer cells and this translocation may facilitate the antioxidant functions of the enzyme.[65] Nuclear localization of GPx1 has also been observed in prostate epithelial cells, but not in prostate tumor-derived cell lines, indicating that this protein may have a novel function in the nucleus of prostate epithelial cells.[66] It has been shown that in response to high endogenous and exogenous reactive oxygen species, SOD1 also rapidly relocates within the nucleus, regulating the expression of a large set of oxidative response genes that are known to provide resistance to oxidative stress, DNA damage repair and relief of replication stress.[67] For this reason, nuclear SOD1 is considered to be an important "genome guardian" for maintaining genomic stability.[68]

Table 9.2 also shows the results obtained from the immunohistochemical investigation extended to other proteins with cytoprotective function and those considered to be biomarkers of the health status of cells and tissues, namely metallothionein (MT) and the tumor suppressor protein p53.

Table 9.2 Presence and localization of cytoprotective proteins in the *P. siculus* liver.[a]

	Treatment		
Protein	Control	Glyphosate (0.05 ng g^{-1} bw)	Glyphosate (0.5 ng g^{-1} bw)
SOD1	Cytoplasm hepatocytes	Hepatocytes cytoplasm and nuclei	Hepatocytes cytoplasm and nuclei
GPx1	Hepatocytes cytoplasm	Hepatocytes cytoplasm and nuclei	Hepatocytes cytoplasm and nuclei
MT	Kupffer cells	Kupffer cells and hepatocytes cytoplasm	Kupffer cells and hepatocytes cytoplasm
p53	n.d.	Kupffer cells and hepatocytes nuclei	Kupffer cells and hepatocytes nuclei

[a]n.d. = not detected.

The former belongs to a family of small ubiquitous proteins, rich in cysteines, which bind essential and non-essential heavy metals and protect cells from the toxicity of metals and free radicals;[69] the latter triggers apoptosis *via* multiple pathways, thus preventing the growth of aberrant cells.[70] In healthy cells, p53 is barely detectable. Conversely, p53 accumulation occurs in most tumoral and non-tumoral cells under various forms of stress.[71] Immunolocalization analysis reveals a sharp increase in the levels of both proteins in the hepatocytes from animals exposed to the herbicide. In the liver of unexposed animals, MT is mainly detected in Kupffer cells, as expected.[36] On the other hand, in livers of GLY-exposed specimens, MT also appears in the cytoplasm of hepatocytes, thus demonstrating that GLY is capable of inducing a cellular response which includes the non-enzymatic antioxidant pathway. It has been demonstrated that in rats the increase of MT synthesis induced by zinc supplementation ameliorated the histopathological changes engendered by exposure to GLY.[72]

Furthermore, the fact that the liver from exposed specimens is fighting against stress signals induced by GLY is also demonstrated by the activation of p53 in Kupffer cells. These cells are specialized macrophages located in the liver and are considered to be critical mediators of both liver damage and repair;[73] p53 activated by GLY-induced stress signals could determine cell fate by specifically regulating cell cycle arrest, DNA repair and apoptosis.[74] Following herbicide exposure, p53 is also detected in the nuclei of hepatocytes, because p53 functions as a transcription factor because its localization in the nucleus is essential in order to exert its function.

It is well documented that many environmental pollutants can disrupt the endocrine system, resulting in an impaired reproductive cycle and fitness.[75,76] In female lizards, as well as in most vertebrates, during the mating season estrogens produced by follicular cells interact with liver cells that expose the ERs inducing them to produce VTG, the nourishing protein of the embryos which, once released into the bloodstream, reaches the ovary where it is internalized in the growing oocytes.[32] The beta form of the estrogen receptor (ERβ) is ubiquitously present in tissues in both males and females, while the alpha form (ERα) is only present in females during the mating season.[77] In males, ERα and VTG genes are present, but silenced; following accidental or experimental estrogenic contamination, male hepatocytes also begin to express and synthesize the corresponding proteins which, for this reason, are recognized as powerful biomarkers of xenoestrogenic contamination in animals.[39,46]

In this context, the possible estrogen-like action of GLY has been evaluated by using *in situ* hybridization and real-time PCR analyses. The obtained results are reported in Table 9.3.

In the liver of control males, transcripts for ERα or VTG are absent, whereas ERβ transcripts are present, as expected. In GLY-exposed animals, livers show the presence of both ERα and VTG transcripts, regardless of the GLY concentration. Following GLY contamination, the transcript level for ERβ shows a dose-dependent pattern; indeed, the expression of ERβ is

Table 9.3 Presence of transcripts for biomarkers of estrogenic contamination in *P. siculus* hepatocytes.

	Treatment		
Transcript	Control	Glyphosate (0.05 ng g^{-1} bw)	Glyphosate (0.5 ng g^{-1} bw)
VTG	n.d.	+	+
ERα	n.d.	+	+
ERβ	++	+	+++

down-regulated in animals exposed to the low concentration, and up-regulated in animals exposed to the higher concentration. The same trend was also found in the liver of females.

These data demonstrate that pure GLY exhibits a clear estrogenic activity, even at the low concentration tested; the herbicide can activate the transcription of VTG and ERα genes by exerting the stimulatory effects through the ER-dependent pathway, interfering with the physiological estrogenic signaling. Since the possible mechanism of GLY–ER interaction is still unknown, it could be hypothesized that GLY stimulates an indirect estrogenic activation pathway that leads, as a last step, to the activation of estrogen-responsive genes. Nuclear SOD1 has also been reported to interact with ERα, and this association appears to be necessary for effective activation of estrogen-responsive genes.[78] Hence, in light of what has been observed in the liver of lizards, it is possible to hypothesize that GLY determines the translocation of SOD1 in the nucleus, which, in turn, triggers the activation of estrogen-responsive genes, such as VTG.

Regarding the GLY-induced increase in ERβ expression, it is known that ERα and ERβ have different functions. ERα is strictly associated with the synthesis of VTG, while ERβ is involved in processes such as cellular and organelle repair, in particular mitochondria, from ROS and hydrogen peroxide damage, and promotes cell proliferation.[79,80] The increase in ERβ induced by the higher exposure to GLY can be considered as a defense mechanism against the adverse effects of the herbicide on cells.

9.6 Conclusion

The results described in this chapter indicate that the oral administration of the active ingredient GLY, even at a low concentration and for a relatively short period of time, induces histological and molecular changes in the liver of the common field lizard *P. siculus*. The analysis demonstrates a severe liver condition, with fibrotic formations that become scarring on increased GLY exposure. However, hepatic cells enhance their defense mechanism by activating both enzymatic and non-enzymatic antioxidant systems, as well as the "guardian of the genome", namely p53. Finally, the results corroborate that the herbicide GLY can be classified as an endocrine disruptor, exhibiting a clear xenoestrogenic effect, at least in the species we employed as the biotic matrix.

These findings show that the effects of GLY are biologically relevant and could endanger the viability and survival of lizard populations, also unbalancing the trophic cascade. Further research is needed to measure potential genotoxic and reproductive impacts of GLY exposure on reptiles and other wildlife species. In the meantime, a more controlled and less intensive use of GLY in areas devoted to crop production would be useful.

Acknowledgements

The research is in the framework of the COST Action CA18221 PERIAMAR (Pesticide RIsk AssessMent for Amphibians and Reptiles).

References

1. H. Maeda and N. Dudareva, The shikimate pathway and aromatic amino acid biosynthesis in plants, *Annu. Rev. Plant Biol.*, 2012, **63**, 73–105.
2. T. Funke, H. Han, M. L. Healy-Fried, M. Fischer and E. Schönbrunn, Molecular basis for the herbicide resistance of Roundup ready crops, *Proc. Natl. Acad. Sci. U. S. A.*, 2006, **103**, 13010–13015.
3. J. V. Tarazona, D. Court-Marques, M. Tiramani, H. Reich, R. Pfeil, F. Istace and F. Crivellente, Glyphosate toxicity and carcinogenicity: a review of the scientific basis of the European Union assessment and its differences with IARC, *Arch. Toxicol.*, 2017, **91**, 2723–2743.
4. C. J. Portier, B. Armstrong, B. C. Baguley, X. Baur, I. Belyaev, R. Bellé, F. Belpoggi, A. Biggeri, M. C. Bosland, P. Bruzzi, L. T. Budnik, M. D. Bugge, K. Burns, G. M. Calaf, D. O. Carpenter, H. M. Carpenter, L. López-Carrillo, R. Clapp, P. L. Cocco, D. Consonni, P. Comba, E. Craft, M. A. Dalvie, D. Davis, P. A. Demers, J. DeWitt, F. Forastiere, J. H. Freedman, L. Fritschi, C. Gaus, J. M. Gohlke, M. Goldberg, E. Greiser, J. Hansen, L. Hardell, M. Hauptmann, W. Huang, J. Huff, M. O. James, C. W. Jameson, A. Kortenkamp, A. Kopp-Schneider, H. Kromhout, M. L. Larramendy, P. J. Landrigan, L. H. Lash, D. Leszczynski, C. F. Lynch, C. Magnani, D. Mandrioli, F. L. Martin, E. Merler, P. Michelozzi, L. Miligi, A. B. Miller, D. Mirabelli, F. E. Mirer, S. Naidoo, M. J. Perry, M. G. Petronio, R. Pirastu, R. J. Portier, K. S. Ramos, L. W. Robertson, T. Rodriguez, M. Röösli, M. K Ross, D. Roy, I. Rusyn, P. Saldiva, J. Sass, K. Savolainen, P. T. J. Scheepers, C. Sergi, E. K. Silbergeld, B. W. Stewart, P. Sutton, F. Tateo, B. Terracini, H. W. Thielmann, D. B. Thomas, H. Vainio, J. E. Vena, P. Vineis, E. Weiderpass, D. D. Weisenburger, T. J. Woodruff, T. Yorifuji, I. J. Yu, P. Zambon, H. Zeeb and S. F. Zhou, Differences in the carcinogenic evaluation of glyphosate between the International Agency for Research on Cancer (IARC) and the European Food Safety Authority (EFSA), *J. Epidemiol. Community Health*, 2016, **70**, 741–745.
5. S. O. Duke and S. B. Powles, Glyphosate: a once-in-a-century herbicide, *Pest Manag. Sci.*, 2008, **64**, 319–325.

6. C. Huhn, More and enhanced glyphosate analysis is needed, *Anal. Bioanal. Chem.*, 2018, **410**, 3041–3045.
7. USEPA, United States Environmental Protection Agency, *Glyphosate issue paper: evaluation of carcinogenic potential*, US Government Printing Office, Washington, 2016, p. 227.
8. S. Singh, V. Kumar, S. Datta, A. B. Wani, D. S. Dhanjal, R. Romero and J. Singh, Glyphosate uptake, translocation, resistance emergence in crops, analytical monitoring, toxicity and degradation: A review, *Environ. Chem. Lett.*, 2020, **18**, 663–702.
9. K. Z. Guyton, D. Loomis, Y. Grosse, F. El Ghissassi, L. Benbrahim-Tallaa, N. Guha, C. Scoccianti, H. Mattock and K. Straif, Carcinogenicity of tetrachlorvinphos, parathion, malathion, diazinon, and glyphosate. International Agency for Research on Cancer Monograph Working Group, IARC, Lyon, France, *Lancet Oncol.*, 2015, **16**, 490–491.
10. International Agency for Research on Cancer, Some Organophosphate Insecticides and Herbicides, IARC Monographs on the Evaluation of Carcinogenic Risks to Humans, *International Agency for Research on Cancer*, Lyon, France, 2017, vol. 112, pp. 1–452.
11. L. Niemann, C. Sieke, R. Pfeil and R. Solecki, A critical review of glyphosate findings in human urine samples and comparison with the exposure of operators and consumers, *J. Verbrauch. Leb.*, 2015, **10**, 3–12.
12. R. Mesnage and M. N. Antoniou, Facts and fallacies in the debate on glyphosate toxicity, *Front. Public Health.*, 2017, **5**, 1–7.
13. S. Singh, V. Kumar, J. P. K. Gill, S. Datta, S. Singh, V. Dhaka, D. Kapoor, A. B. Wani, D. S. Dhanjal, M. Kumar, S. L. Harikumar and J. Singh, Herbicide glyphosate: toxicity and microbial degradation, *Int. J. Environ. Res. Public Health*, 2020, **17**, 1–18.
14. J. P. Myers, M. N. Antoniou, B. Blumberg, L. Carroll, T. Colborn and L. G. Everett, Concerns over use of glyphosate-based herbicides and risks associated with exposures: a consensus statement, *Environ. Health*, 2016, **15**, 1–13.
15. A. H. C. Van Bruggen, M. M. He, K. Shin, V. Mai, K. C. Jeong, M. R. Finckh and J. G. Morris Jr, Environmental and health effects of the herbicide glyphosate, *Sci. Total Environ.*, 2018, **616–617**, 255–268.
16. M. Verderame and R. Scudiero, Glyphosate: toxic or not toxic, this is the question, *Eur. J. Histochem.*, 2018, **62/S1**, 28–32.
17. C. S. Carranza, M. E. Aluffi, N. Benito, K. Magnoli, C. L. Barberis and C. E. Magnoli, Effect of *in vitro* glyphosate on *Fusarium* spp. growth and disease severity in maize, *J. Sci. Food Agric.*, 2019, **99**, 5064–5072.
18. B. L. Dabney and R. Patiño, Low-dose stimulation of growth of the harmful alga, *Prymnesium parvum*, by glyphosate and glyphosate-based herbicides, *Harmful Algae*, 2018, **80**, 130–139.
19. S. J. Gholami-Seyedkolaei, A. Mirvaghefi, H. Farahmand and A. A. Kosari, Effect of a glyphosate-based herbicide in *Cyprinus carpio*: assessment of acetylcholinesterase activity, hematological responses and serum biochemical parameters, *Ecotoxicol. Environ. Saf.*, 2013, **98**, 135–141.

20. T. M. U. Webster and E. M. Santos, Global transcriptomic profiling demonstrates induction of oxidative stress and of compensatory cellular stress responses in brown trout exposed to glyphosate and Roundup, *BMC Genomics*, 2015, **16**, 1–14.
21. N. Wagner, W. Reichenbecher, H. Teichmann, B. Tappeser and S. Lötters, Questions concerning the potential impact of glyphosate-based herbicides on amphibians, *Environ. Toxicol. Chem.*, 2013, **32**, 1688–1700.
22. J. S. Herek, L. Vargas, S. A. Rinas Trindade, C. F. Rutkoski and N. Macagnan, P. A Hartmann and M. T. Hartmann, Genotoxic effects of glyphosate on *Physalaemus* tadpoles, *Environ. Toxicol. Pharmacol.*, 2021, **81**, 103516.
23. M. A. Latorre, E. C. López González, A. Larriera, G. L. Poletta and P. A. Siroski, Effects of *in vivo* exposure to Roundup® on immune system of *Caiman latirostris*, *J. Immunotoxicol.*, 2013, **10**, 349–354.
24. P. A. Siroski, G. L. Poletta, M. A. Latorre, M. E. Merchant, H. H. Ortega and M. D. Mudry, Immunotoxicity of commercial-mixed glyphosate in broad snouted caiman (*Caiman latirostris*), *Chem. - Biol. Interact.*, 2016, **244**, 64–70.
25. A. P. Mestre, P. S. Amavet, A. I. Vanzetti, M. S. Moleón, M. V. Parachú Marcó, G. L. Poletta and P. A. Siroski, Effects of cypermethrin (pyrethroid), glyphosate and chlorpyrifos (organophosphorus) on the endocrine and immune system of Salvator merianae (Argentine tegu), *Ecotoxicol. Environ. Saf.*, 2019, **169**, 61–67.
26. A. G. Oliveira, L. F. Telles, R. A. Hess, G. A. Mahecha and C. A. Oliveira, Effects of the herbicide Roundup on the epididymal region of drakes *Anas platyrhynchos*, *Reprod. Toxicol.*, 2007, **23**, 182–191.
27. S. Ruuskanen, M. J. Rainio, C. Gómez-Gallego, O. Selenius, S. Salminen, M. C. Collado, K. Saikkonen, I. Saloniemi and M. Helander, Glyphosate-based herbicides influence antioxidants, reproductive hormones and gut microbiome but not reproduction: A long-term experiment in an avian model, *Environ. Pollut.*, 2020, **266**, 115108.
28. J. A. Heinemann, S. Z. Agapito-Tenfen and J. A. Carman, A comparative evaluation of the regulation of GM crops or products containing dsRNA and suggested improvements to risk assessments, *Environ. Int.*, 2013, **55**, 43–55.
29. J. Gibbons, D. Scott, T. Ryan, K. Buhlmann, T. Tuberville, B. Metts, J. Greene, T. Mills, Y. Leidena and S. Poppy, The global decline of reptiles, déjà vu amphibians, *Bioscience*, 2000, **50**, 653–666.
30. S. Weir, S. Yu, L. Talent, J. Maul, T. Anderson and C. Salice, Improving reptile ecological risk assessment: oral and dermal toxicity of pesticides to a common lizard species (*Sceloporus occidentalis*), *Environ. Toxicol. Chem.*, 2015, **34**, 1778–1786.
31. M. E. Ortiz-Santaliestra, J. P. Maia, A. Egea-Serrano and I. Lopes, Validity of fish, birds and mammals as surrogates for amphibians and reptiles in pesticide toxicity assessment, *Ecotoxicology*, 2018, **27**, 819–833.

32. M. Verderame and R. Scudiero, Health status of the lizard *Podarcis siculus* (Rafinesque-Schmaltz, 1810) subject to different anthropogenic pressures, *C. R. Biol.*, 2019, **342**, 81–89.
33. R. Sindaco, G. Doria, E. Razzetti and F. Bernini, *Atlante degli Anfibi e dei Rettili d'Italia/Atlas of Italian Amphibians and Reptiles*, Societas Herpetologica Italica – Edizioni Polistampa, Firenze, 2006, p. 773.
34. C. Corti, M. Capula, L. Luiselli, E. Razzetti and R. Sindaco, Fauna d'Italia, Reptilia, *Calderini, Bologna*, 2011, 869.
35. F. Trinchella, M. Riggio, S. Filosa, M. G. Volpe, E. Parisi and R. Scudiero, Cadmium distribution and metallothionein expression in lizard tissues following acute and chronic cadmium intoxication, *Comp. Biochem. Physiol. C Toxicol. Pharmacol.*, 2006, **144**, 272–278.
36. P. Simoniello, S. Filosa, M. Riggio, R. Scudiero, S. Tammaro, F. Trinchella and C. M. Motta, Responses to cadmium intoxication in the liver of the wall lizard *Podarcis sicula, Comp. Biochem. Physiol. C Toxicol. Pharmacol.*, 2010, **151**, 194–203.
37. P. Simoniello, C. M. Motta, R. Scudiero, F. Trinchella and S. Filosa, Cadmium-induced teratogenicity in lizard embryos: Correlation with metallothionein gene expression, *Comp Biochem. Physiol. C Toxicol. Pharmacol.*, 2011, **153**, 119–127.
38. P. Simoniello, S. Filosa, R. Scudiero, F. Trinchella and C. M. Motta, Cadmium impairment of reproduction in the female wall lizard, *Podarcis sicula, Environ. Toxicol.*, 2013, **28**, 553–562.
39. M. Verderame, E. Limatola and R. Scudiero, Estrogenic contamination by manure fertilizer in organic farming: a case study with the lizard *Podarcis sicula, Ecotoxicology*, 2016, **25**, 105–114.
40. R. Scudiero, M. Verderame, C. M. Motta and P. Simoniello, Unravelling the role of metallothionein on development, reproduction and detoxification in the wall lizard *Podarcis sicula, Int. J. Mol. Sci.*, 2011, **18**, 1–12.
41. R. Scudiero, C. M. Motta and P. Simoniello, Impact of environmental stressors on gene expression in the embryo of the Italian wall lizards, *Appl. Sci.*, 2021, **11**, 1–12.
42. S. Buono, L. Cristiano, B. D'Angelo, A. Cimini and R. Putti, PPARalpha mediates the effects of the pesticide methyl thiophanate on liver of the lizard *Podarcis sicula, Comp. Biochem. Physiol. C Toxicol. Pharmacol.*, 2007, **15**, 306–314.
43. P. Simoniello, F. Trinchella, S. Filosa, R. Scudiero, D. Magnani, T. Theil and C. M. Motta, Cadmium contaminated soil affects retinogenesis in lizard embryos, *J. Exp. Zool.*, 2014, **321A**, 207–219.
44. P. Simoniello, M. G. Esposito, F. Trinchella, C. M. Motta and R. Scudiero, Alterations in brain morphology and HSP70 expression in lizard embryos exposed to thermal stress, *C. R. Biol.*, 2016, **339**, 380–390.
45. R. Scudiero, M. Verderame, C. M. Motta, V. Migliaccio and P. Simoniello, HSP70 localization in *Podarcis siculus* embryos under natural thermal regime and following a non-lethal cold shock, *CR Biol.*, 2019, **342**, 299–308.

46. M. Verderame, M. Prisco, P. Andreuccetti, F. Aniello and E. Limatola, Experimentally nonylphenol-polluted diet induces the expression of silent genes VTG and ERα in the liver of male lizard *Podarcis sicula*, *Environ. Pollut.*, 2011, **159**, 1101–1107.
47. Y. Hao, Y. Zhang, H. Ni, J. Gao, Y. Yang, W. Xu and L. Tao, Evaluation of the cytotoxic effects of glyphosate herbicides in human liver, lung, and nerve, *J. Environ. Sci. Health*, 2019, **54**, 737–744.
48. J. M. Pérez-Iglesias, L. Franco-Belussi, L. Moreno, S. Tripole, C. de Oliveira and G. S. Natale, Effects of glyphosate on hepatic tissue evaluating melanomacrophages and erythrocytes responses in Neotropical anuran *Leptodactylus latinasus*, *Environ. Sci. Pollut. Res. Int.*, 2016, **23**, 9852–9861.
49. N. C. Bach, D. J. G. Marino, G. S. Natale and G. M. Somoza, Effects of glyphosate and its commercial formulation, Roundup® Ultramax, on liver histology of tadpoles of the neotropical frog, *Leptodactylus latrans* (Amphibia: Anura), *Chemosphere*, 2018, **202**, 289–297.
50. A. K. Roy and B. Chatterjee, Sexual dimorphism in the liver, *Annu. Rev. Physiol.*, 1983, **45**, 37–50.
51. S. Cağlar and D. Kolankaya, The effect of sub-acute and sub-chronic exposure of rats to the glyphosate-based herbicide Roundup, *Environ. Toxicol. Pharmacol.*, 2008, **25**, 57–62.
52. C. Gasnier, C. Dumont, N. Benachour, E. Clair, M. C. Chagnon and G. E. Séralini, Glyphosate-based herbicides are toxic and endocrine disruptors in human cell lines, *Toxicology*, 2009, **262**, 184–191.
53. United States Environmental Protection Agency USEPA (2018), Glyphosate Registration Review Docket #EPA-HQ-OPP-2009-0361.
54. C. Segnani, C. Ippolito, L. Antonioli, C. Plegrini, C. Blandizzi, A. Dolfi and N. Bernardini, Histochemical detection of collagen fibers by Sirius Red/Fast Green is more sensitive than van Gieson or Sirius Red alone in normal and inflamed rat colon, *PLoS One*, 2015, **10**, 1–10.
55. M. Parola and M. Pinzani, Liver fibrosis: Pathophysiology, pathogenetic targets and clinical issues, *Mol. Aspects Med.*, 2019, **65**, 37–55.
56. T. Kisseleva and D. Brenner, Molecular and cellular mechanisms of liver fibrosis and its regression, *Nat. Rev. Gastroenterol. Hepatol.*, 2021, **18**, 151–166.
57. M. Rozanowska, T. Sarna, E. J. Land and T. G. Truscott, Free radical scavenging properties of melanin interaction of eu- and pheomelanin models with reducing and oxidizing radicals, *Free Radic. Biol. Med.*, 1999, **26**, 518–525.
58. N. S. Loumbourdis, Liver histopathologic alterations in the frog *Rana ridibunda* from a small river of Northern Greece, *Arch. Environ. Contam. Toxicol.*, 2007, **53**, 418–425.
59. N. C. Steinel and D. I. Bolnick, Melanomacrophage centers as a histological indicator of immune function in fish and other poikilotherms, *Front. Immunol.*, 2017, **8**, 1–8.
60. I. A. Badroo, H. P. Nandurkar and A. H. Khanday, Toxicological impacts of herbicide paraquat dichloride on histological profile (gills, liver, and

kidney) of freshwater fish Channa punctatus (Bloch), *Environ. Sci Pollut. Res. Int.*, 2020, **27**, 39054–39067.
61. T. Jung, N. Bader and T. Grune, Lipofuscin: formation, distribution, and metabolic consequences, *Ann. N. Y. Acad. Sci.*, 2007, **1119**, 97–111.
62. S. S. Ellingwood and A. J. Cheng, Biochemical and clinical aspects of glycogen storage diseases, *Endocrinol.*, 2018, **238**, R131–R141.
63. I. N. Zelko, T. J. Mariani and R. J. Folz, Superoxide dismutase multigene family: a comparison of the CuZn-SOD (SOD1), Mn-SOD (SOD2), and EC-SOD (SOD3) gene structures, evolution, and expression, *Free Rad. Biol. Med.*, 2002, **33**, 337–349.
64. R. Margis, C. Dunand, F. K. Teixeira and M. Margis-Pinheiro, Glutathione peroxidase family – an evolutionary overview, *FEBS J.*, 2008, **275**, 3959–3970.
65. E. Lubos, J. Loscalzo and D. E. Handy, Glutathione peroxidase-1 in health and disease: from molecular mechanisms to therapeutic opportunities, *Antioxid. Redox Signal.*, 2011, **15**, 1957–1997.
66. D. N. Ekoue, E. Ansong, L. K. Hong, L. Nonn, V. Macias, R. Deaton, R. Rupnow, P. H. Gann, A. Kajdacsy-Balla and A. M. Diamond, GPX1 Localizes to the nucleus in prostate epithelium and its levels are not associated with prostate cancer recurrence, *Antioxidants*, 2018, **7**, 167.
67. C. K. Tsang, Y. Liu, J. Thomas, Y. Zhang and X. F. Zheng, Superoxide dismutase 1 acts as a nuclear transcription factor to regulate oxidative stress resistance, *Nat. Commun.*, 2014, **5**, 3446.
68. E. Inoue, K. Tano, H. Yoshii, J. Nakamura, S. Tada, M. Watanabe, M. Seki and T. Enomoto, SOD1 is essential for the viability of DT40 cells and nuclear SOD1 functions as a guardian of genomic DNA, *J. Nucl. Acids*, 2010, 795946.
69. P. Coyle, J. C. Philcox, L. C. Carey and A. M. Role, Metallothionein: the multipurpose protein, *Cell. Mol. Life Sci.*, 2002, **59**, 627–647.
70. J. Chen, The cell-cycle arrest and apoptotic functions of p53 in tumor initiation and progression, *Cold Spring Harb. Perspect. Med.*, 2016, **6**, 1–15.
71. A. Hafner, M. L. Bulyk, A. Jambhekar and G. Lahav, The multiple mechanisms that regulate p53 activity and cell fate, *Nat. Rev. Mol. Cell Biol.*, 2019, **20**, 199–210.
72. E. V. Tizhe, N. D. Ibrahim, M. Y. Fatihu, I. O. Igbokwe, B. D. George, S. F. Ambali and J. M. Shallangwa, Serum biochemical assessment of hepatic and renal functions of rats during oral exposure to glyphosate with zinc, *Comp. Clin. Path.*, 2014, **23**, 1043–1050.
73. L. J. Dixon, M. Barnes, H. Tang, M. T. Pritchard and L. E. Nagy, Kupffer cells in the liver, *Comp. Physiol.*, 2013, **3**, 785–797.
74. A. J. Levine and M. Oren, The first 30 years of p53: growing ever more complex, *Nat. Rev. Cancer*, 2019, **9**, 749–758.
75. F. Maqbool, S. Mostafalou, H. Bahadar and M. Abdollahi, Review of endocrine disorders associated with environmental toxicants and possible involved mechanisms, *Life Sci.*, 2016, **145**, 265–273.

76. T. Encarnação, A. A. Pais, M. G. Campos and H. D. Burrows, Endocrine disrupting chemicals: Impact on human health, wildlife and the environment, *Sci. Prog.*, 2019, **102**, 3–42.
77. M. Verderame and E. Limatola, Molecular identification of estrogen receptors (ERα and ERβ) and their differential expression during VTG synthesis in the liver of lizard *Podarcis sicula*, *Gen. Comp. Endocrinol.*, 2010, **168**, 231–238.
78. A. K. Rao, Y. S. Ziegler, I. X. McLeod, J. R. Yates and A. M. Nardulli, Effects of Cu/Zn superoxide dismutase on estrogen responsiveness and oxidative stress in human breast cancer cells, *Mol. Endocrinol.*, 2008, **22**, 1113–1124.
79. I. Paterni, C. Granchi, J. A. Katzenellenbogen and F. Minutolo, Estrogen receptors alpha (ERα) and beta (ERβ): subtype-selective ligands and clinical potential, *Steroids*, 2014, **90**, 13–29.
80. K. M. Eyster, The estrogen receptors: An overview from different perspectives, *Methods Mol. Biol.*, 2016, **1366**, 1–10.

CHAPTER 10

What Is Caiman latirostris *Teaching Us About Endocrine Disruptors?*

M. DURANDO,*[a,c] G. H. GALOPPO,[b,c,d] Y. E. TAVALIERI,[b,d] M. V. ZANARDI[a,c] AND M. MUÑOZ-DE-TORO[b,d]

[a] Laboratorio de Endocrinología y Carcinogénesis, Instituto de Salud y Ambiente del Litoral (ISAL, UNL-CONICET), Facultad de Bioquímica y Ciencias Biológicas, Universidad Nacional del Litoral (UNL), Santa Fe, Argentina; [b] Laboratorio de Eco-Fisiopatología, Instituto de Salud y Ambiente del Litoral (ISAL, UNL-CONICET), Facultad de Bioquímica y Ciencias Biológicas, Universidad Nacional del Litoral (UNL), Santa Fe, Argentina; [c] Cátedra de Fisiología Humana, Facultad de Bioquímica y Ciencias Biológicas, Universidad Nacional del Litoral (UNL), Santa Fe, Argentina; [d] Cátedra de Patología Humana, Facultad de Bioquímica y Ciencias Biológicas, Universidad Nacional del Litoral (UNL), Santa Fe, Argentina
*Email: mdurando@fbcb.unl.edu.ar

10.1 Ecophysiological Characteristics of the Broad-snouted Caiman (*Caiman latirostris*)

The broad-snouted caiman (*Caiman latirostris*) (see Figure 10.1) is a South American crocodilian species broadly distributed in tropical and subtropical wetlands of Argentina, Brazil, Bolivia, Paraguay and Uruguay. In comparison with other crocodilian species that inhabit tropical regions, *C. latirostris*

Issues in Toxicology No. 45
Bird and Reptile Species in Environmental Risk Assessment Strategies
Edited by Guillermo Eli Liwszyc and Marcelo L. Larramendy

Published by the Royal Society of Chemistry, www.rsc.org

Figure 10.1 Ten day-old hatchling of a broad-snouted caiman (*Caiman latirostris*). Photograph: German Hugo Galoppo.

inhabits the southernmost latitudes. This characteristic makes this caiman species be considered as cold-tolerant.[1]

10.1.1 Sex Determination

As with all crocodilian species reported so far, *C. latirostris* displays temperature-dependent sex determination (TSD). Eggs incubated at 29–31 °C produce 100% of female caimans (for which these temperatures are called female-producing temperatures (FPTs)), whereas eggs incubated at 33–34 °C produce 100% of male caimans (for which these temperatures are called male-producing temperatures (MPTs)). Nevertheless, eggs incubated at intermediate temperatures or temperatures higher than 34.5 °C produce variable percentages of both sexes.[2–4] The period of embryonic sex determination has been named the thermosensitive window.[5] Although temperature is a major factor in sex determination, it has been demonstrated that hormones or hormonally active compounds also play a determinant role in this process. In fact, exposure to exogenous 17β-estradiol (E_2) during the thermosensitive window overrides the effects of temperature, producing phenotypic females when eggs are incubated at the MPT.[4,5] This effect has been defined as sex reversal or hormone-induced sex determination (HSD).

10.1.2 Reproduction

Caiman latirostris is an oviparous species with seasonal reproduction. Since the reproductive season of *C. latirostris* depends on the latitude and environmental factors such as temperature, rainfall and sunlight hours, the period in which this species is reproductively active shows geographical variations. In Brazil, the nesting season spreads from August to January, in Uruguay from January to March[7] and in Argentina from December to March.[7,8] During each reproductive season, an adult female lays about 18–50 eggs per clutch.[7]

10.1.3 Hatchling Growth and Development

At birth, the total length of hatchlings is about 23 cm.[4] In wild environments, sexual maturity is considered to be reached between 6 and 10 years after birth, when the animals reach a total length between 120 and 140 cm.[9] To the best of our knowledge, the timing of puberty in *C. latirostris* has not yet been studied. For other crocodilian species, it has been reported that, in wild environments, the time of puberty onset is highly variable among individuals, being affected by several physiological and environmental cues.[10] However, the temporal pattern of development in caimans raised under controlled conditions can differ from that observed in wild environments.[11,12]

10.1.4 Feeding Habits, Social Behavior and Longevity

Caiman latirostris feeds mainly on a variety of aquatic insects, crustaceans and other invertebrates, particularly snails, but its diet can also include other reptiles and mammals. As caimans grow, the diet characterized by small invertebrates can shift to a diet characterized by vertebrates, which makes *C. latirostris* an apex predator in its environment.[13]

The movement and home ranges of adult caimans depend on the availability of resources for survival, the reproductive activity, anthropogenic disturbances and flooding events.[14] Despite the different factors affecting movement, the relative restricted home ranges reported by Marques *et al.*[14] makes *C. latirostris* a site-fidelity species.

Regarding its longevity, although *C. latirostris* has been described as a long-lived species, there are no reports indicating the average survival age in wild environments. However, a 26 year-old specimen raised in captivity and 26 year-old wild-caught specimens (age estimated by the snout-to-vent length) have been reported.[15] Additionally, a 58 year-old specimen of *C. latirostris* has been reported in captivity,[16] supporting the concept that *C. latirostris* is a long-lived species with a human-like lifespan.

10.1.5 Sexual Dimorphism

Possibly, the most evident sexual dimorphism observed in *C. latirostris* relies on body size. Adult female *C. latirostris* specimens are characteristically smaller than males.[9,15] Besides, the craniofacial features of adult wild-caught specimens are also sexually dimorphic.[17] This sexual dimorphism, important for sex recognition during the reproductive season, seems to be achieved at advanced age, since no sex-related differences in craniofacial features have been found in juvenile caimans.[18] However, at the juvenile stage, sexual dimorphism is evident in external genitalia.[19] Accordingly, these morphometric characteristics constitute an important tool for classifying caiman specimens as females or males in fieldwork.

Sexual dimorphism also includes differential characteristics at biochemical, molecular and histofunctional levels, which can be altered by exposure to hormonally active pollutants.[20,21] An example of this situation is the induction of vitellogenin in male caimans after postnatal exposure to estrogenic compounds.[22] Vitellogenin is a large phospholipoglycoprotein normally synthesized in the liver of adult female caimans in response to estrogenic stimulation and normally undetectable in the plasma of male caimans. Since vitellogenin can be induced by exposure to xenoestrogens, the presence of this protein in the blood of male caimans would be a suitable biomarker of exposure to exogenous estrogenic compounds.[22] At the molecular level, sexually dimorphic expression of genes related to sexual differentiation (such as *amh* and *sox-9*)[6] and sexually dimorphic expression of estrogen receptor alpha (ERα) in the cerebral cortex have been reported in 10 day-old caimans.[23] At the histological level, male and female gonads can be differentiated at embryonic developmental stage 27.[24] Moreover, subtle sexual dimorphism in the histological compartmentalization of the thyroid gland has been reported in juvenile caimans and sexually dimorphic responses have been observed when female or male caimans are exposed to several environmental pollutants.[21]

10.2 Endocrine-disrupting Chemicals

The increasing urbanization, the expansion of areas destined to agriculture and the continuous industrialization of productive processes have led to the introduction of a large variety of man-made chemicals into the environment. These chemicals can induce adverse effects both in humans and in wildlife.[25] When such effects include the endocrine system, these chemicals are referred to as endocrine-disrupting chemicals (EDCs).[26] An EDC has been defined as "an exogenous substance or mixture that alters function(s) of the endocrine system and consequently causes adverse health effects in an intact organism, or its progeny, or (sub) populations".[27] To consider a chemical as an EDC, the European Food Safety Authority (EFSA) has proposed three criteria, namely that it should (I) exhibit endocrine activity, (II) generate adverse effects in an intact organism or (sub)population and (III) produce a causal relationship between the chemical and the effect on the endocrine activity.[28]

EDCs can be found in food, cosmetics, plastic bottles, detergents, flame retardants, pharmaceutical products and pesticides, among others.[29] Exposure to EDCs can occur through drinking water, food intake, dermal absorption and inhalation of gases as well as air particles.[29] Since wildlife and humans are constantly exposed to EDCs, these chemicals constitute a real threat to human and wildlife health.[30]

The effects of EDCs depend on both the level of exposure and the developmental stage at which the exposure occurs.[31] Regarding the level of exposure, many EDCs exhibit non-monotonic dose–response curves. Thus, the principle of traditional toxicology that "the dose makes the poison" may not apply to EDCs.[32] Regarding the developmental stage at which the exposure occurs, although EDCs may alter physiological responses during the whole life of an individual, the prenatal and neonatal stages constitute a critically sensitive window of vulnerability and the consequences can be observed later in life.[25,32–35]

10.2.1 Mechanism of Action of EDCs

Some EDCs bind to hormone receptors, acting either as agonists or as antagonists, and may thus alter the action of endogenous hormones.[35,36] Exposure to a single EDC at the no-observed-adverse-effect level (NOAEL) can be innocuous, but when combined with another EDC could have negative effects in the exposed organism.[33,35] The most studied EDCs are those that mimic E_2. EDCs that interfere with ER signaling generate biological responses *via* both classical and non-classical pathways.[25,37]

The effects of EDC exposure are typically categorized as either organizational or activational. Organizational effects are typically permanent and are the result of EDC exposure during the prenatal and/or neonatal stages. Activational effects are typically transitory and are the result of exposure occurring during adulthood.[38] Epigenetic changes that can be inherited by subsequent generations can be induced by EDC exposure during the prenatal and/or neonatal period, whereas non-epigenetic changes have been reported when the exposure occurs during adulthood (for review, see ref. 33).

10.2.2 Types of EDCs

Following the classification proposed by the World Health Organization (WHO), EDCs can be classified based on their physico-chemical properties or application areas in: (I) persistent and bioaccumulative chemicals (*e.g.*, organochlorine compounds (OCCs)), (II) less persistent and less bioaccumulative chemicals (*e.g.*, bisphenol A (BPA)), (III) pesticides, pharmaceuticals and personal care product ingredients (*e.g.*, atrazine (ATZ)) and (IV) other chemicals that include metals, natural hormones and phytoestrogens.[39]

Here, we will focus on some of the EDCs most studied by our research group.

- OCCs are characterized by their persistence in the environment, capacity to be transported over long distances and high toxicity and lipophilicity.[40,41] OCCs include organochlorine pesticides (OCPs) and polychlorinated biphenyls (PCBs). Among OCPs, endosulfan is a pesticide that has been extensively used worldwide in cotton, corn, soy, wheat, sunflower and sorghum crops against several species of insects and arachnids. Due to the proved deleterious effects in both wildlife and humans,[42] in the USA, its use was banned by the US Environmental Protection Agency (US EPA) in 2010,[43] whereas in Argentina, endosulfan was banned in 2013.[44] Unfortunately, due to its high efficiency and low cost, in other countries such as India, China, Brazil, South Korea and Israel, endosulfan has not yet been banned and is still in use.[45–48] In fact, endosulfan has been detected not only in environmental abiotic matrices (soil and water)[49–51] but also in biological matrices (serum, eggs and tissues) of wild species,[52–56] including *C. latirostris* eggs.[57] In addition, endosulfan has been found to be an ER agonist and androgen receptor (AR) antagonist.[25,58–60]

- BPA is one of the most studied EDCs because it is one of the highest-volume chemicals produced worldwide and a chemical used in many everyday products.[61] BPA is used mainly in the production of polycarbonate plastic and epoxy resins. The use of BPA-based polycarbonates in baby bottles is banned in some countries,[39] including Argentina.[62] BPA has been detected in many environmental matrices, such as ground and surface waters, river sediments, soil and air.[63] Besides, BPA has been observed in biological matrices such as tissues of marine and freshwater aquatic species like fish and amphibians (reviewed in ref. 64). BPA is an estrogenic chemical that binds to both ERα and ERβ. Additionally, in *Xenopus laevis*, it has been demonstrated to act as an antagonist of thyroid hormone receptor and to have anti-androgenic activity.[61,65,66]
- ATZ is a triazine herbicide used worldwide.[67] ATZ has been detected in several environmental matrices, such as soil, surface water and ground water at concentrations higher than the acceptable limit.[67] Regarding biological matrices, ATZ has been found in tissues of aquatic turtles, earthworms, bivalves and microalgae.[68–71] Because of its potential risk of contaminating drinking water, the commercial use of ATZ has been banned in several countries.[67] ATZ acts as an anti-androgen[72] and as an estrogen agonist that induces the activity of aromatase, leading to an excess of estrogen production.[25] In addition, ATZ alters thyroid homeostasis, which is critical in amphibian metamorphosis.[73] In this sense, several studies have found that exposure to ATZ can lead to a decrease in the metamorphic rate and a delay in this process.[73–76]

In the following sections, readers will be introduced to the employment of *C. latirostris* as a sentinel organism to detect and monitor EDC exposure, highlighting the main effects of EDCs on the reproductive system and thyroid function.

10.3 *Caiman latirostris* as a Sentinel of Environmental Pollution

The use of animal sentinel systems to detect and monitor pollutants in a given environment has been applied for several years.[77] An animal sentinel system is defined as "a system in which data on animals exposed to contaminants in the environment are regularly and systematically collected and analyzed to identify potential health hazards to other animals or humans".[78]

Species used as sentinels must have measurable responses to the pollutants aimed to be monitored, which may range from molecular to population levels. To qualify as a suitable species, when considering pollutant monitoring, a potential sentinel must meet, according to the previously reported criteria,[77,79] the following aspects:

- high habitat fidelity and, preferably, territorial;
- representativeness toward its environment (apex predators fulfill this last quality more prominently as they bioaccumulate local pollutants during large periods of time);
- wide and known distribution range in the study area, availability for capture all year round and a well-known biology and ecology;
- easy identification, wide recognition by the general public[80] and interest in its preservation;
- sufficient population size and easy sampling to ensure practicability and reproducibility;
- body size large enough to proportionate adequate samples of tissues and fluids to study several biological markers;
- greater susceptibility or greater exposure to a particular pollutant than humans in the same environment;
- bioaccumulation of the studied pollutant at measurable concentrations or enough sensitivity to the pollutant to show measurable changes in its physiology or behavior; and
- longevity of several years and sufficient resistance to pollutants.

The ecophysiological characteristics of *C. latirostris* (described in Section 10.1) make this species a potential sentinel organism to study EDC pollution in wetlands. Besides, *C. latirostris* has been shown to be a sensitive species to contamination with EDCs.[4,20,22,81]

10.4 EDCs and Their Effects on Reproductive Features

It is known that exposure to EDCs can affect the development of embryos in several species, depending on the exposure time. If the endocrine system is perturbed, the normal hormonal signals that guide proper development could be temporarily or permanently altered, with dramatic consequences for life.[38]

Our research group is interested in understanding how EDCs can modify the reproductive functions. Thus, we have studied the impact of EDCs on caiman reproduction after natural and experimental exposures.

10.4.1 Natural Exposure

Caimans are susceptible to be exposed to pollutants, mainly OCCs that persist in the environment and in the food chain. Additionally, features such as feeding habits and longevity allow caimans to accumulate high concentrations of OCCs in fat tissues, as shown for other crocodilian species.[82–84]

Exposure to OCCs can also occur before hatching by maternal transfer into the yolk or from the nest material through the eggshell.[85] After hatching, there are three main sources of contamination: the remaining yolk, the aquatic environment and the diet. Considering these last concepts, we performed two studies.[57,86]

In the first study, we evaluated the presence and concentrations of OCCs in *C. latirostris* eggs laid in clutches from areas with different degrees of anthropogenic intervention.[57] At least 26 OCCs were measured and classified as OCPs and PCBs. As expected, the concentrations of OCCs in eggs collected from areas with high degrees of anthropogenic disturbances were significantly different from those in eggs from areas with low degrees of anthropogenic intervention. Indeed, eggs collected from zones characterized by intensive agricultural practices showed the highest OCP levels. Additionally, the species of OCPs in neighboring nests were similar. These results show not only that these anthropogenic activities are important sources of OCC pollution but also that OCC pollution follows a spatial pattern. On the other hand, OCCs were detected even in areas with low anthropogenic influence considered to be pristine, and PCBs, which are chemicals considered to be present mainly in urban habitats due to their use in electrical transformers, were found in 93% of the eggs.[57] These findings alert about the long-transport range of OCCs and show their persistence in the environment. It is remarkable that, even when the use of dichlorodiphenyltrichloroethane (DDT) in Argentina was banned in 1991, the most frequently detected OCCs in our study were dichlorodiphenyldichloroethylene (*p,p′*-DDE) and oxychlordane (OCLD), which are the most persistent metabolites of DDT and chlordane, respectively. This finding supports the importance of persistence as a criterion in the assessment of OCC pollution. Although, as mentioned earlier, chlordane and DDT were banned in 1991, they were still used extensively both in agricultural activities and as household insecticides until 1998.[87,88] Although the use and disposal of PCBs in Argentina are also restricted,[89,90] the replacement of PCB transformers has been in process since 2007.[91]

It has been reported that eggs of turtles exposed to PCBs incubated at the MPT produce female animals.[92] In contrast, in our studies in *C. latirostris*, we found that natural exposure to OCCs (including PCBs) did not modify the phenotypic sex expected according to the incubation temperature.[57] This discrepancy could be attributed to different factors. Firstly, turtle species may be more sensitive to OCC exposure than caimans. Secondly, the incubation temperature used could account for a relatively low-sensitivity stage. This is in agreement with the results published by Milnes *et al.*,[93] which show that developing crocodilians are more sensitive to OCC exposure when incubated at more intermediate temperatures than the FPT or MPT. Thirdly, the concentrations of OCCs found may be below the NOAEL, since they are lower than those reported for reference sites.[94,95]

Although we found no correlation between the OCC burden and hatching success or hatchling survival, we did find a significant negative association between the concentration of OCCs – mainly OCPs – and clutch size.[57]

This finding caught our attention because it was in agreement with that described for alligators.[95] Specifically, studies in alligators and turtles have shown a strong correlation between the concentration of OCPs in eggs and the burden of OCPs in maternal tissue.[96,97] Taking into account these findings in alligators and turtles, we can speculate that the reproductive performance of the mother is affected by the exposure to contaminants, which means that the higher the OCP concentration in the eggs, the higher the OCP burden in the maternal tissue. Therefore, we can assume that if the mother has a high level of contamination, the number of the eggs laid is reduced. Moreover, the exposure of the developing embryo to EDCs at the concentrations detected could alter the reproductive performance in adulthood and drastically interfere with population stability.

Since caimans are non-migratory and have short dispersal distance and a small home range, caiman eggs could be sufficiently sensitive to monitor local contamination with OCCs, supporting the role of *C. latirostris* as a non-conventional vertebrate species in environmental risk assessment.

In the second study,[86] we explored the relationship between the levels of OCCs in eggs and the eggshell features. Based on early reports describing a dramatic avian eggshell thinning as a consequence of exposure to DDT metabolites,[98] many researchers have focused on the negative impact of OCCs and other contaminants on the reproductive failures of wildlife oviparous species.[99–101]

The caiman eggshell is composed of an internal flexible fibrous membrane and an external calcareous rigid and breakable porous shell.[102] The external porous shell allows the exchange of water and gases between the embryo and the environment. Thus, the survival of the embryo is highly dependent on the eggshell features.

Unlike observations in the eggshell of birds,[100] we failed to find an association between eggshell thickness and the total egg burden of OCCs or DDT metabolites. However, we found an inverse correlation between the OCC burden and the porosity of the eggshell[87] (see Figure 10.2). The low porosity was associated with a low caiman survival at 10 days of age.[86] This finding suggests that 10 day-old hatchlings from embryos exposed to OCCs have a reduced survival as a consequence of hypoxia, due to the OCC-mediated reduced porosity. These results resemble those obtained in alligators, chickens and turtles where hypoxic conditions threaten the life of the hatchlings.[103–106]

10.4.2 Experimental Exposure

The environmental presence of EDCs has prompted researchers to focus their studies not only on laboratory species but also on wildlife species.[107–110] At our lab, for more than 20 years, we have been studying the impact of BPA, endosulfan and ATZ on the reproductive system of male and female caimans at different developmental stages. A summary of the results found is shown in Table 10.1.

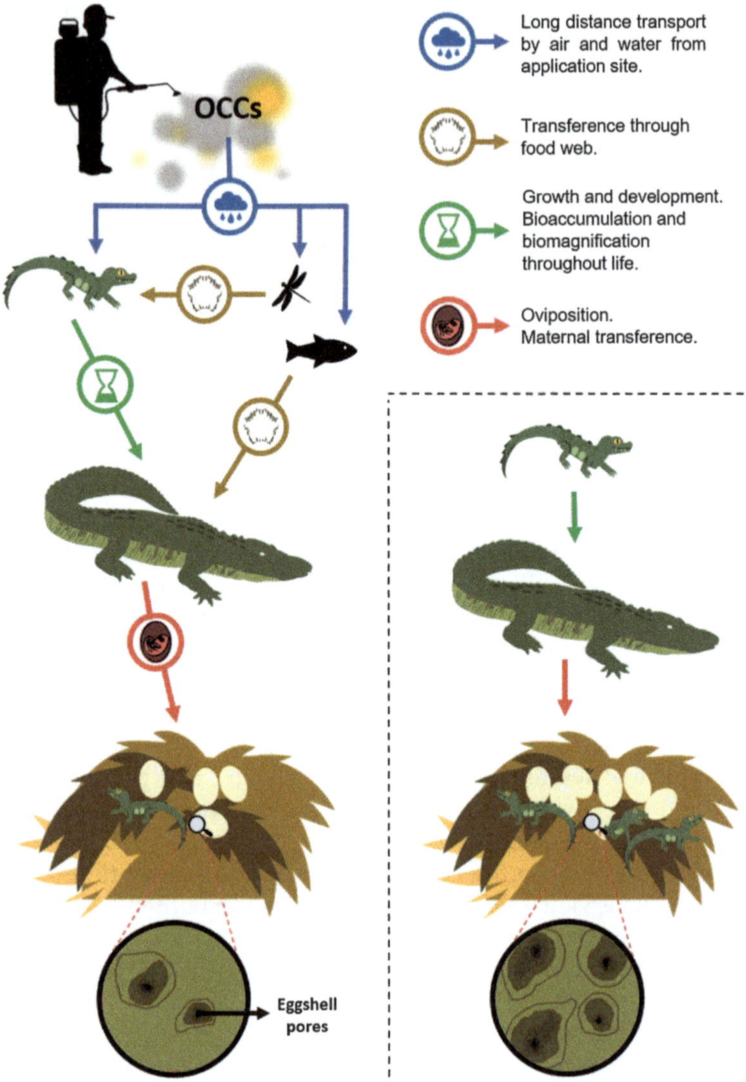

Figure 10.2 Natural exposure to organochlorine compounds (OCCs). The left panel represents the exposure paths of OCCs and their effects on the eggs and offspring of exposed caimans. Alterations due to the OCC burden include decreased clutch size,[53] decreased eggshell porosity and, as a consequence, decreased hatchling survival.[80] The right panel represents the growth, development and nesting of caimans living in zones with low anthropogenic sources of OCCs. The blue arrows symbolize long-distance transport of OCCs, by air and water, from application sites to areas inhabited by non-target species. The yellowish-brown arrows symbolize the transference of OCCs through the food web. The green arrows symbolize the growth and development as well as the bioaccumulation and biomagnification of OCCs in the adipose tissue of caimans living in high burdened zones. The red arrows symbolize the nesting and oviposition as well as the maternal transference of OCCs to the eggs, mainly in the lipid-rich yolk.

Table 10.1 Effects of *in ovo* exposure to endocrine-disrupting chemicals (EDCs) including bisphenol A (BPA), endosulfan (END) and atrazine (ATZ) on the reproductive system of *Caiman latirostris*.[a]

EDC	Doses/ incubation temperature	Time of endpoint evaluation	Main reported effects	Ref.
BPA	1.4 and 140 ppm/30 °C and 33 °C	Neonatal: 10 day-old Late postnatal: 90 day-old Juvenile	BPA 1.4 ppm: males exhibited disrupted seminiferous tubules (with empty lumens). Ten day-old caimans showed a decreased relative tubular area in testes. Juvenile caimans exhibited an increased testicular intratubular proliferation. Females exhibited altered ovarian follicular dynamics. Juvenile females showed an increased incidence of ovarian multioocyte follicles. BPA 140 ppm overrode the temperature effect on sex determination: females were obtained at the MPT. These females exhibited ovaries with delayed development.	3, 5, 10 and 105
END	0.02; 2 and 20 ppm/ 30 °C and 33 °C	At birth Neonatal: 10 day-old Late postnatal: 90 day-old Juvenile	END 2 and 20 ppm led to higher egg weight loss and hatchlings with a lower weight, both in females and in males. END 2 and 20 ppm decreased the serum testosterone levels of neonatal females. END 0.02 and 20 ppm caused an altered testicular histoarchitecture characterized by tortuous seminiferous tubules with empty tubular lumens, in neonates and juveniles. END 2 and 20 ppm increased both the proliferation and the apoptotic index in testes of neonates, although END 20 ppm decreased the intratubular proliferation in testes of juveniles. END 20 ppm modified the distribution pattern of ERα in testes. It also increased the mRNA expression of *amh*, *sox9* and *sf-1* (genes involved in sex determination, gonadal differentiation and steroidogenesis) in the GAM complexes of neonates.	5, 10, 77, 105 and 112

Table 10.1 (*Continued*)

EDC	Doses/ incubation temperature	Time of endpoint evaluation	Main reported effects	Ref.
ATZ	0.2 ppm/ 30 °C and 33 °C	At birth Neonatal: 10 day-old Juvenile	ATZ led to higher egg weight loss and hatchlings with a lower weight, without differences between females and males. ATZ disrupted the testicular histoarchitecture and reduced the percentage of tubular perimeter length occupied by desmin-positive cells. Females exhibited altered ovarian follicular dynamics and decreased levels of serum. ATZ slowed down phallus growth in males. Males exposed to ATZ exhibited decreased expression of androgen receptor (AR), named a demasculinization effect. Females showed increased AR expression. The phallus of both males and females exhibited increased estrogen receptor alpha (ERα) expression due to ATZ exposure.	77, 78, 105 and 112

[a]All treatments were administered with a single dose of the xenobiotic applied on the eggshell at stage 20 of embryo development.

10.4.2.1 In Ovo *Exposure*

In ovo exposure was performed to assess the effects of BPA, ATZ and endosulfan on different target organs of endocrine disruption. Our results demonstrated some organizational effects, as those shown in Figure 10.3. Exposure to 2 ppm ATZ and 20 ppm endosulfan increased egg weight loss and decreased hatchling weight at birth,[81] whereas exposure to BPA 140 ppm overrode the temperature effect on sex determination. Since female caimans were obtained at the MPT, this last result agrees with the effect observed after exposure to a high dose of E_2, which is the positive control of endocrine disruption.[4,6,111] Our results also demonstrated that a dose of BPA 100-fold larger than that of E_2 evoked a comparable effect. In fact, our *in vivo* system was more sensitive than an *in vitro* system, in which an E_2/BPA ratio of 1/10 000 was suggested.[112–117] Moreover, all the EDCs evaluated caused histomorphological changes in caiman male and female gonads from 10 days old to the juvenile stage.[4,6,11,111,118] The most significant results were associated with disrupted seminiferous tubules with empty lumens and altered ovarian follicular dynamics, in males and females, respectively.

Several pieces of evidence from alligators living in Lake Apopka (Florida, USA) have shown that males exhibited reduced phallus (external genitalia)

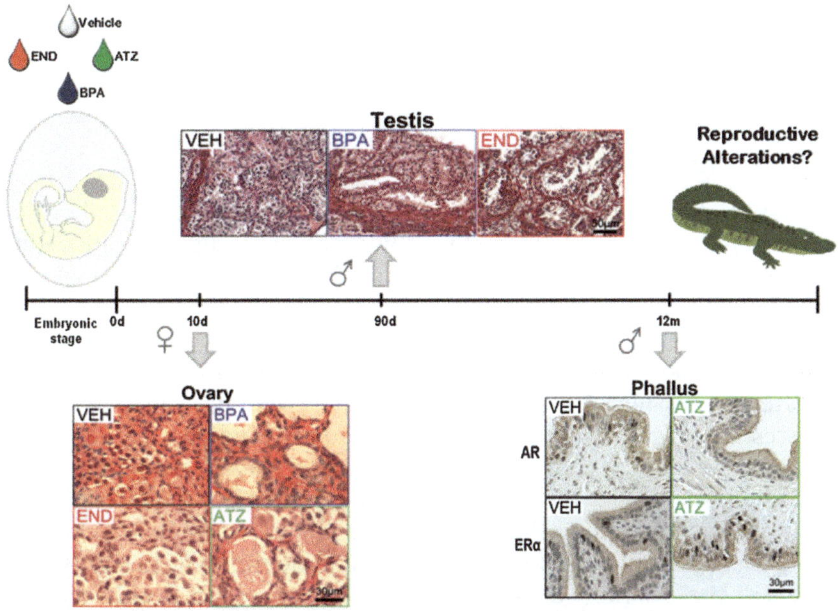

Figure 10.3 Alterations in the reproductive system of caimans after prenatal exposure to atrazine (ATZ), bisphenol A (BPA) or endosulfan (END). Representative images of histomorphological alterations in the ovary, testis and phallus are shown. ATZ, BPA and endosulfan increased the presence of type-3 follicles in ovaries of 10 day-old caimans;[105] BPA and endosulfan altered the morphology of seminiferous tubules in the testis of 90 day-old caimans;[5,10] ATZ increased the expression of estrogen receptor alpha (ERα) and decreased the expression of androgen receptor (AR) in the phallus of prepubertal juvenile male caimans.[78] For a detailed description, please see text.

size and decreased concentrations of serum testosterone, compared to male alligators living in non-contaminated environments.[119,120] In agreement with this, we found changes in the phallus of juvenile male and female caimans experimentally exposed to ATZ (see Table 10.1). Interestingly, we showed an apparent slower growth of the phallus compared to the body growth. By analyzing microscopic features, we demonstrated a differential response to ATZ between males and females. In males, ATZ induced a decreased protein expression of AR, evidencing its demasculinization effect, whereas in females, ATZ increased the expression of AR. Besides, we detected an increased expression of ERα in all ATZ-exposed caimans.[20]

Since E_2 is considered to be the positive control to study EDCs with estrogenic or anti-estrogenic activity, we also studied the reproductive features of female caimans after exposure to E_2 or BPA. In the following, we will summarize the main results obtained. Ten day-old HSD females (obtained after E_2 or BPA exposure) exhibited a delayed development of their gonads[4] and showed a high mRNA expression of *sox9*,[6] a typical result obtained in males.[121] Regarding juvenile caimans, the phallus size and histological features of TSD and HSD females were similar; however, the expression of AR and ERα in HSD females resembled that observed in TSD males.[19]

10.4.2.2 *Early Postnatal Exposure*

Our results after neonatal exposure to BPA and E_2 showed increased oviduct luminal epithelial height, probably as a result of epithelial stratification, which is in turn induced by increased proliferative activity.[122] Females exposed to BPA and E_2 also showed changes in the collagen remodeling and in the subepithelial area of α-smooth muscle actin (α-SMA)-positive cells. These changes were accompanied by an upregulated expression of AR and a deregulated expression of ERα in the luminal epithelium.[122] Taken together, these results suggest that neonatal exposure to EDCs induces changes at the oviductal level, which could lead to a precocious oviduct development and differentiation.[122] With these results in mind, we further tried to clarify the mechanisms behind these changes. In a recent work, we found that BPA and E_2 modified the protein expression of FoxA2, β-catenin, Wnt-5a and Wnt-7a, as well as the desmin/α-SMA ratio, all proteins associated with oviductal development and differentiation.[123]

10.5 EDCs and Their Effects on the Thyroid Histofunctional Characteristics

The phenomenon of thyroid disruption is a relatively new field of study compared to the hypothalamic–pituitary–gonadal axis disruption. The experimental exposure to OCCs has been found to cause thyroid disruption in fishes, birds, reptiles and mammals.[39,124,125] Additionally, the natural exposure of wildlife to mixtures of environmental pollutants has shown thyroid-disrupting effects. These effects include altered circulating levels of thyroid hormones in alligators, polar bears, fishes and birds,[126–129] and histomorphological or histofunctional alterations in alligators and birds.[129,130] Finally, changes in thyroid hormone activity, as a consequence of modifications of deiodinase activity, in polar bears and birds have been reported as well.[129,131] Similarly, lizards from sites known to be contaminated with herbicides have been found to present a more active thyroid gland, with significantly larger follicular area, and a far greater abundance of reabsorption vacuoles.[132]

By using models of experimental exposure, we also detected that some EDCs could cause long-lasting effects on the thyroid gland. Prepubertal juvenile caimans prenatally exposed to endosulfan showed increased stromal compartments along with decreased colloid production/storage.[133] Our results also showed that ATZ caused thyroid enlargement, colloid depletion, increased follicular epithelial height and increased hyperplastic follicles, all of which are considered to be signs of excessive stimulation of thyroid-stimulating hormone.[21] Interestingly, preliminary results suggest that more subtle ATZ-induced thyroid alterations involving cell proliferation are present at earlier developmental stages, with females being more sensitive than males.[134] This result confirms the sex-related sensitivity to thyroid disruption related to ATZ exposure and suggests that exposure to environmentally relevant doses to this pesticide induces organizational effects that become evident as caimans grow and mature. Thyroid disruption could, in turn, have potential long-lasting effects on caiman growth, development, metabolism and reproduction.

10.6 Conclusions

The impact of environmental pollution on different species is increasing day by day. Our role as scientists is both to provide more evidence of the consequences of this pollution and to establish a network with control agencies to improve human social or occupational behavior.

In this chapter, we summarized the most relevant results obtained after two decades of work. We highlight the alterations related to male and female reproductive systems, because they could have negative consequences for the species population dynamics. We also showed several pieces of evidence that underpin how anthropogenic intervention is affecting large areas considered as pristine in the near past and demonstrate the possible use of *C. latirostris* as a sentinel species due to its ecophysiological characteristics. Last but not least, we present the changes detected in both the thyroid gland and the hormonal milieu.

Finally, while sex determination, gonadal histoarchitecture and sex hormone profiles, among others, are potential biomarkers of prenatal and early postnatal exposure to EDCs (organizational effects), the induction of vitellogenin in male caimans represents a more sensitive biochemical endpoint of postnatal exposure to estrogenic compounds (activational effect).

All these results warn us about the high potency of chemicals that are daily discharged into wetland ecosystems, which are home to *C. latirostris* and other wild species.

Acknowledgements

YET, MVZ and MD work at the Consejo Nacional de Investigaciones Científicas y Técnicas (CONICET, Argentina): the former two as fellows and the latter as a researcher. GHG and MMT are researchers of the Universidad Nacional del Litoral (UNL, Argentina). This work was supported by grants from the Agencia Nacional de Promoción de la Investigación, el Desarrollo Tecnológico y la Innovación (Agencia I + D + I; PICT2016 No. 0656) and the Universidad Nacional del Litoral (UNL), Argentina.

References

1. C. I. Piña, Un estudio del efecto de las temperaturas de incubación en la determinación sexual y el primer año de crecimiento del yacaré overo, Caiman latirostris (Daudin, 1802), Facultad de Ciencias Exactas, Físicas y Naturales, Universidad Nacional de Córdoba, 2001. Available at https://catalogo.biblio.unc.edu.ar/Record/exactas.18549.
2. M. V. Parachú-Marcó, P. Leiva, J. L. Iungman, M. S. Simoncini and C. I. Piña, New evidence characterizing temperature-dependent sex determination in broad-snouted caiman, *Caiman latirostris*, *Herpet. Conserv. Biol.*, 2017, **12**, 78–84.
3. C. I. Piña, A. Larriera and M. R. Cabrera, Effect of incubation temperature on incubation period, sex ratio, hatching success, and survivorship in *Caiman latirostris* (Crocodylia, Alligatoridae), *J. Herpetol.*, 2003, **37**, 199–202.
4. C. Stoker, F. Rey, H. Rodriguez, J. G. Ramos, P. Sirosky, A. Larriera, E. H. Luque and M. Muñoz-de-Toro, Sex reversal effects on *Caiman latirostris* exposed to environmentally relevant doses of the xenoestrogen bisphenol A, *Gen. Comp. Endocrinol.*, 2003, **133**, 287–296.
5. C. A. Smith and J. M. P. Joss, Gonadal sex differentiation in *Alligator mississippiensis*, a species with temperature-dependent sex determination, *Cell Tissue Res.*, 1993, **273**, 149–162.
6. M. Durando, L. Cocito, H. A. Rodríguez, J. Varayoud, J. G. Ramos, E. H. Luque and M. Muñoz-de-Toro, Neonatal expression of *amh, sox9* and *sf-1* mRNA in *Caiman latirostris* and effects of *in ovo* exposure to endocrine disrupting chemicals, *Gen. Comp. Endocrinol.*, 2013, **191**, 31–38.
7. L. M. Verdade, A. Larriera and C. I. Piña, in *Crocodiles: Status Survey and Conservation Action Plan*, ed. S. C. Manolis and C. Stevenson, Crocodile Specialist Group: Darwin, 2010, pp. 18–22.
8. M. S. Simoncini, C. I. Piña, F. B. Cruz and A. Larriera, Climatic effects on the reproductive biology of *Caiman latirostris* (Crocodylia: Alligatoridae), *Amphibia-Reptilia*, 2011, **32**, 305–314.
9. P. A. Micucci and T. Waller, in *La conservación y el manejo de caimanes y cocodrilos de América Latina, Vol. I*, ed. A. Larriera and L. M. Verdade, Fundación Banco Bica, Santo Tomé, Santa Fe, Argentina, 1995, pp. 81–112.
10. A. A. Rooney, D. A. Crain, A. R. Woodward and L. J. Guillette Jr, Seasonal variation in plasma sex steroid concentrations in juvenile American alligators, *Gen. Comp. Endocrinol.*, 2003, **135**, 25–34.
11. M. Durando, G. Canesini, L. L. Cocito, G. H. Galoppo, M. A. Zayas, E. H. Luque and M. Muñoz-de-Toro, Histomorphological changes in testes of broad-snouted caimans (*Caiman latirostris*) associated with *in ovo* exposure to endocrine-disrupting chemicals, *J. Exp. Zool. A Ecol. Genet. Physiol.*, 2016, **325**, 84–96.
12. M. V. Parachú-Marcó, C. I. Piña, M. Simoncini and A. Larriera, Effects of incubation and rearing temperatures on *Caiman latirostris* growth, *Zool. Stud.*, 2010, **49**, 367–373.

13. C. Borteiro, F. Gutierrez, M. Tedrosa and F. Kolenc, Food habits of the broad-snouted caiman (*Caiman latirostris*: Crocodylia, Alligatoridae) in northwestern Uruguay, *Stud. Neotrop. Fauna Environ.*, 2008, **44**, 31–36.
14. T. S. Marques, L. A. B. Bassetti, N. R. F. Lara, T. C. G. Portelinha and C. I. Piña, Home range and movement pattern of the broad-snouted caiman (*Caiman latirostris*) in a silviculture dominated landscape, *South Amer, J. Herpetol.*, 2020, **16**, 16–25.
15. N. Carl and J. Darlington, Extended parental care in broad-snouted caiman *Caiman latirostris* (Daudin, 1801) (Crocodylia, Alligatoridae), *Herpetol. Notes*, 2017, **10**, 127–129.
16. R. Weigl, Longevity of crocodilians in captivity, *Int. Zoo News*, 2014, **61**, 363–373.
17. L. M. Verdade, Cranial sexual dimorphism in captive adult broad-snouted caiman (*Caiman latirostris*), *Amphibia-Reptilia*, 2003, **24**, 92–99.
18. M. A. Zayas, Exposición natural y experimental a compuestos agroindustriales: efectos sobre variables bioquímicas y parámetros dimórficos de relevancia en la reproducción de *Caiman latirostris*, Facultad de Bioquímica y Ciencias Biológicas, Universidad Nacional del Litoral, 2014. Available at https://bibliotecavirtual.unl.edu.ar:8443/handle/11185/523.
19. Y. E. Tavalieri, G. H. Galoppo, G. Canesini, J. C. Truter, J. G. Ramos, E. H. Luque and M. Muñoz-de-Toro, The external genitalia in juvenile *Caiman latirostris* differ in hormone sex determinate-female from temperature sex determinate-female, *Gen. Comp. Endocrinol.*, 2019, **273**, 236–248.
20. Y. E. Tavalieri, G. H. Galoppo, G. Canesini, E. H. Luque and M. Muñoz-de-Toro, Effects of agricultural pesticides on the reproductive system of aquatic wildlife species, with crocodilians as sentinel species, *Mol. Cell. Endocrinol.*, 2020, **518**, 110918.
21. G. H. Galoppo, Y. E. Tavalieri, G. Schierano-Marotti, M. R. Osti, E. H. Luque and M. Muñoz-de-Toro, Long-term effects of *in ovo* exposure to an environmentally relevant dose of atrazine on the thyroid gland of *Caiman latirostris*, *Environ. Res.*, 2020, **186**, 109410.
22. F. Rey, J. G. Ramos, C. Stoker, L. E. Bussmann, E. H. Luque and M. Muñoz-de-Toro, Vitellogenin detection in *Caiman latirostris* (Crocodylia: Alligatoridae): a tool to assess environmental estrogen exposure in wildlife, *J. Comp. Physiol., B*, 2006, **176**, 243–251.
23. J. Varayoud, L. Monje, G. S. Moreno-Piovano, G. H. Galoppo, E. H. Luque, M. Muñoz-de-Toro and J. G. Ramos, Sexually dimorphic expression of receptor-alpha in the cerebral cortex of neonatal *Caiman latirostris* (Crocodylia: Alligatoridae), *Gen. Comp. Endocrinol.*, 2012, **179**, 205–213.
24. G. Canesini, C. Stoker, G. H. Galoppo, M. L. Durando, M. V. Tschopp, E. H. Luque, M. Muñoz-de-Toro and J. G. Ramos, Temperature- vs. estrogen-induced sex determination in *Caiman latirostris* embryos:

Both females, but with different expression patterns of key molecules involved in ovarian development, *Gen. Comp. Endocrinol.*, 2018, **259**, 176–188.
25. E. H. Luque, M. Muñoz-de-Toro and J. G. Ramos, Estrogenic Agonist, in *Encyclopedia of Reproduction*, ed. M. Skinner, Academic Press, Oxford, England, 2018, vol. 2, pp. 753–759.
26. T. Colborn and C. Clement, *Chemically-induced alterations in sexual and functional development: the wildlife/human connection*, Princeton Scientific Publishing Co., Inc., Princeton, USA, 1992.
27. International Programme on Chemical Safety Global Assessment of the State-of-the-Science of Endocrine Disruptors. Available: https://apps.who.int/iris/bitstream/handle/10665/67357/WHO_PCS_EDC_02.2.pdf?sequence=1&isAllowed=y.
28. European Food Safety Authority (EFSA) Scientific Committee, Scientific opinion on the hazard assessment of endocrine disruptors: scientific criteria for identification of endocrine disruptors and appropriateness of existing test methods for assessing effects mediated by these substances on human health and the environment: Identification/Characterisation of EDs: scientific criteria and test methods, *EFSA J.*, 2013, **11**, 3132.
29. B. Yilmaz, H. Terekeci, S. Sandal and F. Kelestimur, Endocrine disrupting chemicals: exposure, effects on human health, mechanism of action, models for testing and strategies for prevention, *Rev. Endocr. Metab. Disord.*, 2020, **21**, 127–147.
30. T. K. Kasonga, M. Coetzee, I. Kamika, V. M. Ngole-Jeme and M. N. B. Momba, Endocrine-disruptive chemicals as contaminants of emerging concern in wastewater and surface water: a review, *J. Environ. Manage.*, 2021, **277**, 111485.
31. A. Bergman, J. J. Heindel, S. Jobling, K. A. Kidd and R. T. Zoeller, *Interorganization programme for the sound management of chemicals, State of the science of endocrine disrupting chemicals*, United Nations Environment Programme and World Health Organization, 2012.
32. L. N. Vandenberg, T. Colborn, T. B. Hayes, J. J. Heindel, D. R. Jacobs Jr, D. H. Lee, T. Shioda, A. M. Soto, F. S. V. Saal, W. V. Welshons, R. T. Zoeller and J. P. Myers, Hormones and endocrine disrupting chemicals: low-dose effects and nonmonotonic dose responses, *Endocr. Rev.*, 2012, **33**, 378–455.
33. A. C. Gore, V. A. Chappell, S. E. Fenton, J. A. Flaws, A. Nadal, G. S. Prins, J. Toppari and R. T. Zoeller, EDC-2: The Endocrine Society's second scientific statement on endocrine-disrupting chemicals, *Endocr. Rev.*, 2015, **36**, E1–150.
34. L. Gross, The toxic origins of disease, *PLoS Biol.*, 2007, **5**, 1392–1398.
35. R. Lauretta, A. Sansone, M. Sansone, F. Romanelli and M. Appetecchia, Endocrine disrupting chemicals: effects on endocrine glands, *Front. Endocrinol.*, 2019, **10**, 178.

36. C. Sonnenschein and A. M. Soto, An updated review of environmental estrogen and androgen mimics and antagonists, *J. Steroid Biochem. Mol. Biol.*, 1998, **65**, 143–150.
37. E. K. Shanle and W. Xu, Endocrine disrupting chemicals targeting estrogen receptor signaling: identification and mechanisms of action, *Chem. Res. Toxicol.*, 2011, **24**, 6–19.
38. L. J. Guillette Jr, D. A. Crain, A. A. Rooney and D. B. Pickford, Organization versus activation: the role of endocrine-disrupting contaminants (EDCs) during embryonic development in wildlife, *Environ. Health Perspect.*, 1995, **103**, 157–164.
39. World Health Organization (WHO), in *State of the Science of endocrine Disrupting Chemicals – 2012. An assessment of the state of the science of endocrine disruptors prepared by a group of experts for the United Nations Environment Programme (UNEP) and WHO*, ed. WHO/UNEP, 2013, pp. 189–258. Available: https://apps.who.int/iris/handle/10665/78102
40. T. Encarnação, A. A. Pais, M. G. Campos and H. D. Burrows, Endocrine disrupting chemicals: impact on human health, wildlife and the environment, *Sci. Prog.*, 2019, **102**, 3–42.
41. C. J. Martyniuk, A. C. Mehinto and N. D. Denslow, Organochlorine pesticides: agrochemicals with potent endocrine-disrupting properties in fish, *Mol. Cell. End.*, 2020, **507**, 110764.
42. N. Lubick, Environment. Endosulfan's exit: U.S. EPA pesticide review leads to a ban, *Science*, 2010, **328**, 1466.
43. United States Environmental Protection Agency (U.S. EPA), Office of Pesticide Programs, Endosulfan phase-out pesticides. US EPA, 2010. Available: https://archive.epa.gov/pesticides/reregistration/web/html/endosulfan-agreement.html.
44. Servicio Nacional de Sanidad y Calidad Agroalimentaria (SENASA), Res. 511/11. Available: https://www.cac.com.ar/data/documentos/44_511%2011.pdf.
45. Y. Fang, Z. Nie, Q. Die, Y. Tian, F. Liu, J. He and Q. Huang, Spatial distribution, transport dynamics, and health risks of endosulfan at a contaminated site, *Environ. Pollut.*, 2016, **216**, 538–547.
46. K. Ghosh, B. Chatterjee, A. G. Jayaprasad and S. R. Kanade, The persistent organochlorine pesticide endosulfan modulates multiple epigenetic regulators with oncogenic potential in MCF-7 cells, *Sci. Total Environ*, 2018, **624**, 1612–1622.
47. J. Nasir, X. Wang, B. Xu, C. Wang, D. R. Joswiak, S. Rehman, A. Lodhi, S. Shafiq and R. Jilani, Selected organochlorine pesticides and polychlorinated biphenyls in urban atmosphere of Pakistan: concentration, spatial variation and sources, *Environ. Sci. Technol.*, 2014, **48**, 2610–2618.
48. I. C. Yadav, N. L. Devi, J. Li, G. Zhang and P. R. Shakya, Occurrence, profile and spatial distribution of organochlorines pesticides in soil of Nepal: implication for source apportionment and health risk assessment, *Sci. Total Environ*, 2016, **573**, 1598–1606.

49. M. L. Ballesteros, K. S. Miglioranza, M. Gonzalez, G. Fillmann, D. A. Wunderlin and M. A. Bistoni, Multimatrix measurement of persistent organic pollutants in Mar Chiquita, a continental saline shallow lake, *Sci. Total Environ*, 2014, **490**, 73–80.
50. S. I. Grondona, M. Gonzalez, D. E. Martínez, H. E. Massone and K. S. Miglioranza, Endosulfan leaching from Typic Argiudolls in soybean tillage areas and groundwater pollution implications, *Sci. Total Environ*, 2014, **484**, 146–153.
51. L. Regaldo, M. F. Gutierrez, U. Reno, V. Fernandez, S. Gervasio, M. R. Repetti and A. M. Gagneten, Water and sediment quality assessment in the Colastiné Corralito stream system (Santa Fe, Argentina): impact of industry and agriculture on aquatic ecosystems, *Environ. Sci. Pollut. Res. Int.*, 2018, **25**, 6951–6968.
52. J. A. Araujo-León, G. J. Mena-Rejón, E. B. Canché-Pool and H. A. Ruiz-Piña, Biomonitoring organochlorine pesticides in *Didelphis virginiana* from Yucatan, Mexico by GC-ECD, *Bull. Environ. Contam. Toxicol.*, 2019, **102**, 836–842.
53. M. F. Barni, M. Gonzalez and K. S. Miglioranza, Assessment of persistent organic pollutants accumulation and lipid peroxidation in two reproductive stages of wild silverside (*Odontesthes bonariensis*), *Ecotoxicol. Environ. Saf.*, 2014, **99**, 45–53.
54. A. De Roma, C. Rossini, C. Riverso, G. Galiero and M. Esposito, Endosulfan poisoning in canids and felids in the Calabria region of southern Italy, *J. Vet. Diagn. Invest.*, 2017, **29**, 122–125.
55. A. T. Kuvarega and P. Taru, Accumulation of endosulfan in wild rat, *Rattus norvegicus* as a result of application to soya bean in Mazoe (Zimbabwe), *Environ. Monit. Assess.*, 2007, **125**, 333–345.
56. M. G. Paulino, T. P. Benze, H. Sadauskas-Henrique, M. M. Sakuragui, J. B. Fernandes and M. N. Fernandes, The impact of organochlorines and metals on wild fish living in a tropical hydroelectric reservoir: bioaccumulation and histopathological biomarkers, *Sci. Total Environ.*, 2014, **497–498**, 293–306.
57. C. Stoker, M. R. Repetti, S. R. García, M. A. Zayas, G. H. Galoppo, H. R. Beldoménico, E. H. Luque and M. Muñoz-de-Toro, Organochlorine compound residues in the eggs of broad-snouted caimans (*Caiman latirostris*) and correlation with measures of reproductive performance, *Chemosphere*, 2011, **84**, 311–317.
58. L. Kass, A. L. Gomez and G. A. Altamirano, Relationship between agrochemical compounds and mammary gland development and breast cancer, *Mol. Cell. Endocrinol.*, 2020, **508**, 110789.
59. A. M. Soto, K. L. Chung and C. Sonnenschein, The pesticides endosulfan, toxaphene, and dieldrin have estrogenic effects on human estrogen-sensitive cells, *Environ. Health Perspect.*, 1994, **102**, 380–383.
60. J. Varayoud, L. Monje, T. Bernhardt, M. Muñoz-de-Toro, E. H. Luque and J. G. Ramos, Endosulfan modulates estrogen-dependent genes like

a non-uterotrophic dose of 17beta-estradiol, *Reprod. Toxicol.*, 2008, **26**, 138–145.
61. F. S. Vom Saal and L. N. Vandenberg, Update on the health effects of bisphenol A: overwhelming evidence of harm, *Endocrinology*, 2021, **162**, bqaa171.
62. Administración Nacional de Medicamentos, Alimentos y Tecnología Médica (ANMAT), Resolución 1207/2012. Available: http://www.anmat.gov.ar/boletin_anmat/BO/ Disposicion_1207-2012.pdf.
63. A. Careghini, A. F. Mastorgio, S. Saponaro and E. Sezenna, Bisphenol A, nonylphenols, benzophenones, and benzotriazoles in soils, groundwater, surface water, sediments, and food: A review, *Environ. Sci. Pollut. Res. Int.*, 2015, **22**, 5711–5741.
64. J. Corrales, L. A. Kristofco, W. B. Steele, B. S. Yates, C. S. Breed, E. S. Williams and B. W. Brooks, Global assessment of bisphenol A in the environment: review and analysis of its occurrence and bioaccumulation, *Dose Response*, 2015, **13**, 1559325815598308.
65. R. A. Heimeier, B. Das, D. R. Buchholz and Y. B. Shi, The xenoestrogen bisphenol A inhibits postembryonic vertebrate development by antagonizing gene regulation by thyroid hormone, *Endocrinology*, 2009, **150**, 2964–2973.
66. J. Li, Y. Li, M. Zhu, S. Song and Z. Qin, A multiwell-based assay for screening thyroid hormone signaling disruptors using *thibz* expression as a sensitive endpoint in *Xenopus laevis*, *Molecules*, 2022, **27**, 798.
67. M. E. Taverna, C. A. Busatto, M. R. Lescano, V. V. Nicolau, C. S. Zalazar, G. R. Meira and D. A. Estenoz, Microparticles based on ionic and organosolv lignins for the controlled release of atrazine, *J. Hazard. Mater.*, 2018, **359**, 139–147.
68. D. L. Douros, K. F. Gaines and J. M. Novak, Atrazine and glyphosate dynamics in a lotic ecosystem: the common snapping turtle as a sentinel species, *Environ. Monit. Assess.*, 2015, **187**, 114.
69. F. Wang, R. Ji, Z. Jiang and W. Chen, Species-dependent effects of biochar amendment on bioaccumulation of atrazine in earthworms, *Environ. Pollut.*, 2014, **186**, 241–247.
70. A. E. Jacomini, W. E. Paiva Avelar, A. S. Martinez and P. S. Bonato, Bioaccumulation of atrazine in freshwater bivalves *Anodontites trapesialis* (Lamarck, 1819) and *Corbicula fluminea* (Muller, 1774), *Arch. Environ. Contam. Toxicol.*, 2006, **51**, 387–391.
71. A. N. Kabra, M. Ji, J. Choi, J. R. Kim, S. P. Govindwar and B. Jeon, Toxicity of atrazine and its bioaccumulation and biodegradation in a green microalga, *Chlamydomonas mexicana*, *Environ. Sci. Pollut. Res. Int.*, 2014, **21**, 12270–12278.
72. T. B. Hayes, L. L. Anderson, V. R. Beasley, S. R. de Solla, T. Iguchi, H. Ingraham, P. Kestemont, J. Kniewald, Z. Kniewald, V. S. Langlois, E. H. Luque, K. A. McCoy, M. Muñoz-de-Toro, T. Oka, C. A. Oliveira, F. Orton, S. Ruby, M. Suzawa, L. E. Tavera-Mendoza, V. L. Trudeau, A. B. Victor-Costa and E. Willingham, Demasculinization and

feminization of male gonads by atrazine: consistent effects across vertebrate classes, *J. Steroid Biochem. Mol. Biol.*, 2011, **127**, 64–73.
73. V. S. Langlois, A. C. Carew, B. D. Pauli, M. G. Wade, G. M. Cooke and V. L. Trudeau, Low levels of the herbicide atrazine alter sex ratios and reduce metamorphic success in *Rana pipiens* tadpoles raised in outdoor mesocosms, *Environ. Health Perspect.*, 2010, **118**, 552–557.
74. T. D. Hoskins and M. D. Boone, Atrazine feminizes sex ratio in Blanchard's cricket frogs (*Acris blanchardi*) at concentrations as low as 0.1 μg/L, *Environ. Toxicol. Chem.*, 2018, **37**, 427–435.
75. L. Sai, Q. Wu, B. Qu, C. Bo, G. Yu, Q. Jia, L. Xie, Y. Li, Q. Guo, J. C. Ng and C. Peng, Assessing atrazine-induced toxicities in *Bufo bufo gargarizans* Cantor, *Bull. Environ. Contam. Toxicol.*, 2015, **94**, 152–157.
76. R. M. Zaya, Z. Amini, A. S. Whitaker, S. L. Kohler and C. F. Ide, Atrazine exposure affects growth, body condition and liver health in *Xenopus laevis* tadpoles, *Aquat. Toxicol.*, 2011, **104**, 243–253.
77. J. C. Amiard and C. Amiard-Triquet, in *Ecological Biomarkers: Indicators of Ecotoxicological Effects*, ed. C. Amiard-Triquet, J. C. Amiard and P. S. Rainbow, CRC Press, Boca Raton, Florida, 2013, pp. 155–185.
78. National Academy of Sciences, Committee on Animals as Monitors of Environmental Hazards, Board on Environmental Studies, Toxicology National Research Council, in *Animals as Sentinels of Environmental Health Hazards*, ed. National Academy of Sciences, National Academies Press, Washington D.C. USA, 1991, pp. 33–35.
79. B. Berthet, in *Aquatic Ecotoxicology. Advancing Tools for Dealing with Emerging Risks*, ed. C. Amiard-Triquet, J. C. Amiard and C. Mouneyrac, Academic Press, London, 2015, pp. 205–227.
80. R. M. Seston, M. J. Zwiernik, T. B. Fredricks, S. J. Coefield, D. L. Tazelaar, D. W. Hamman, J. D. Paulson and J. P. Giesy, Utilizing the great blue heron (Ardea herodias) in ecological risk assessments of bioaccumulative contaminants, *Environ. Monit. Assess.*, 2009, **157**, 199–210.
81. P. M. Beldoménico, F. Rey, W. S. Prado, J. C. Villarreal, M. Muñoz-de-Toro and E. H. Luque, In ovum exposure to pesticides increases the egg weight loss and decreases hatchlings weight of *Caiman latirostris* (Crocodylia: Alligatoridae), *Ecotoxicol. Environ. Saf.*, 2007, **68**, 246–251.
82. T. A. Bargar, C. Sills-McMurry, R. L. Dickerson, W. E. Rhodes and G. P. Cobb, Relative distribution of polychlorinated biphenyls among tissues of neonatal American alligators (*Alligator mississippiensis*), *Arch. Environ. Contam. Toxicol.*, 1999, **37**, 364–368.
83. M. Yoshikane, W. R. Kay, Y. Shibata, M. Inoue, T. Yanai, R. Kamata, J. S. Edmonds and M. Morita, Very high concentrations of DDE and toxaphene residues in crocodiles from the Ord River, Western Australia: an investigation into possible endocrine disruption, *J. Environ. Monit.*, 2006, **8**, 649–661.
84. M. S. Humphries, J. G. Myburgh, R. Campbell, A. Buah-Kwofie and X. Combrink, Organochlorine pesticide bioaccumulation in wild Nile crocodile (*Crocodylus niloticus*) fat tissues: environmental influences on

changing residue levels and contaminant profiles, *Sci. Total Environ*, 2021, **753**, 142068.
85. H. J. Hamlin and L. J. Guillette Jr, Embryos as targets of endocrine disrupting contaminants in wildlife, *Birth Defects Res. C Embryo Today*, 2011, **93**, 19–33.
86. C. Stoker, M. A. Zayas, M. A. Ferreira, M. Durando, G. H. Galoppo, H. A. Rodríguez, M. R. Repetti, H. R. Beldoménico, E. G. Caldini, E. H. Luque and M. Muñoz-de-Toro, The eggshell features and clutch viability of the broad-snouted caiman (*Caiman latirostris*) are associated with the egg burden of organochlorine compounds, *Ecotoxicol. Environ. Saf.*, 2013, **98**, 191–195.
87. Administración Nacional de Medicamentos, Alimentos y Tecnología Médica (ANMAT), 1998. ANMAT No. 7292/1998. Available: http://www.anmat.gov.ar/Legislacion/Domisanitarios/Disposicion_ANMAT_7292-1998.pdf.
88. Ley Nacional N 17.751, Prohibición de plaguicidas en base a dieldrin, heptacloro ehidrocarburos clorados, Boletin oficial de la República Argentina N 21449, publication date 06/03/1968, Buenos Aires, Argentina.
89. Ley Nacional N 24.051/91, Régimen de desechos peligrosos, Boletín oficial de la República Argentina N 27.307, año C, publication date 01/17/1992, Buenos Aires Argentina. Available: http://www2.medioambiente.gov.ar/sian/pan/Leyesdecretos/Ley24051.htm.
90. Ley Nacional N 25.670, Presupuestos mínimos para la gestión y eliminación de los PCBs, Boletin oficial de la República Argentina N 30.029, año CX, publication date 11/19/2002, Buenos Aires, Argentina.
91. Decreto N 657/07 Provincia de Entre Ríos, Disposición de medidas y destino defondos para las obras necesarias para la eliminación de PCB's de los equipos y transformadores afectados al servicio, ubicados en el territorio provincial. Available: http://www.entrerios.gov.ar/wsdecreto/decreto.php?codigo=32903.
92. J. M. Bergeron, D. Crews and J. A. McLachlan, PCBs as environmental estrogens: turtle sex determination as a biomarker of environmental contamination, *Environ. Health Perspect.*, 1994, **102**, 780–781.
93. M. R. Milnes, T. A. Bryan, J. G. Medina, M. P. Gunderson and L. J. Guillette Jr, Developmental alterations as a result of *in ovo* exposure to the pesticide metabolite p,p0-DDE in *Alligator mississippiensis*, *Gen. Comp. Endocrinol.*, 2005, **144**, 257–263.
94. G. P. Cobb, P. D. Houlis and T. A. Bargar, Polychlorinated biphenyl occurrence in American alligators (*Alligator mississippiensis*) from Louisiana and South Carolina, *Environ. Pollut.*, 2002, **118**, 1–4.
95. R. H. Rauschenberger, M. S. Sepúlveda, J. J. Wiebe, N. J. Szabo and T. S. Gross, Predicting maternal body burdens of organochlorine pesticides from eggs and evidence of maternal transfer in *Alligator mississippiensis*, *Environ. Toxicol. Chem.*, 2004, **23**, 2906–2915.

96. R. H. Rauschenberger, J. J. Wiebe, M. S. Sepúlveda, J. E. Scarborough and T. S. Gross, Parental exposure to pesticides and poor clutch viability in American alligators, *Environ. Sci. Technol.*, 2007, **41**, 5559–5563.
97. R. de Solla, K. J. Fernie and S. Ashpole, Snapping turtles (*Chelydra serpentina*) as bioindicators in Canadian areas of concern in the Great Lakes Basin. II. Changes in hatching success and hatchling deformities in relation to persistent organic pollutants, *Environ. Pollut.*, 2008, **153**, 529–536.
98. R. Carson, *Silent spring*, Houghton Mifflin, Boston, MA, 1962.
99. L. J. Blus, T. G. Lamont and B. S. Neely Jr, Effects of organochlorine residues on eggshell thickness, reproduction, and population status of brown pelicans (Pelecanus occidentalis) in South Carolina and Florida, 1969-76, *Pestic. Monit. J.*, 1979, **12**, 172–184.
100. D. M. Fry, Reproductive effects in birds exposed to pesticides and industrial chemicals, *Environ. Health Perspect.*, 1995, **103**, 165–171.
101. H. J. Hamlin and L. J. Guillette Jr, Birth defects in wildlife: the role of environmental contaminants as inducers of reproductive and developmental dysfunction, *Syst. Biol. Reprod. Med.*, 2010, **56**, 113–121.
102. A. Paz, A. A. Yanosky, C. Mercolli and A. Pisanó, Structure of eggshells from two sympatric species of caimans (*Caiman latirostris* and *Caiman yacare*) (Crocodylia: Alligatorinae), *Anim. Biol.*, 1995, **4**, 71–77.
103. D. C. Deeming and M. W. J. Ferguson, Effects of incubation temperature on growth and development of embryos of *Alligator mississippiensis*, *J. Comp. Physiol., B*, 1989, **159**, 183–193.
104. Y. Handrich and H. Girard, Gas diffusive conductance of sea-level hen eggs incubated at 2900 m altitude, *Respir. Physiol.*, 1985, **60**, 237–252.
105. Y. C. Kam, Physiological effects of hypoxia on metabolism and growth of turtle embryos, *Respir. Physiol.*, 1993, **92**, 127–138.
106. J. Metcalfe, M. K. Stock and R. L. Ingermann, in *Respiration and Metabolism of Embryonic Vertebrates*, ed. R. S. Seymour, Junk Press, Boston, 1984, pp. 205–219.
107. J. T. Ackerman and C. A. Eagles-Smith, Agricultural wetlands as potential hotspots for mercury bioaccumulation: experimental evidence using caged fish, *Environ. Sci. Technol.*, 2010, **44**, 1451–1457.
108. W. Kloas, R. Urbatzka, R. Opitz, S. Wuertz, T. Behrends, B. Hermelink, F. Hofmann, O. Jagnytsch, H. Kroupova, C. Lorenz, N. Neumann, C. Pietsch, A. Trubiroha, C. Van Ballegooy, C. Wiedemann and I. Lutz, Endocrine disruption in aquatic vertebrates, *Ann. N. Y. Acad. Sci.*, 2009, **1163**, 187e200.
109. J. P. Sumpter and A. C. Johnson, Lessons from endocrine disruption and their application to other issues concerning trace organics in the aquatic environment, *Environ. Sci. Technol.*, 2005, **39**, 4321–4332.
110. J. P. Sumpter, Protecting aquatic organisms from chemicals: the harsh realities, *Philos. Trans. A: Math. Phys. Eng. Sci.*, 2009, **367**, 3877–3894.
111. C. Stoker, P. M. Beldoménico, V. L. Bosquiazzo, M. A. Zayas, F. Rey, H. Rodríguez, M. Muñoz-de-Toro and E. H. Luque, Developmental

exposure to endocrine disruptor chemicals alters follicular dynamics and steroid levels in *Caiman latirostris*, *Gen. Comp. Endocrinol.*, 2008, **156**, 603–612.
112. H. R. Andersen, A. M. Andersson, S. F. Arnold, H. Autrup, M. Barfoed, N. A. Beresford, P. Bjerregaard, L. B. Christiansen, B. Gissel, R. Hummel, E. Bonefel Jorgensen, B. Korsgaard, R. Le Guevel, H. Leffers, J. McLachlan, A. Moller, J. B. Nielsen, N. Olea, A. Oles-Karasko, F. Pakdel, K. L. Pedersen, P. Perez, E. Skakkeboek, C. Sonnenschein, A. M. Soto, J. P. Sumpter, S. M. Thorpe and P. Grandjean, Comparison of short-term estrogenicity tests for identification of hormone-disrupting chemicals, *Environ. Health Perspect.*, 1999, **107**, 89–108.
113. A. V. Krishnan, P. Stathis, S. F. Permuth, L. Tokes and D. Feldman, Bisphenol-A: an estrogenic substance is released from polycarbonate flasks during autoclaving, *Endocrinology*, 1993, **132**, 2279–2286.
114. G. G. J. M. Kuiper, B. Carlsson, K. Grandien, E. Enmark, J. Haggblad, S. Nilsson and J. Gustafsson, Comparison of the ligand binding specificity and transcript tissue distribution of estrogen receptors alpha and beta, *Endocrinology*, 1997, **138**, 863–870.
115. H. Leffers, M. Naesby, B. Vendelbo, N. E. Skakkebaek and M. Jorgensen, Oestrogenic potencies of zeranol, oestradiol, diethylstilboestrol, bisphenol A and genistein: implications for exposure assessment of potential endocrine disrupters, *Hum. Reprod.*, 2001, **16**, 1037–1045.
116. N. Olea, R. Pulgar, P. Perez, M. F. Olea-Serrano, A. Rivas, A. Novillo Fertrell, V. Pedraza, A. M. Soto and C. Sonnenschein, Estrogenicity of resin-based composites and sealants used in dentistry, *Environ. Health Perspect.*, 1996, **104**, 298–305.
117. R. Steinmetz, N. G. Brown, D. L. Allen, R. M. Bigsby and N. Ben-Jonathan, The environmental estrogen bisphenol A stimulates prolactin release in vitro and in vivo, *Endocrinology*, 1997, **138**, 1780–1786.
118. F. Rey, M. González, M. A. Zayas, C. Stoker, M. Durando, E. H. Luque and M. Muñoz-de-Toro, Prenatal exposure to pesticides disrupts testicular histoarchitecture and alters testosterone levels in male *Caiman latirostris*, *Gen. Comp. Endocrinol.*, 2009, **162**, 286–292.
119. L. J. Guillette Jr and E. A. Guillette, Environmental contaminants and reproductive abnormalities in wildlife: implications for public health?, *Toxicol. Ind. Health*, 1996, **12**, 537–550.
120. L. J. Guillette, A. R. Woodward, D. A. Crain, D. B. Pickford, A. A. Rooney and H. F. Percival, Plasma steroid concentrations and male phallus size in juvenile alligators from seven Florida lakes, *Gen. Comp. Endocrinol.*, 1999, **116**, 356–372.
121. B. C. Morrish and A. H. Sinclair, Vertebrate sex determination: many means to an end, *Reproduction*, 2002, **124**, 447–457.
122. G. H. Galoppo, G. Canesini, Y. E. Tavalieri, C. Stoker, L. Kass, E. H. Luque and M. Muñoz-de-Toro, Bisphenol A disrupts the temporal

pattern of histofunctional changes in the female reproductive tract of *Caiman latirostris*, *Gen. Comp. Endocrinol.*, 2017, **254**, 75–85.
123. Y. E. Tavalieri, R. Alarcón, M. V. Tschopp, G. Canesini, E. H. Luque, M. Muñoz-de-Toro and G. H. Galoppo, Exposure to xenoestrogens alters the expression of key morphoregulatory proteins of oviduct adenogenesis in the broad-snouted caiman (*Caiman latirostris*), *Aquat. Toxicol.*, 2021, **235**, 105817.
124. V. Calsolaro, G. Pasqualetti, F. Niccolai, N. Caraccio and F. Monzani, Thyroid disrupting chemicals, *Int. J. Mol. Sci.*, 2017, **18**, 2583.
125. A. A. Thambirajah, M. G. Wade, J. Verreault, N. Buisine, V. A. Alves, V. S. Langlois and C. C. Helbing, Disruption by stealth – Interference of endocrine disrupting chemicals on hormonal crosstalk with thyroid axis function in humans and other animals, *Environ. Res.*, 2022, **203**, 111906.
126. D. A. Crain, L. J. Guillette, D. B. Pickford, H. F. Percival and A. R. Woodward, Sex-steroid and thyroid hormone concentrations in juvenile alligators (*Alligator mississippiensis*) from contaminated and reference lakes in Florida, USA, *Environ. Toxicol. Chem.*, 1998, **17**, 446–452.
127. E. G. Sormo, I. Jussi, M. Jussi, M. Braathen, J. U. Skaare and B. M. Jenssen, Thyroid hormone status in gray seal (*Halichoerus grypus*) pups from the Baltic Sea and the Atlantic Ocean in relation to organochlorine pollutants, *Environ. Toxicol. Chem.*, 2005, **24**, 610–616.
128. N. K. Brar, C. Waggoner, J. A. Reyes, R. Fairey and K. M. Kelley, Evidence for thyroid endocrine disruption in wild fish in San Francisco Bay, California, USA. Relationships to contaminant exposures, *Aquat. Toxicol.*, 2010, **96**, 203–215.
129. G. J. Mayne, C. A. Bishop, P. A. Martin, H. J. Boermans and B. Hunter, Thyroid function in nestling tree swallows and eastern bluebirds exposed to non-persistent pesticides and p,p'-DDE in apple orchards of southern Ontario, Canada, *Ecotoxicology*, 2005, **14**, 381–396.
130. E. A. Hewitt, D. A. Crain, M. P. Gunderson and L. J. Guillette Jr, Thyroid status in juvenile alligators (*Alligator mississippiensis*) from contaminated and reference sites on Lake Okeechobee, Florida, USA, *Chemosphere*, 2002, **47**, 1129–1135.
131. K. M. Gabrielsen, J. S. Krokstad, G. D. Villanger, D. A. Blair, M. J. Obregon, C. Sonne, R. Dietz, R. J. Letcher and B. M. Jenssen, Thyroid hormones and deiodinase activity in plasma and tissues in relation to high levels of organohalogen contaminants in East Greenland polar bears (*Ursus maritimus*), *Environ. Res.*, 2015, **136**, 413–423.
132. R. C. Bicho, M. J. Amaral, A. M. Faustino, D. M. Power, A. Rema, M. A. Carretero, A. M. Soares and R. M. Mann, Thyroid disruption in the lizard *Podarcis bocagei* exposed to a mixture of herbicides: a field study, *Ecotoxicology*, 2013, **22**, 156–165.

133. G. Schierano-Marotti, M. Osti, G. Galoppo, G. Canesini, L. Ibarra Bouzada, E. H. Luque and M. Muñoz-de-Toro, Histoarquitectura funcional de la glándula tiroides del yacaré overo (*Caiman latirostris*). Efectos de la exposición *in ovo* a endosulfán, *Scientia Interfluvius*, 2015, p. 111.
134. A. Almada, Y. E. Tavalieri, E. H. Luque, M. Muñoz-de-Toro and G. H. Galoppo, *In ovo* exposure to atrazine alters thyroid histomorphology in late postnatal broad-snouted caiman (*Caiman latirostris*), XXIII Jornadas Anuales de la Sociedad Argentina de Biología, *Biocell*, 2021, in press.

CHAPTER 11

The Broad-snouted Caiman (Caiman latirostris): *A Model Species for Environmental Pesticide Contamination Assessment Through Molecular Markers*

L. M. ODETTI,[a,b] M. F. SIMONIELLO,[a] P. A. SIROSKI[c,d] AND G. L. POLETTA*[a,b,d]

[a] Cátedra de Toxicología, Farmacología y Bioquímica Legal, Facultad de Bioquímica y Ciencias Biológicas, Universidad Nacional del Litoral, Santa Fe, Argentina; [b] Consejo Nacional de Investigaciones Científicas y Técnicas (CONICET), Ciudad Autónoma de Buenos Aires, Argentina; [c] Laboratorio de Ecología Molecular Aplicada, Instituto de Ciencias Veterinarias del Litoral (ICiVet-Litoral-UNL/CONICET), Esperanza, Santa Fe, Argentina; [d] Proyecto Yacaré, Laboratorio de Zoología Aplicada: Anexo Vertebrados, Facultad de Humanidades y Ciencias, Universidad Nacional del Litoral/Ministerio de Ambiente y Cambio Climático, Gobierno de Santa Fe, Santa Fe, Argentina

*Email: gisepoletta@hotmail.com

Issues in Toxicology No. 45
Bird and Reptile Species in Environmental Risk Assessment Strategies
Edited by Guillermo Eli Liwszyc and Marcelo L. Larramendy

Published by the Royal Society of Chemistry, www.rsc.org

11.1 *Caiman latirostris*: Life History Characteristics and Population Situation in Argentina

11.1.1 Geographical Distribution and Sustainable Use Programs

Reptiles are relevant models in evolutionary history and have great ecological, sociological and commercial importance. They are transcendental in natural ecosystems as prey, predators, herbivores, seed dispersers and bioindicators of environmental health.[1]

Caiman latirostris is a crocodilian species that inhabits the north-central region of Argentina and it constitutes a valuable commercial and ecological resource in South America. The geographical distribution of this species includes the northeastern part of Argentina, southeastern Bolivia, Paraguay, Brazil and northern Uruguay.[2] In particular, in Argentina, *C. latirostris* is distributed in Formosa, Santa Fe, Misiones, Corrientes, Entre Ríos, Chaco, Santiago del Estero, Salta and Jujuy provinces. It is an ectothermic species with a well-defined seasonal cycle that depends on climatic factors (external temperature and hydrological regime). In Argentina, breeding takes place between November and January, with an incubation period of approximately 65–75 days at 31 °C. At hatching, the young weigh about 40 g and measure approximately 22 cm long. The size of adult males is in the order of 2.60 m in length and about 80 kg in weight (see Figure 11.1).[3]

Since the year 1990, the wild populations of *C. latirostris* of the Province of Santa Fe have been managed through a program based on the "ranching" technique (harvesting eggs in nature, captive breeding and returning to nature), known as "Proyecto Yacaré", which allowed a significant recovery of natural populations of this species.[3] From then on, the sustainable commercial use of *C. latirostris* through the passing of Provincial Law No. 11.820 has become an alternative of maximum interest to the Province of Santa Fe without any risk for the species conservation. Ranching programs are based on the critical factor represented by embryonic and perinatal high mortality, characteristic of these species. The implementation of this management methodology prevents the loss of eggs produced in nature and the great

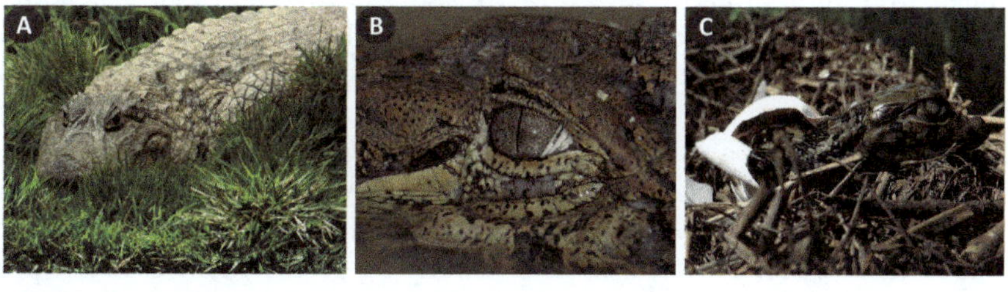

Figure 11.1 (A and B) *Caiman latirostris* (broad-snouted caiman) adult (male approx. 2.53 m). (C) *C. latirostris* hatchling coming out from the egg. Pictures from Pablo A. Siroski.

mortality of individuals under the first year of life, with the reintroduction into the nature of a number of animals higher than those that would have survived naturally. These programs actively involve local people in the identification and preservation of nesting areas, which has beneficial effects not only for the species under management but also for all the taxa that share its habitat.[3] As a result of the implementation of these programs in Argentina, natural populations of caimans have recovered from an extremely endangered situation, returning to stability since 1997.[2,3]

11.1.2 Environmental Problems Associated With Agricultural Expansion and the Use of Pesticides

The increase in the use of pesticides worldwide has generated great concern about environmental damage, contributing to a large extent to soil, water and air pollution.[4–6] Argentina has never been exempted from this situation. The use of pesticides has increased exponentially in the country since the 1990s in correlation with the scale of the areas planted with soybean.[7]

Pesticides are probably one of the most studied environmental contaminants. The vast majority of applied pesticides drift into the surrounding environment (see Figure 11.2). Non-specific application methods such as aerial spraying from airplanes or direct application as well as the lack of training, technical services, regulations and control on the safe use of pesticides are the main contributors to the accumulation of pesticides in the countryside.[8] The fate of a pesticide in the environment is subject to a variety of physical, chemical and biological processes that can affect its degradation as well as its interactions with environmental components. In turn, irrigation conditions and the number of applications are factors that also influence its final fate.[9]

Depending on their chemical properties, pesticides can leach into ground and surface water, persist in soil or enter into organisms and influence the health of different species, in some cases bioaccumulating in the food chain.[10] In this sense, different works have reported the presence of pesticide residues in surface water and sediments. Aparicio *et al.*[11] studied the environmental fate of glyphosate (GLY) and quantified its concentration in agricultural watersheds in Argentina. This work reported that the GLY levels in surface water ranged from 0.5 to 7.6 $\mu g\,L^{-1}$, while the soil concentrations ranged from 35 to 1502 $\mu g\,kg^{-1}$ in GLY and 299 to 2256 $\mu g\,kg^{-1}$ in AMPA (aminomethylphosphonic acid). Primost *et al.*[12] determined the presence of GLY and AMPA in soil and water in the agroecosystems of the Mesopotamian (Argentina) region. The maximum concentration levels reported in soil samples were 8.105 and 38.939 $\mu g\,kg^{-1}$ for GLY and AMPA, respectively. Maximum values of 1.8 μg GLY per L and 1.9 μg AMPA per L were recorded in water samples. Hunt *et al.*[13] measured the concentrations of 17 insecticides in sediments collected from streams belonging to the main soybean producing regions of South America (Argentina, Paraguay and Brazil) during high application periods of the pesticides. The concentration ranges found were 1.2–7.4 $ng\,g^{-1}$ chlorpyrifos (CPF); 0.9–8.3 $ng\,g^{-1}$ cypermethrin (CP);

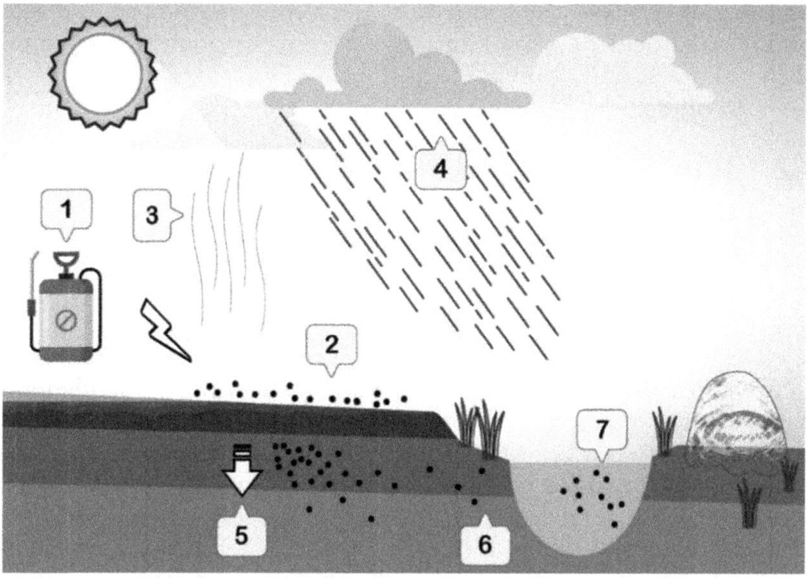

Figure 11.2 Possible ways of pesticide movement through the hydrological cycle. 1 – Application of the pesticide; 2 – absorption by the crop; 3 – evaporation (vaporized into the atmosphere); 4 – deposition by rain; 5 – infiltration; 6 – adherence to soil particles; and 7 – contaminated water.

0.42–16.6 ng g^{-1} lambda-cyhalothrin and 0.49–2.1 ng g^{-1} endosulfan (ES). Etchegoyen *et al.*[14] determined the presence of CPF, CP and ES in the surface water and bottom sediments of the main tributaries and the main watercourse of the Paraguay and Paraná rivers. The ranges of total pesticide concentrations measured in water and sediments were 0.004–6.62 μg L^{-1} and 0.16–221.3 μg kg^{-1} dry weight, respectively. In turn, CP residues have been widely detected in water and sediment samples from streams and rivers draining major agricultural districts.[15,16]

One of the most serious consequences of agricultural intensification is the impact on the environment, which has led to habitat destruction through deforestation and the degradation of the remaining ecosystems ("patches") due to the massive use of agrochemicals.[17] This has great effects on many species of native fauna and flora, many of which represent invaluable ecological components as well as resources of great economic importance for the region.[18]

In this context, one of the main problems affecting populations of *C. latirostris* during the past 25 years is habitat loss due to anthropogenic pressure, tending to assign more and more land to agricultural production.[19] It has been estimated that more than 50% of the transgenic soybean sown over the past decade spread over areas that corresponded previously to native forests, leaving untouched fragmented patches as remnants. Natural populations living in these areas are also constantly exposed to agrochemicals used for pest control, such as herbicides and insecticides, most of the time applied as complex mixtures.[20] Moreover, the period of the year of maximum application of pesticides coincides with the breeding season of *C. latirostris*, implying a particularly significant risk of contamination for developing embryos and neonates of the species.[19]

11.2 Why Use *C. latirostris* as a Sentinel of Pesticide Contamination?

11.2.1 Evidence of Pesticide Effects on the Broad-snouted Caiman

The biological impact of xenobiotics is the result of a cascade of events that begins with their introduction into the environment, their assimilation by exposed species and their interaction with sensitive components in affected organisms, which can be evident at the genetic, immunological and biochemical levels, a marked effect on populations. Sensitive early biomarkers to be applied in natural biota are essential to properly assess the damage risk. Taking into account the wide use of pesticide formulations in current agricultural activities, early detection of their effects through an integral approach that combines the analysis of biomarkers of different endpoints should be considered of paramount importance.

In this context, and for more than 15 years, we have been studying the possible effects of pesticides widely used in agricultural activities in Argentina on caimans. Biomarkers of genotoxicity, oxidative stress, immunotoxicity and biochemical–enzymatic alterations have been adapted, in order to use them together in a battery of tests to evaluate xenobiotic exposure in caiman blood (see Figure 11.3). To date, there have been no reports on the application of many of these techniques on blood for any crocodilian and even in other reptile species, which represents a huge advantage for the study of wild populations, as it allows constant biomonitoring without causing any damage to the animals.[21–23] Hence, genotoxicity and reduced growth were demonstrated for caimans after embryonic and postnatal exposure to Roundup® (66.2% GLY).[19,24] Furthermore, a mixture of GLY, CPF and CP formulations commonly used in agricultural practices produced a greater genotoxic effect on neonates, in addition to inducing biochemical–enzymatic alterations. We also observed differences in the growth parameters in different nests during the first year of life (3 and 12 months) after exposure during embryonic development to GLY and/or a CP + ES + GLY mixture.[25] In turn, LópezGonzález *et al.*[23] and Burella *et al.*[26] reported genotoxicity, oxidative damage and alteration in the activity of superoxide dismutase (SOD) and catalase (CAT) enzymes in erythrocytes of *C. latirostris* exposed to sublethal concentrations of two GLY commercial formulations: Panzer Gold® and Roundup® Full II (500–1000 μg per egg); the ES-containing formulation Galgofan® and the CP formulation Atanor® (1–1 000 μg per egg) by topical application through the eggshell at the beginning of the incubation period. Subsequently, the effects of three binary mixtures of CP, CPF and GLY formulations were studied

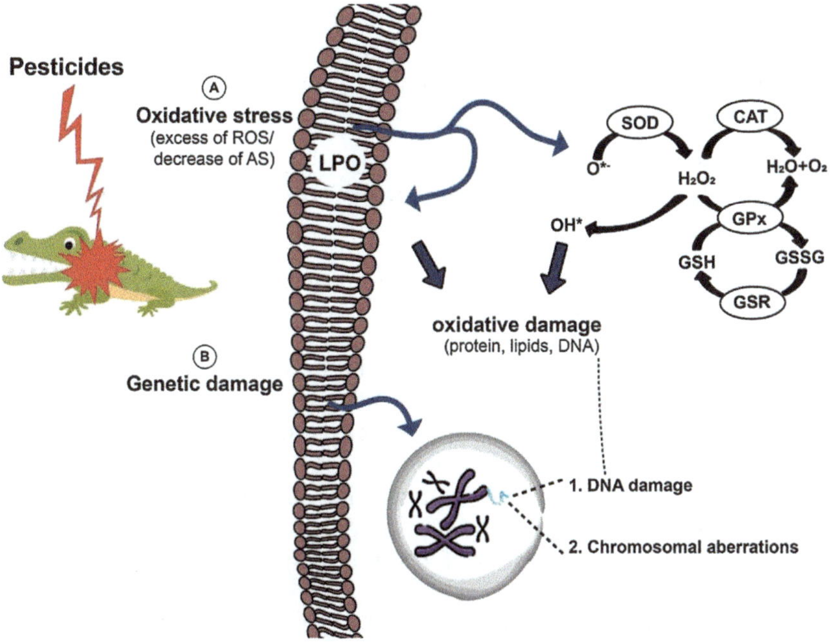

Figure 11.3 Illustration of possible genetic damage and oxidative stress caused by pesticide exposure. (A) Imbalance between reactive oxygen species and antioxidant defense systems within a cell. (B) Induction of genetic alterations including DNA damage and chromosomal aberrations.

reporting a significant increase in the frequency of micronuclei (FMN).[27] In turn, Odetti *et al.*[28] evaluated the effects of the same formulations separately and when exposed as binary and ternary mixtures during the embryonic period, simulating the exposure conditions that occur on nest materials in natural environments by spraying recommended concentrations for soybean crops. They demonstrated that exposure to pesticides under these conditions generates an increase in the FMN, DNA damage and oxidative DNA damage for all the formulations studied and their mixtures. Recently, genotoxicity, immunotoxicity and oxidative stress were demonstrated in caiman hatchlings (20 days old) exposed *in vivo* to sublethal concentrations of ES-, CP- and CPF-based formulations, two GLY-based formulations and two different ternary complex mixtures of them.[29]

Subsequently, in order to simulate the actual environmental conditions of exposure, we conducted a semi-natural experiment in which caimans were maintained in outdoor enclosures exposed to a single pesticide and/or mixtures of pesticides, following concentrations and application schedules as recommended for soybean crops. These semi-natural conditions of exposure generate an increase in the FMN and DNA damage with an important contribution of base oxidation not only in all the studied formulations, but also in their mixtures.[30]

As previously shown most of the biomarkers used so far to identify contamination stress in broad-snouted caimans are based on alterations in biochemical parameters and/or cytological or histological abnormalities that can be correlated with pesticide exposure.[19,23–31]

11.2.2 Identification and Development of Gene Expression Markers

11.2.2.1 Importance of the Technique and Advantages Over Other Tools

Rapid progress in recent years in the genomics field has provided tools that can contribute to a better understanding of how chemicals can impact upon human health and the ecosystem. Critical parameters for cellular homeostasis are under continuous vigilance by a set of protein factors that act as cellular sensors. Cells respond to alterations in these parameters by re-adjusting metabolism, usually by activating the *de novo* synthesis of specific proteins in order to compensate for stress damage, from the mRNAs that code for these proteins.[32]

Based on this principle, a new family of biomarkers was developed for the analysis of the gene transcription related to cellular stress, which can be used to detect exposure to xenobiotics, acting, in turn, as early warning sensors. Alterations in mRNA expression patterns offer new insights into the role of genes in the context of toxicity.[33] These changes are immediate and generally more sensitive than the majority of the endpoints traditionally used in toxicology, making them relevant markers under various stress conditions. In this sense, altered gene expression can be used to detect biological toxicity and/or monitor the impact of different xenobiotics in natural environments.[34] Several studies in fish and mammals have demonstrated their usefulness as markers for biomonitoring and ecotoxicology tests.[35–37]

Nowadays, quantification of specific gene expression is mainly done through quantitative real-time polymerase chain reaction (qPCR). It allows the accumulation of the amplification product to be detected as the reaction progresses, that is to say in "real time".[32] It has been one of the most promising approaches to perform gene expression analyses and become increasingly popular compared to other methods because it has higher sensitivity, greater speed and a broader dynamic quantification range.[38] Advances in RNA technologies have had a very strong impact on molecular ecology studies, providing new tools for understanding the response of organisms to environmental stress.[39] Appropriate RNA sampling, preservation and extraction techniques may vary depending on downstream applications, the availability of laboratory equipment and the size and nature of the organisms or tissues analyzed.[40] In this sense, to evaluate any changes in gene modulation, an optimal RNA extraction procedure is essential, minimizing the activity of RNases to prevent their degradation.

11.2.2.2 *Optimization of Protocols for the Determination of the Expression Levels in* C. latirostris

Analysis of gene expression patterns in response to the negative impact of different xenobiotics can provide a more complete scenario of the origin of damage at both the molecular and cellular levels. Responses to contaminants may not only manifest in the form of changes in one or a small number of genes but may also involve a cascade of genetic interactions.[33,41] It has been shown that gene expression is altered directly or indirectly as a result of exposure to xenobiotics. In particular, several works have shown that pesticides alter the expression levels of genes encoding for antioxidant systems, as well as those involved in biotransformation processes, cell cycle regulators and damage repair, among others.[34,42–44] Specifically, SOD and CAT are two enzymes that contribute to the first line of antioxidant defense. When free radical production exceeds the level that natural antioxidant defense mechanisms can cope with, a cellular oxidative environment is created.[45] The alteration in the response of the antioxidant system can modify the expression of the genes that encode these enzymes. Although many studies indicate that cells have the means to detect reactive oxygen species (ROS), the mechanism of action of xenobiotics and ROS regulators that activate the expression of certain genes has not been fully elucidated yet.[34,46,47] The transcriptional pathway that responds to ROS in eukaryotes is currently being deciphered. A large number of biological responses and cell signaling pathways are involved in oxidative stress induced by toxicological effects. During the process, significant changes in gene expression and stimulation or inhibition of signal transduction are observed.[48] It has been reported that many fish and mammalian species exhibited altered expression of *cat* and *sod* genes as a response to exposure to pesticides (atrazine, organochlorines, CP and indoxacarb) in different tissues such as the liver, kidneys, ovaries and gills.[34,44,49,50] The cytochrome p450 (CYP) superfamily has a large number of genes that encode these enzymes and are present in all domains of life. Generally, they act as oxidizing agents in a variety of metabolic and biosynthetic reactions, including detoxification and regulatory processes.[51] The CYP1, CYP2 and CYP3 families have been shown to be responsible for a variety of functions of xenobiotic metabolism, mainly in the liver. The presence and expression levels of these enzymes influence the tissue's ability to metabolize and/or detoxify xenobiotics.[52] On the other hand, glutathione *S*-transferases (GSTs) are among the major antioxidant enzymes and first-line indicators of antioxidant status. In biotransformation phase II, GSTs play a central role through the detoxification and excretion of xenobiotics such as herbicides and endogenous compounds that are conjugated to glutathione.[53] Finally, apoptosis, or programmed cell death, is a highly regulated process by which an organism eliminates unwanted cells without eliciting an inflammatory response.[54] Apoptotic signaling alterations are related to modifications of genes such as *p53*, *bcl2* and *bax*, among others.[55]

The study of gene expression profiles in reptiles has only recently begun, with few papers published. Waits *et al.*[56] developed and annotated blood transcriptomes from six reptile species (three snakes, two lizards and one turtle). They demonstrated that similar ecologically relevant gene groups (insulin and insulin-like signaling, electron transport chain, oxidative stress and glucocorticoid receptors) transcribed in both blood and the liver. In turn, Capraro *et al.*[57] examined gene expression in the *Pogona vitticeps* using mRNA-seq and label-free quantitative mass spectrometry in brain, heart and skeletal muscle samples from animals at late hibernation, and at 2 days and 2 months post-arousal. They identified differentially expressed genes in all tissues between hibernation and post-arousal and stress response pathways were induced in all tissues during hibernation. They also reported evidence for gene expression regulation at transcription, translation and post-translation. Héritier *et al.*[58] selected *Trachemys scripta elegans* exposed to high concentrations of GLY and contaminated water from a local watercourse in France, to measure *cat* and *sod* gene expression in the liver, showing statistically significant increases in exposed animals. In the same work, the authors standardized the qPCR technique in blood of the endangered freshwater endemic species *Mauremys leprosa* for the analysis of expression of the antioxidant enzyme genes to conduct ecotoxicological studies without euthanizing specimens. Then, they reported significant variations in gene expression of antioxidant enzyme genes in blood of turtles exposed to degraded waters at different points along a river in the eastern Pyrenees (France).[59] Recently, Morao *et al.*[60] evaluated the possible impacts of metal contamination in *Chelonia mydas* by gene expression analysis performed in blood, targeting genes involved in detoxification/quenching and metal transport (*mt*, *mtf* and *fer*), as well as in oxidative stress responses (*cat*, *sod*, *gr*, *tdx*, *txrd*, *selp* and *gclc*). Sujiwattanarat *et al.*[61] were the first to report the molecular cloning and characterization of the copper, zinc-superoxide dismutase gene (Cu-, Zn-*sod*) of the Siamese crocodile (*Crocodylus siamensis*) among crocodilian species, and the analysis of its expression levels in different tissues (liver, pancreas, lung, kidney, heart and whole blood) by qPCR. More recently, other authors carried out studies in different crocodilian species with samples from different tissues or organs (gonads, heart, kidney, skeletal muscle, red and white blood cells) of *Alligator mississippiensis* and *Alligator sinensis*. In these works, identification and determination of the relative mRNA abundance of androgen receptors, glucocorticoid receptors, heat shock proteins (HSPA2, A8, A9, C1 and C3) and proto-oncogenes were carried out.[62–64] In the same way in *Crocodylus niloticus*, a transcriptome study of adipose tissue of healthy animals was performed by RNAseq and *de novo* assembly followed by differential gene expression analysis.[65]

Specifically, in *C. latirostris*, in order to analyze gene expression levels as potential biomarkers of pesticide exposure, a protocol for RNA isolation and preservation from blood samples was adapted.[66] In particular, peripheral blood is an important and less invasive sample compared to solid or semi-solid tissues for assessing the health of wildlife species exposed to environmental stressors. In addition, this tissue allows molecular responses to be associated with broader physiological responses.[64] At the same time, the possibility of extracting RNA from blood is relevant for threatened or endangered species where generally a limited amount of material can be obtained and culling of animals should be avoided. Blood is an especially challenging sample, particularly in non-mammalian vertebrates that possess nucleated red blood cells and thus a higher DNA and nuclease content, which makes RNA degradation extremely rapid.[67] In this context, the presence of the genes related to antioxidant defense pathways (*cat* and *sod*), damage repair (*p53*), biotransformation (*p450* and *gst*) and cell cycle regulators (*bcl2* and *bax*) was identified in peripheral blood by the conventional PCR technique (see Figure 11.4).[68] Then, we molecularly characterized the Cu, Zn-*sod* gene in *C. latirostris* and determined its tissue-specific expression pattern, including the liver, gonads, the spleen, the heart and whole blood.[69]

Studies have shown that toxic compounds, including pesticides, have diverse effects on modulating gene expression levels, including alterations in the expression of detoxifying enzymes, antioxidant enzymes and DNA repair enzymes, among others.[41,60] To date, no information existed on alterations in *cat* and *sod* gene expression levels as a response to pesticide exposure in any crocodilian species. Bearing this in mind, we quantified the expression levels of these genes in the blood of *C. latirostris* exposed to GLY-, CP- and CPF-based pesticide formulations and their mixtures. We conducted two different experiments simulating the exposure conditions that occur in natural environments, and under the same conditions where we previously evaluated genotoxicity, oxidative damage to lipids and DNA and antioxidant enzyme activity.[28,30]

In the first experiment, the expression levels of *cat* as well as Cu- and Zn-*sod* genes were quantified after embryonic exposure to CP, CPF, GLY and their binary and ternary mixtures by spraying them on the nest material in contact with the eggs, at an early stage of embryonic development (within 5 days after laying).[28] One hundred and eighty eggs from five nests (36 eggs per nest) were distributed into nine experimental groups with three replicas for each group. The treatments were defined as follows: negative control (NC) treated with distilled water; vehicle control (VC) treated with ethanol; three groups exposed to CP (Atanor®, 0.12%), CPF (Lorsban®, 0.8%) and GLY (Roundup® Full II, 2%) formulations separately; three groups exposed to the binary mixtures (BM1: GLY + CP; BM2: CPF + CP; BM3: CPF + GLY) and one group exposed to the ternary mixture of the three formulations (TM: CP + CPF + GLY). Each replicate was placed in a plastic container separately, using vermiculite as the substrate and covering them with vegetal material similar to the nesting material, free of any exogenous substance. The concentrations applied correspond to those recommended for field application in soybean crops, considering the surface of the incubation container (0.034 m^2) as the area of spraying. Ethanol was used as the vehicle for pesticide formulations insoluble in water such as CPF, CP and all their mixtures. Expression levels were determined for each experimental group (see Figure 11.5). The results obtained showed that the gene expression of *sod* and *cat* was significantly

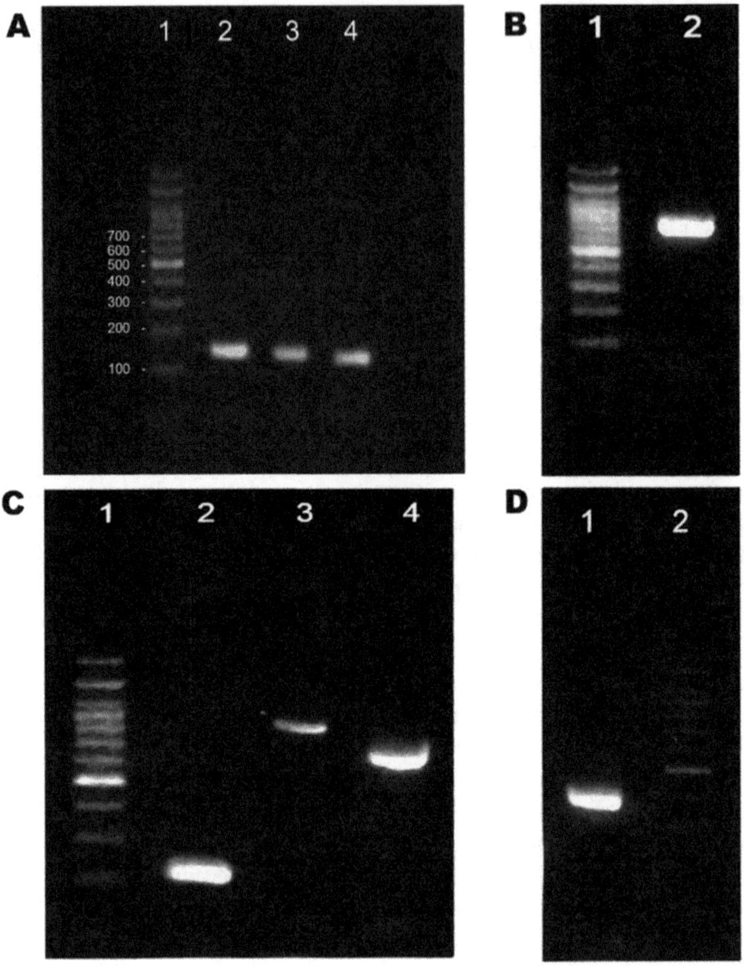

Figure 11.4 Agarose gel (2%) showing the specific PCR products for each gene. (A) Lane 1: 100 bp molecular size marker (Plus DNA – TransGen Biotech); lane 2: β-actin – 138 bp; lane 3: *cat* – 135 bp and lane 4: *sod* – 124 bp. (B) Lane 1: 100 bp molecular size marker (Plus DNA – TransGen Biotech) and lane 2: *gst* – 838 bp. (C) Lane 1: 100 bp molecular size marker (Plus DNA – TransGen Biotech); lane 2: *p53* – 217 bp; lane 3: *bcl2* – 842 bp and lane 4: *p450* – 674 bp. (D) Lane 1: *bax* – 406 bp and lane 2: 100 bp molecular size marker (Plus DNA – TransGen Biotech).

lower in BM3 in relation to the VC ($p = 0.0357$ in both cases, Mann–Whitney *U*-test) and the same was observed for *sod* in BM1 and BM2 ($p = 0.0357$ and $p = 0.0286$, respectively; Mann–Whitney *U*-test). No statistically significant effect was observed for any of the compounds separately. In the comparison of the mixtures with the individual compounds, we found a lower expression of both genes in BM3 compared to GLY ($p_{cat} = 0.0071$ and $p_{sod} = 0.0014$, respectively; Mann–Whitney *U*-test).[70]

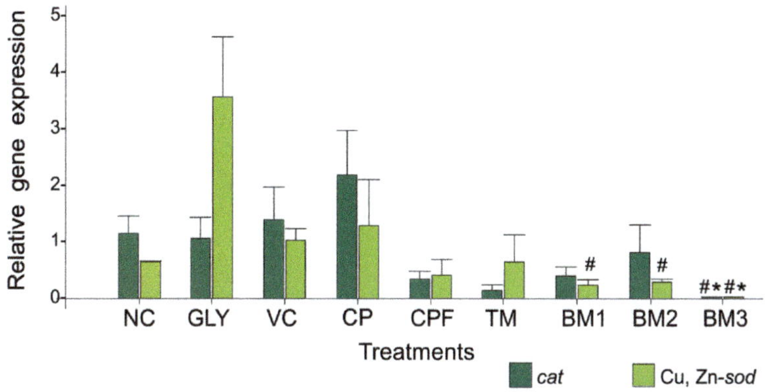

Figure 11.5 Cu-, Zn-*sod* and *cat* gene expression in *C. latirostris* hatchlings exposed to different pesticide formulations and their mixtures during embryonic development. NC: negative control; VC: vehicle control; GLY: glyphosate formulation at 2%; CPF: chlorpyrifos formulation at 0.8%; CP: cypermethrin formulation at 0.12%; TM: ternary mixture at field concentrations: CP + CPF + GLY; BM1: GLY + CP binary mixture; BM2: CPF + CP binary mixture and BM3: CPF + GLY binary mixture. #Statistically significant compared to the VC; *statistically significant compared to GLY.

The downexpression of *sod* and *cat* genes is possibly due to the inhibitory regulation of their expression directly caused by the toxic action of xenobiotics.[71] The mechanisms by which xenobiotics could directly inhibit antioxidant defenses are still unclear and depend on the type of compound. The possible accumulation of ROS, which depends on the intensity and duration of the applied agent, as well as the susceptibility of the species, may increase or decrease antioxidant activity.[34] In our previous work, there were no significant differences in CAT and SOD enzymatic activities under the same exposure conditions, but still significant DNA damage and oxidation of purines and pyrimidines were found.[28] In contrast, some authors have suggested that mRNA levels are representative of cellular activity at a particular time, whereas protein activity might be regulated at the post-translational level, enzymatically and not by transcriptional control.[72,73]

In the second study, 20 day-old caimans were exposed to the same pesticide formulations and concentrations as those used in the previous study but under semi-natural (*ex situ*) conditions. A total of one hundred specimens were equally distributed in 5 enclosures, with each containing 20 animals. The treatments were defined as follows: NC treated with tap water; three groups exposed to CP (Atanor®, 0.12%), CPF (Lorsban®, 0.8%) and GLY (Roundup® Full II, 2%) formulations, and one group exposed to the ternary mixture of the three formulations (Mx3: CP + CPF + GLY). The enclosures were outdoors, with each one separated from the others by a sufficient distance to avoid cross-contamination (more than 50 m). They were approximately 4.5 m^2 in size, closed at the top with small wire meshes to prevent the entry of predators. Each enclosure had an artificial water body, a feeder and a cover to generate partial shade. Throughout the experiment, the animals were fed every 2 days with a mixture consisting of 50% ground chicken head and 50% dry reptile pellets supplied *ad libitum*.[8] The experiments were carried out during the warm months, with a duration of 75 days. All caimans were acclimatized for 10 days in each enclosure under natural conditions of temperature, rainfall, humidity and photoperiod, previous to the beginning of the experiment.

After acclimatization and immediately prior to spraying with the pesticides, the animals were removed from the enclosures to avoid direct spraying on them and were placed back 24 h later. The control group was also removed to maintain the same conditions in all groups. Applications were made using a hand-held sprayer, over the entire surface of the enclosure from a height of 60 cm. GLY was applied twice, with a difference of one month, considering that it is applied generally at pre-sowing for weed desiccation and in post-emergence to control weeds before harvest.[73] Insecticides, such as CP and CPF, are usually applied before soybean flowering, and when a punctual pest appears during the crop-growing season,[74] they were applied once. After sub-chronic exposure, animals were bled and different biomarkers evaluated. The results showed a significant overexpression of the *cat* gene in CP-exposed specimens with respect to the NC ($p = 0.0286$, Mann–Whitney *U*-test) but not in the other groups. When Mx3 was compared with the compounds separately, we found differences with respect to CP in both genes, with the expression in the mixture being significantly lower ($p_{sod} = 0.0100$ and $p_{cat} = 0.0004$; Mann–Whitney *U*-test) (see Figure 11.6).[70]

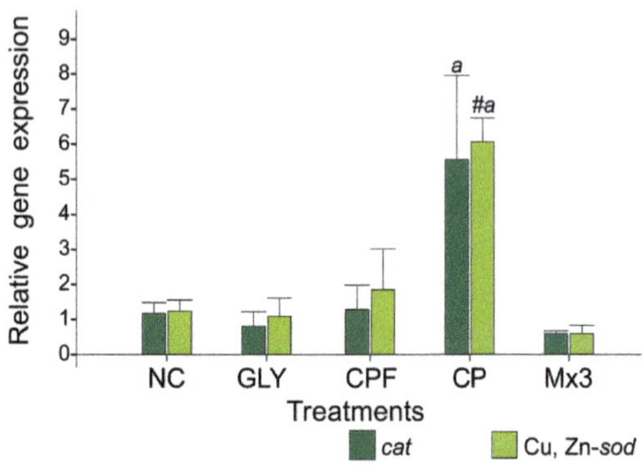

Figure 11.6 Cu-, Zn-*sod* and *cat* gene expressions in *C. latirostris* neonates exposed to different pesticide formulations and their mixture under semi-natural conditions. NC: negative control; GLY: 2% glyphosate formulation; CPF: 0.8% chlorpyrifos formulation; CP: 0.12% cypermethrin formulation and Mx3: mixture of the three formulations at field concentrations: CP + CPF + GLY. $^{\#}p < 0.05$ statistically significant compared to NC. The letter "a" indicates a significant difference between the individual compounds and the constituent mixture: statistically significant compared to Mx3.

This result suggests that CP induces the antioxidant system by *de novo* synthesis of the two main enzymes (CAT and SOD) acting in the first line of response to oxidative stress. This could signify a greater toxicity effect of CP compared to the other pesticides. This significant result does not correlate with the results obtained on enzyme activity. However, increased expression of the *cat* gene may not always mean increased enzyme activity, since gene expression can be controlled by post-transcriptional regulation.[49] The lack of correlation between antioxidant enzyme activities and expression of their encoding genes could be attributed to the delay effect between transcription and translation or to post-translational modifications.[75] Moreover, in agreement with these results, different works on different species (zebrafish, Swiss albino mice, among others) have reported an overexpression of *cat* genes upon exposure to CP.[75–77]

11.3 Future Perspectives

Considering the wide use of pesticide formulations in current agricultural environments, the early detection of their effects by means of sensitive markers provides valuable tools for the medium- and long-term conservation of wild species. This enables the identification of environmental problems before biological systems are seriously affected and is considered to be of paramount importance for the sustainability of natural environments.

We believe that it is necessary to continue investigating the possible toxicity mechanisms underlying pesticide exposures that may affect the physiological responses of caiman populations, especially concerning molecular regulation of gene expression under toxicity stress. These findings could contribute to predicting the environmental risk of real situations on wildlife species, which is especially important for their conservation as natural resources.

Acknowledgements

This study was supported by the Agencia Nacional de Promoción Científica y Tecnológica (PICT 2016-2020 to GLP), Consejo Nacional de Investigaciones Científicas y Técnicas (PIP 11220130100478CO to GLP) and Proyecto Yacaré – Yacarés Santafesinos (Gob. Sta. Fe/MAyCC).

References

1. J. Alföldi, F. Di Palma, M. Grabherr, C. Williams, L. Kong, E. Mauceli, P. Russell, C. B. Lowe, R. E. Glor, J. D. Jaffe, D. A. Ray, S. Boissinot, A. M. Shedlock, C. Botka, T. A. Castoe, J. K. Colbourne, M. K. Fujita, R. Godinez Moreno, B. F. ten Hallers, D. Haussler, A. Heger, D. Heiman, D. E. Janes, J. Johnson, P. J. de Jong, M. Y. Koriabine, M. Lara, P. A. Novick, C. L. Organ, S. E. Peach, S. Poe, D. D. Pollock, K. de Queiroz, T. Sanger, S. Searle, J. D. Smith, Z. Smith, R. Swofford, J. Turner-Maier, J. Wade, S. Young, A. Zadissa, S. V. Edwards, T. C. Glenn, C. J. Schneider,

J. B. Losos, E. S. Lander, M. Breen, C. P. Ponting and K. Lindblad-Toh, The genome of the green anole lizard and a comparative analysis with birds and mammals, *Nature*, 2011, **477**, 587.

2. P. Siroski, L. A. B. Bassetti, C. Piña and A. Larriera, *Caiman latirostris*, The IUCN Red List of Threatened Species 2020, https://www.iucnredlist.org/species/46585/3009813 (accessed April 2022).
3. A. Larriera, A. Imhof and P. Siroski, in *Contribución al conocimiento del Genero Caiman de Suramérica*, ed. J. Castroviejo, J. Ayarzaguena and A. Velasco, Public. Asoc. Amigos de Doña Ana, Sevilla, 2008, vol. 18, pp. 139–179.
4. E. N. Papadakis, Z. Vryzas, A. Kotopoulou, K. Kintzikoglou, K. C. Makris and E. Papadopoulou-Mourkidou, A pesticide monitoring survey in rivers and lakes of northern Greece and its human and ecotoxicological risk assessment, *Ecotoxicol. Environ. Saf.*, 2015, **116**, 1–9.
5. M. F. Silva-Barnia, M. Gonzaleza, F. Waniab, Y. D. Leib and K. S. B. Miglioranza, Spatial and temporal distribution of pesticides and PCBs in the atmosphere using XAD-resin based passive samplers: A case study in the Quequén Grande River watershed, Argentina, *Atmos. Pollut. Res.*, 2017, **9**, 238–245.
6. M. K. Uddin, Agrochemicals and Environmental Risks, *Environ. Policy Law*, 2018, **48**, 91–96.
7. Food and Agriculture Organization of the United Nations, http://www.fao.org/faostat/es/#data (accessed April 2020).
8. B. Bonicelli, E. Cotteux and J. P. Douzals, Pesticides applications assessment in banana crop conditions: a methodological approach, *Environ. Sci. Pollut. Res.*, 2015, **22**, 17255–17259.
9. V. I. Lushchak, T. M. Matviishyn, V. V. Husak, J. M. Storey and K. B. Storey, Pesticide toxicity: a mechanistic approach, *EXCLI J.*, 2018, **17**, 1101–1136.
10. R. Van der Oost, J. Beyer and N. P. Vermeulen, Fish bioaccumulation and biomarkers in environmental risk assessment: a review, *Environ. Toxicol. Pharmacol.*, 2003, **13**, 57–149.
11. V. C. Aparicio, E. de Geronimo, D. Marino, J. Primost, P. Carriquiriborde and J. L. Costa, Environmental fate of glyphosate and aminomethylphosphonic acid in surface waters and soil of agricultural basins, *Chemosphere*, 2013, **93**, 1866–1873.
12. J. E. Primost, D. Marino, V. Aparicio, J. L. Acosta and P. Carriquiborde, Glyphosate and AMPA, "pseudo-persistent" pollutants under real-world agricultural management practices in the Mesopotamic Pampas, *Environ. Pollut.*, 2017, **229**, 771–779.
13. L. Hunt, C. Bonetto, V. Resh, D. V. F. Buss, S. Fanelli, N. Marrochi and M. Lydy, Insecticide concentrations in stream sediments of soy production regions of South America, *Sci. Total Environ.*, 2016, **547**, 114–124.
14. M. A. Etchegoyen, A. E. Ronco, P. Almada, M. Abelando and D. J. Marino, Occurrence and fate of pesticides in the Argentine stretch of the Paraguay-Paraná basin, *Environ. Monit. Assess.*, 2017, **189**, 63.

15. D. Marino and A. Ronco, Cypermethrin and chlorpyrifos concentration levels in surface water bodies of the Pampa Ondulada, Argentina, *Bull. Environ. Contam. Toxicol.*, 2005, **75**, 820–826.
16. Z. Vryzas, C. Alexoudis, G. Vassiliou, K. Galanis and E. Papadopoulou-Mourkidou, Determination and aquatic risk assessment of pesticide residues in riparian drainage canals in northeastern Greece, *Ecotoxicol. Environ. Saf.*, 2011, **74**, 174–181.
17. L. Gaona, F. Bedmar, V. Gianelli, A. V. Faberi and H. Angelini, Estimating the risk of groundwater contamination and environmental impact of pesticides in an agricultural basin in Argentina, *Int. J. Sci. Technol.*, 2019, **16**, 6657–6670.
18. S. Stehle and R. Schulz, Agricultural insecticides threaten surface waters at the global scale, *Proc. Natl. Acad. Sci. U. S. A.*, 2015, **112**, 5750–5755.
19. G. L. Poletta, A. Larriera, E. Kleinsorge and M. D. Mudry, Genotoxicity of the herbicide formulation Roundup® (glyphosate) in broad-snouted caiman (*Caiman latirostris*) evidenced by the comet assay and the micronucleus test, *Mutat. Res.*, 2009, **672**, 95–102.
20. CASAFE: Cámara de Sanidad Agropecuaria y Fertilizantes de Argentina, https://www.casafe.org/informe-industria-de-fitosanitarios-en-argentina (accessed November 2020).
21. G. L. Poletta, A. Larriera, E. Kleinsorge and M. D. Mudry, *Caiman latirostris* (broad snouted caiman) as a sentinel organism for genotoxic monitoring: Basal values determination of micronucleus and comet assay, *Mutat. Res.*, 2008, **650**, 202–209.
22. G. L. Poletta, M. F. Simoniello and M. D. Mudry, Biomarkers of oxidative damage and antioxidant defense capacity in *Caiman latirostris* blood, *Comp. Biochem. Physiol., Part C: Toxicol. Pharmacol.*, 2016, **179**, 29–36.
23. E. C. López González, A. Larriera, P. A. Siroski and G. L. Poletta, Micronuclei and other nuclear abnormalities on *Caiman latirostris* (Broad-snouted caiman) hatchling after embryonic exposure to different pesticide formulations, *Ecotoxicol. Environ. Saf.*, 2017, **136**, 84–91.
24. E. C. López González, M. A. Latorre, A. Larriera, P. A. Siroski and G. L. Poletta, Induction of micronuclei in broad snouted caiman (*Caiman latirostris*) hatchlings exposed *in vivo* to Roundup® (glyphosate) concentrations used in agriculture, *Pestic. Biochem. Physiol.*, 2013, **105**, 131–134.
25. G. L. Poletta, E. Kleinsorge, A. Paonessa, M. D. Mudry, A. Larriera and P. A. Siroski, Genetic, enzymatic and developmental alterations observed in *Caiman latirostris* exposed *in ovo* to pesticide formulations and mixtures in an experiment simulating environmental exposure, *Ecotoxicol. Environ. Saf.*, 2011, **74**, 852–859.
26. P. M. Burella, L. M. Odetti, M. F. Simoniello and G. L. Poletta, Oxidative damage and antioxidant defense in *Caiman latirostris* (Broad snouted caiman) exposed *in ovo* to pesticide formulations, *Ecotoxicol. Environ. Saf.*, 2018, **161**, 437–443.

27. E. C. López González, P. A. Siroski and G. L. Poletta, Genotoxicity induced by widely used pesticide binary mixtures on *Caiman latirostris* (Broad-snouted caiman), *Chemosphere*, 2019, **232**, 337–344.
28. L. M. Odetti, E. C. López Gonzalez, M. L. Romito, M. F. Simoniello and G. L. Poletta, Genotoxicity and oxidative stress in *Caiman latirostris* hatchling exposure to pesticide formulations and their mixtures during incubation period, *Ecotoxicol. Environ. Saf.*, 2020, **193**, 110312.
29. E. C. López González, L. M. Romito, M. A. Latorre, P. A. Siroski and G. L. Poletta, Biomarkers of genotoxicity, immunotoxicity and oxidative stress on *Caiman latirostris* (Broad-snouted caiman) hatchlings exposed to pesticide formulations and mixtures widely used in agriculture, *Environ. Adv.*, 2021, **5**, 100114.
30. E. C. López González, L. M. Odetti, M. A. Latorre, O. B. Ávila, L. E. Contini, P. A. Siroski and G. L. Poletta, A comprehensive approach using multiple biomarkers to detect damage induced by pesticides in broad-snouted caiman (*Caiman latirostris*) under *ex situ* conditions, *Heliyon*, 2022, **8**, e08667.
31. P. Siroski, G. L. Poletta, M. A. Latorre, M. E. Merchant, H. H. Ortega and M. D. Mudry, Immunotoxicity of commercial-mixed glyphosate in broad snouted caiman (*Caiman latirostris*), *Chem. – Biol. Interact.*, 2016, **244**, 64–70.
32. S. A. Bustin, V. Benes, J. A. Garson, J. Hellemans, J. Huggett, M. Kubista, M. Reinhold, T. Nolan, M. V. Pfaffl, G. L. Shipley, J. Vandesompele and C. T. Wittwer, The MIQE Guidelines: minimum Information for publication of quantitative real-time PCR experiment, *Clin. Chem.*, 2009, **55**, 611–622.
33. M. J. Aardema and J. T MacGregor, Toxicology and genetic toxicology in the new era of "toxicogenomics": impact of "-omics" technologies, *Mutat. Res.*, 2003, **499**, 13–25.
34. M. Ghelichpoura, A. T. Mirghaeda, S. H. Hoseinifar, M. Khalilic, M. Youseifd and A. Perez-Jimenez, Expression of immune, antioxidant and stress related genes indifferent organs of common carp exposed to indoxacarb, *Aquat. Toxicol.*, 2019, **208**, 208–216.
35. Z. Zhang, Q. Liu, J. Cai, J. Yang, Q. Shen and S. Xu, Chlorpyrifos exposure in common carp (*Cyprinus carpio L.*) leads to oxidative stress and immune responses, *Fish Shellfish Immunol.*, 2017, **67**, 604–611.
36. S. Özdemira, S. Altunb and H. Arslanc, Imidacloprid exposure cause the histopathological changes, activation of TNF-α, iNOS, 8-OHdG biomarkers, and alteration of caspase 3, iNOS, CYP1A, MT1 gene expression levels in common carp (*Cyprinus carpio* L.), *Toxicol. Rep.*, 2018, **5**, 125–133.
37. R. Anjithaa, A. Antonya, O. Shilpaa, K. P. Anupamaa, S. Mallikarjunaiahb and H. P. Gurushankara, Malathion induced cancer-linked gene expression in human lymphocytes, *Environ. Res.*, 2020, **182**, 109131.

38. D. G. Ginzinger, Gene quantification using real-time quantitative PCR: an emerging technology hits the mainstream, *Exp. Hematol.*, 2002, **6**, 503–512.
39. R. Ekblom and J. Galindo, Applications of next-generation sequencing in molecular ecology of non-model organisms, *Heredity*, 2011, **107**, 1–15.
40. P. Gayral, L. Weinert, Y. Chiari, G. Tsagkogeorga, M. Ballenghien and N. Galtier, Next-generation sequencing of transcriptomes: a guide to RNA isolation in nonmodel animals, *Mol. Ecol. Resour.*, 2011, **11**, 650–661.
41. D. L. Sheader, T. D. Williams, B. P. Lyonsand and J. K. Chipman, Oxidative stress response of European flounder (*Platichthys flesus*) to cadmium determined by a custom cDNA microarray, *Mar. Environ. Res.*, 2006, **62**, 33–44.
42. M. Zhao, Y. Zhang, C. Wang, Z. Fu, W. Liu and J. Gan, Induction of macrophage apoptosis by an organochlorine insecticide acetofenate, *Chem. Res. Toxicol.*, 2009, **22**, 504–510.
43. Z. Zhang, Q. Liu, J. Cai, J. Yang, Q. Shen and S. Xu, Chlorpyrifos exposure in common carp (*Cyprinus carpio L.*) leads to oxidative stress and immune responses, *Fish Shellfish Immunol.*, 2017, **67**, 604–611.
44. E. V. Paravani, M. F. Simoniello, G. L. Poletta and V. H. Casco, Cypermethrin induction of DNA damage and oxidative stress in zebrafish gill cells, *Ecotoxicol. Environ. Saf.*, 2019, **173**, 1–7.
45. O. M. Ighodaroa and O. A. Akinloyeb, First line defense antioxidants-superoxide dismutase (SOD), catalase (CAT) and glutathione peroxidase (GPX): Their fundamental role in the entire antioxidant defense grid, *Alexandria J. Med.*, 2018, **54**, 287–293.
46. P. A. Olsvik, T. Kristensen, R. Waagbø, B. O. Rosseland, K. E. Tollefsen, G. Baeverfjord and M. H. Berntssen, mRNA expression of antioxidant enzymes (SOD, CAT and GSH-Px) and lipid peroxidative stress in liver of Atlantic salmon (Salmo salar) exposed to hyperoxic water during smoltification, Comp. Biochem, *Comp. Biochem. Physiol., Part C: Toxicol. Pharmacol.*, 2005, **141**, 314–323.
47. T. Liu, L. Li, Z. Zhang and Y. Wang, Transcriptional inhibition of the catalase gene in phosphine-induced oxidative stress in *Drosophila melanogaster*, *Pestic. Biochem. Physiol.*, 2015, **124**, 1–7.
48. J. G. Scandalios, Oxidative stress: molecular perception and transduction of signals triggering antioxidant gene defenses, *Braz. J. Med. Biol. Res.*, 2005, **38**, 995–1014.
49. Y. Jin, X. Zhang, L. Shu, L. Chen, L. Sun, H. Qian, W. Liu and Z. Fu, Oxidative stress response and gene expression with atrazine exposure in adult female zebrafish (*Danio rerio*), *Chemosphere*, 2010, **78**, 846–852.
50. M. Karaca, L. Varihs, K. Korkmaz, O. Özaydın, F. Percind and H. Orhan, Organochlorine pesticides and antioxidant enzymes are inversely correlated with liver enzyme gene expression in *Cyprinus carpio*, *Toxicol. Lett.*, 2014, **230**, 198–207.

51. F. A. Yamamoto, G. F. Diamante, M. S. Santana, D. R. Santos, D. R. Bombardeli, C. C. Martins, C. A. Oliveira Ribeiro and D. Schlenk, Alterations of cytochrome P_{450} and the occurrence of persistent organic pollutants in tilapia caged in the reservoirs of the Iguaçu River, *Environ. Pollut.*, 2018, **240**, 670–682.
52. Z. Siroká and J. Drastichová, Biochemical markers of aquatic environment contamination cytochrome P_{450} in fish. A review, *Acta Vet. Brno*, 2004, **73**, 123–132.
53. H. Y Yang and T. H Lee, Antioxidant enzymes as redox-based biomarkers: a brief review, *Rep. Biochem. Mol.*, 2015, **48**, 200–208.
54. Y. Jin, S. Zheng and Z. Fu, Embryonic exposure to cypermethrin induces apoptosis and immunotoxicity in zebrafish (*Danio rerio*), *Fish Shellfish Immunol.*, 2011, **30**, 1049–1054.
55. M. Zhao, Y. Zhang, C. Wang, Z. Fu, W. Liu and J. Gan, Induction of macrophage apoptosis by an organochlorine insecticide acetofenate, *Chem. Res. Toxicol.*, 2009, **22**, 504–510.
56. D. S. Waits, D. Y. Simpson, A. M. Sparkman, A. M. Bronikowski and T. S. Schwartz, The utility of reptile blood transcriptomes in molecular ecology, *Mol. Ecol. Resour.*, 2019, **20**, 308–317.
57. A. Capraro, D. O'Meally, S. A. Waters, H. R. Patel, A. Georges and P. D. Waters, Waking the sleeping dragon: gene expression profiling reveals adaptive strategies of the hibernating reptile *Pogona vitticeps*, *BMC Genomics*, 2019, **20**, 460.
58. L. Héritier, D. Duval, R. Galinier, A.-L. Meistertzheim and O. Verneau, Oxidative stress induced by glyphosate-based herbicide on freshwater turtles, *Environ. Toxicol. Chem.*, 2017, **12**, 3343–3350.
59. L. Héritier, A.-L. Meistertzheimand and O. Verneau, Oxidative stress biomarkers in the Mediterranean pond turtle (*Mauremys leprosa*) reveal contrasted aquatic environments in Southern France, *Chemosphere*, 2017, **183**, 332–338.
60. I. Morao, M. Lemos, R. Féliz, S. Vieira, C. Barata and S. Novais, Stress response markers in the blood of Sao Tomé green sea turtles (*Chelonia mydas*) and their relation with accumulated metal levels, *Environm. Pollut.*, 2022, **293**, 118490.
61. P. Sujiwattanarat, P. Pongsanarakul, Y. Temsiripong, T. Temsiripong, C. Thawornkuno, Y. Uno, S. Unajak, Y. Matsuda, K. Choowongkomon and K. Srikulnath, Molecular cloning and characterization of Siamese crocodile (Crocodylus siamensis) copper, zinc superoxide dismutase (CSI-Sod) gene, *Comp. Biochem. Physiol., Part A: Mol. Integr. Physiol.*, 2016, **191**, 187–195.
62. R. Yatsu, S. Miyagawa, S. Kohno, B. B. Parrott, K. Yamaguchi, Y. Ogino, H. Miyakawa, R. H. Lowers, S. Shigenobu, L. J. Guillette Jr and T. Iguchi, RNA-seq analysis of the gonadal transcriptome during *Alligator mississippiensis* temperature-dependent sex determination and differentiation, *BMC Genomics*, 2016, **17**, 77.

63. H. Sun, X. Zuo, L. Sun, P. Yan, F. Z. H. Xue, E. Li, Y. Zhou, R. Wu and X. Wu, Insights into the seasonal adaptive mechanisms of *Chinese alligator* (*Alligator sinensis*) from transcriptomic analyses, *Aust. J. Zool.*, 2018, **66**, 93–102.
64. S. Kohno, J. Zhu and L. J. Guillette, Stress responses in the chemistry and mRNA abundance of the peripheral blood in the American alligator, *J. Exp. Zool., Part A*, 2019, **333**, 151–163.
65. O. I. Azeez, J. G. Myburgh, A. M. Bosman, J. Featherston, K. P. Sibeko-Matjilla, M. C. Oosthuizen and J. P. Chamunorwa, Next generation sequencing and RNA-seq characterization of adipose tissue in the Nile crocodile (*Crocodylus niloticus*) in South Africa: Possible mechanism(s) of pathogenesis and pathophysiology of pansteatitis, *PLoS One*, 2019, **14**, e0225073.
66. E. C. López González, L. M. Odetti, G. L. Poletta, N. Denslow, K. Kroll, P. A. Siroski and M. V. Parapachu Marcó, Optimizing protocols for high-quality RNA extraction from blood and liver tissues of the Broad-snouted caiman, *Russ. J. Herpetol.*, 2021, **28**, 197–204.
67. Y. Chiari and N. Galtier, RNA extraction from sauropsids blood: evaluation and improvement of methods, *Amphib. Reptiles*, 2011, **32**, 136–139.
68. B. Stringhini, Identificación de isoformas del gen citocromo p450 (cyp) en sangre periférica de *Caiman latirostris* y análisis de su potencialidad como marcador de exposición a xenobióticos, presented in part at Encuentro de Jóvenes Investigadores de la Universidad Nacional del Litoral, Santa Fe, September, 2021.
69. E. V. Paravani, L. M. Odetti, M. F. Simoniello and G. L. Poletta, Molecular analysis and bioinformatic characterization of cooper, zinc-superoxide dismutase (Cu/Zn-*sod*) gene of *Caiman latirostris*, *Mol. Biol. Rep.*, 2020, **47**, 8849–8857.
70. L. M. Odetti, PhD thesis, Universidad Nacional del Litoral, Santa Fe, Argentina, 2021.
71. A. Slaninova, M. Smutna, H. Modra and Z. Svobodova, A review: oxidative stress in fish induced by pesticides, *Neuroendocrinol. Lett.*, 2009, **30**, 2–12.
72. P. M. Craig, C. M. Wood and G. B. McClelland, Oxidative stress response and gene expression with acute copper exposure in zebrafish (*Danio rerio*), *Am. J. Physiol.: Regul., Integr. Comp. Physiol.*, 2007, **293**, R1882–1892.
73. M. Arregui, R. Scotta and D. Sánchez, Improved weed control with broad-leaved herbicides in glyphosate-tolerant soybean (Glycine max), *Crop Prot.*, 2006, **25**, 653–656.
74. F. Spurlock and M. Lee, in *Synthetic pyrethroid use patterns, properties, and environmental effects, in Synthetic Pyrethroids*, ed. J. Gan, F. Spurlock, P. Hendley and D. P. Weston, American Chemical Society, Washington DC, 2008.
75. P. Singh, P. Lata, S. Patel, A. K. Pandey, S. K. Jain, R. Shanker and A. Dhawan, Expression profiling of toxicity pathway genes by real-time

PCR array in cypermethrin-exposed mouse brain, *Toxicol. Mech. Methods*, 2011, **21**, 193–199.

76. Y. Jin, S. Zheng, Y. Pu, L. Shu, L. Sun and W. Liu, Cypermethrin has the potential to induce hepatic oxidative stress, DNA damage and apoptosis in adult zebrafish (*Danio rerio*), *Chemosphere*, 2011, **82**, 398–404.
77. E. V. Paravani, M. F. Simoniello, G. L. Poletta, F. R. Zolessi and V. H. Casco, Cypermethrin: Oxidative stress and genotoxicity in retinal cells of the adult zebrafish, *Mutat. Res., Genet. Toxicol. Environ. Mutagen.*, 2008, **826**, 25.

CHAPTER 12

Epilogue and Final Remarks

GUILLERMO ELI LIWSZYC*[a] AND
MARCELO L. LARRAMENDY[b,c]

[a] Former Guest Scientist at the Department of Pathology, University of Helsinki, Helsinki, Finland; [b] Cátedra de Citología, Facultad de Ciencias Naturales y Museo, Universidad Nacional de La Plata, Calle 64 No. 3, B1904AMA La Plata, Argentina; [c] Consejo Nacional de Investigaciones Científicas y Técnicas (CONICET), Buenos Aires, Argentina
*Email: GuillermoLiwszyc@yahoo.com

The growing impact of anthropogenic activities generates huge quantities of toxic residues that can have direct or indirect detrimental effects on the quality of our environment. The increase in their jeopardizing effects can exert short-, medium- and long-term consequences, affecting not only human health but also environmental health in general. Furthermore, these disturbances are usually capable, in most cases, of reducing the biodiversity of native flora and fauna, which will, in turn, encourage the resistance and emergence of new pests and diseases. This book, *Bird and Reptile Species in Environmental Risk Assessment Strategies*, intends to provide an outline, relevant examples and stimulate practical discussions on the use of these non-conventional groups of vertebrates, bird and reptile species, within the scientific challenges faced by the Ecotoxicology and Risk Assessment academic world. It includes relevant tools which may be of use in the analysis prior to the implementation of decisions leading to actions which will hopefully reduce environmental health risks against environmental factors. Our aim was to compile information from a diversity of sources into a single volume. The rationale is to give some real-life examples in order to widen the concept that

Issues in Toxicology No. 45
Bird and Reptile Species in Environmental Risk Assessment Strategies
Edited by Guillermo Eli Liwszyc and Marcelo L. Larramendy

Published by the Royal Society of Chemistry, www.rsc.org

the use of non-conventional animal models, far from being a scientific conundrum, may instead provide useful avenues to be explored and further developed as a way to provide help and real answers to some of the actual problems the whole planet is dealing with currently. Concomitantly, these concrete examples extend concepts of hazardous factors to living species worldwide that may stimulate new research ideas and trends in their respective important fields.

In recent times, the growing impact of anthropogenic activities has generated huge quantities of toxic residues that can have direct or indirect detrimental effects on the quality of our environment. The increase in their jeopardizing effects can exert short-, medium- and long-term consequences, affecting not only human health but also environmental health in general. Furthermore, these disturbances are usually capable, in most of the cases, of reducing the biodiversity of native flora and fauna, which will, in turn, encourage the resistance and emergence of new pests and diseases.

This book, *Bird and Reptile Species in Environmental Risk Assessment Strategies*, intends to provide an outline and relevant examples and stimulate practical discussions on the use of non-conventional groups of vertebrates, bird and reptile species, within the scientific challenges faced by the Ecotoxicology and Risk Assessment academic world. It is worth underpinning that these two taxa are known under the name Sauropsida, a term that includes land vertebrates which comprise all existing reptiles and birds as well as their fossil ancestors. Birds and reptiles are the two groups of species which evolved from egg-laying amniotes in the Carboniferous period. It should be noted that the other group of species that evolved from amniotes is referred to as Synapsida, a group that gave rise, eventually, to mammals. Living sauropsids include lizards, snakes, turtles, crocodiles and birds as well. On the other hand, extinct sauropsids encompass non-avian vertebrates, *e.g.*, dinosaurs, pterosaurs, plesiosaurs and ichthyosaurs, among many others. In brief, sometime during the middle or lower Carboniferous period, amniotes split into two evolutionary lines. One of these evolved to all kinds of reptiles, known as Sauropsida, an evolutionary group that eventually led to mammals. Whether this assumption is accepted or not, it is more than evident that it is not correct to state that "mammals evolved from reptiles" since both groups of vertebrates derived from early amniotes. In any case, modern reptiles are vastly different from modern mammals. However, it is accepted that both groups have differentially evolved for about 300 million years from early amniotes.

This book endows relevant tools which may be of use in the analysis prior to the implementation of decisions leading to actions which will hopefully reduce environmental health risk against environmental factors that may adversely impact upon human and other living species health or causing ecological imbalances between them and our environment.

This volume comprises three major sections, namely: (1) *Bird and Reptile Species as Experimental Models in Environmental Risk Assessment Studies: General Aspects*; (2) *Bird Species as Experimental Models in Environmental*

Risk Assessment Studies and (3) *Reptile Species as Experimental Models in Environmental Risk Assessment Studies*.

In the first and general section, *i.e.*, *Bird and Reptile Species as Experimental Models in Environmental Risk Assessment Studies: General Aspects*, readers will find a broad-spectrum chapter (Chapter 1). It provides background and introductory information about the concept and implications of this hot topic in our current world, environmental risk assessment, while highlighting the nature of some environmental stressors and a sample of the most versatile and validated methods of analyses. In general terms, the narrative focuses on the worldwide accepted concept that Environmental Risk Assessment is a scientific practice that identifies and evaluates the likelihood of an activity or substance to threaten the environment in general and/or to living organisms, natural habitats and ecosystems in particular. In other words, the term employed describes the overall process/es and/or method/s allowing the identification of hazards and risk factors that have the potentiality to cause harm (hazard identification) as well as to analyse and evaluate the risk associated with that particular hazard (risk analysis and risk evaluation). It also suggests several, if not all, appropriate ways to eliminate or minimize the hazard, or control the risk when this cannot be eliminated (risk control).

The second section, *i.e.*, *Bird Species as Experimental Models in Environmental Risk Assessment Studies*, includes four chapters (Chapters 2–5) in which these vertebrates are used as biotic matrices. Actually, all extant bird species are some of the most successful vertebrates on Earth. Due to their ability to fly, birds have managed to occupy unique ecological niches, which helped them survive the demise of their close relatives, dinosaurs. The class Aves (or commonly known as Birds) is just one of the classes of the bigger kingdom Animalia. Birds share some characteristics with another closely related class, Reptilia, but they also have some features that are unique to these fantastic creatures. If we go through their main traits, some peculiar aspects can be underpinned, since they characterize and define the group, which are as follows. (a) Birds are vertebrates. Thus, they possess an inner skeleton that contains a spine, limbs and skull. (b) Birds are endothermic; they maintain constant body temperature by themselves without depending on outside factors. (c) Birds are bipedal. (d) The upper limbs of birds have evolved into specialized structures, the wings, that allow flight. (e) The body of most birds is spindle-shaped to make the flight more comfortable and more effective. (f) The bones of birds are hollow inside, to lower the overall weight of the body. (g) The birds have a beak instead of a mouth with teeth. (h) The body is covered with unique structures, feathers, while only legs are covered with scales. (i) Birds have a heart with four chambers, which separate oxygen-rich and oxygen-poor blood correctly, helping with the maintenance of a constant temperature. (j) Birds have a complex nervous system and a well-developed brain. It is well known that many birds are recognized to be highly teachable and intelligent. (k) There are two defined sexes in birds, male and female. (l) The fertilization in birds is internal. As a result of fertilization, a closed-off, waterproof egg is formed. (m) The eggs develop outside the body. Birds are characterized to show parental care for their eggs, usually under the care of either both parents or sometimes only one.

In brief, if we want to define birds strictly, we would say they are warm-blooded, bipedal vertebrate animals, with forelimbs evolved into wings, covered in feathers and with a beak instead of a mouth. Because of the constant flow of new data obtained by advances in DNA methodologies, the classification of birds and the relationships between orders and families is under constant revision. Accordingly, the latest classification contains 40 orders of these vertebrates distributed among the six continents. Furthermore, to the best of our knowledge, there are more than 9000 recognized bird species which are spread over the whole planet, from the poles to the equator.

Taking into account the explosive increase in the list of pollutants, the scientific community is facing challenges to protect humans and wildlife from the potentially negative consequences of their effects. The majority, if not all, of the previous efforts made to achieve this goal have revealed the global occurrence of high levels of various pollutants in birds. Laboratory research and epidemiological studies imply that pollutants exert a broad-spectrum of side-effects on birds by interfering with several parameters, including their endocrine, immune and neural systems, reproduction, development and growth. Nevertheless, inconsistent results suggest that the potential effects of pollutant exposure on the physiological parameters in birds are multifactorial, involving simultaneously several and diverse biological processes, species-specific differences, sex and age depending on the types of xenobiotics. Besides, another factor deserving attention is the type of interaction of more than one pollutant by acting simultaneously with another one/s. When the conventional concentration additivity is analysed, several studies have suggested that different classes of pollutants possibly act synergistically or antagonistically based on their concentration. However, risk assessment information is still recorded in a scattered and inadequate manner, because of the lack of enough avian species under study, the limited number of xenobiotics investigated and insufficient geographical representation. For these reasons, our knowledge of xenobiotic-induced deleterious effects on birds remains rudimentary, although mechanistic understanding of their mode of action is progressing. In particular, research on what happens to wild bird populations and their ecosystems under pollutant-induced stress is still unavailable. Accordingly, the aim of this book section is to provide some pieces of practical evidence predicting and tracing the effects of several xenobiotics at different biological organization levels, especially from the molecular, cellular and individual levels to the population, community and ecosystem levels, due to the limited and scattered information available, as mentioned earlier.

Chapter 2 provides the description of over 25 years of research in mercury (Hg) contamination and sub-lethal effects in the Florida Everglades (USA) using local wading bird species as experimental models. The authors assessed various potential indicator tissues and used nestling feathers as an indicator of Hg exposure and Hg risk to breeding. Long-term monitoring of wading bird populations and analysis of museum specimens' Hg revealed temporal and geographical variation in Hg within the study area. The results demonstrate a strong and relevant reduction in the number of nests in colonies exposed to Hg supported by the reduced breeding propensity and increased early failure, before egg-laying, in exposure experiments as well as a strong influence of variation in food availability on failure associated with Hg. Furthermore, the variation of food availability modulated Hg effects and the same amount of Hg in feathers could be associated with no apparent reduction in breeding success if food availability was high or with a reduction higher than expected from toxicity reference values when food availability was low. The evaluation of nestling feathers and other tissue indicators suggests that there is no single, unique best indicator, rather results suggesting specific endpoint–indicator associations.

In Chapter 3 the authors underpinned that seabirds play a key/crucial role in ecosystem functioning and serve as a useful indicator of its health status. The focus of this chapter is on the analysis of the stage of external exposure to risk factor/s. The authors explored different approaches to assess the vulnerability of seabirds to environmental risks, discussed current challenges, proposed a trait-based solution to these challenges and explored additional complicating factors in using non-model species in environmental risk assessments. The trait-based approach is an ecological dimension reduction technique to make sense of datasets with a high species dimension. This approach exists hand-in-hand with additional experimental approaches that investigate the internal response of a specific biological receptor of interest to risk factors given a prescribed level of exposure. When combined, these approaches should provide a more complete picture of the impact of risk factor/s to the native species and their communities. The authors suggested that integrating non-conventional species in environmental risk assessments is one important step towards this goal.

Chapter 4 stresses that the eggs of wild birds represent a biotic matrix for documenting the toxic effects of organochlorine pesticides (OCPs) in natural ecosystems. The authors categorized OCP residues reported so far in the literature based on species and food habits, among other factors. The list of dominant species studied so far includes the Indian white-backed vulture, house sparrow, junglefowl, scaly-breasted munia and red-whiskered bulbul. All egg samples monitored show varying levels of OCP residues. The frequencies of detection of different compounds belonging to OCPs were in the order of total hexachlorocyclohexanes ($\sum$HCHs) > total dicholorodiphenyltrichloroethanes ($\sum$DDTs) > cyclodienes (dieldrin and heptachlor epoxide). Among the $\sum$OCPs reported in the reviewed studies, the residues of DDT and its metabolites represented the major share (60%). Besides, *p,p*′-dicholorodiphenyldichloroethane (*p,p*′-DDE) residues were reported to exceed the threshold levels of toxicity in the eggs of the wild birds. Furthermore, 28% of the eggs of house sparrows analysed for $\sum$OCP residues contained *p,p*′-DDE residues leading to their reproductive impairment. However, the magnitude of residue accumulation does not show any significant differences with respect to species and feeding habits. Inferences from results hence uphold the inevitable need for systemic toxicity studies and real-time biomonitoring programmes towards assessing the harmful effects of toxic contaminants over the biotic components of our environment.

The last chapter of this section, Chapter 5, documents that agricultural intensification affects bird populations through both a reduction in landscape heterogeneity and an increase in the use of pesticides, which negatively influence a variety of bird species. The authors aimed to examine how agricultural intensification and the associated increase in pesticide use can affect farmland birds. Special attention is given to the case of pesticide-treated seeds, and the current guidelines on how to perform the risk assessment of seed treatments. Nevertheless, because a number of registered pesticide seed treatments have been found to pose a risk to farmland birds, current regulatory risk assessment protocols need to be improved so as to provide more accurate predictions of real-life situations and to better protect bird populations.

The third section, *i.e.*, *Reptile Species as Experimental Models in Environmental Risk Assessment Studies*, consists of six chapters (Chapters 6–11) in which these vertebrates, Reptiles, are employed as biotic matrices. It is known that Reptiles (from the late Latin, *neuter of reptilis*, meaning "*creeping*") are cold-blooded (ectothermic) vertebrates of the class Reptilia. They are often found in every kind of habitat, including grasslands, swamps, forests, deserts, oceans and other arid environments. Over 11 500 vertebrate species are currently recognized and classified as reptiles, constituting, thus, the dominant terrestrial vertebrate group. Reptiles have some features that are unique to these fantastic creatures. Among them, several characteristics deserve attention since they typify and define the group, which are as follows. (a) Reptiles are vertebrates. Reptiles show characteristics of other vertebrates such as mammals, birds and some amphibians. All reptiles have spinal cords to support their body structure from the base of the head to the tail. This bony endoskeleton protects the inner tissue and helps in body movement. In the Reptilia class, crocodiles have the most extensive body structures. (b) All reptiles are cold-blooded animals, which are incapable of regulating their own body temperature. Unlike mammals, reptiles use their surroundings to keep them warm, the reason why they prefer to occupy hot/warm habitats. Reptiles tend to bask under the sun on and off to regulate their body temperature. (c) Reptiles have horny/rough scales. Their bodies are covered with scutes, sharing these features with the feet of birds, or horny layers of scales and bony plates. These scales are made of α- and β-keratin and originate from the epidermis, unlike the dermis in mammals. As a result, reptile skin is thinner than in other groups of vertebrates. (d) Reptiles lay eggs. Some snake and lizard species give birth to live young, *i.e.*, viviparous (*e.g.*, Eurasian or common lizard *Lacerta vivipara*). It is worth mentioning that in most reptiles the incubation temperature determines the sex. Temperatures above 30 °C produce all female specimens, whereas temperatures below 25 °C produce all males. (e) Reptiles have one or two developed lungs. Since reptiles have dry and scaly skin, the only way they can breathe would be through their lungs. So first, they breathe through their nasal passage or mouth and pass air into their lungs. All terrestrial reptiles have two well-developed lungs, whereas some giant snakes have only one lung, while the other is atrophied. (f) Reptiles have chemoreceptors. To find their prey, some reptiles use a lining on their nose and roof of the mouth to absorb the chemicals in the air. This phenomenon known as chemoreception is prevalent in snakes and lizards. (g) Reptiles have two different digestion methods. From a dietary perspective, there are carnivorous reptiles (snakes and lizards in the Squamata order) and herbivorous reptiles (*i.e.*, tortoises). Carnivorous reptiles have shorter guts for efficiently breaking down meat but have a slower digestion than mammals. On the other hand, as herbivorous reptiles cannot grind, to speed up digestion, they swallow rocks and pebbles to help crush the food. (h) All reptiles undergo skin-shedding. Reptiles shed their skin as they grow and throughout their lifecycle. Sometimes, improper or incomplete shedding may occur because of mites, incorrect humidity or handling, malnutrition, dermatitis or trauma. As the shedding process begins, the skin appears slightly dull when the separation between old and new skin begins. Shedding may also be necessary for removing parasites or other things that have attached themselves to the skin during growth periods. Turtles and lizards shed their skin in several pieces, whereas snakes shed their skin in one piece, a phenomenon that varies from species to species.

Chapter 6 describes the usefulness of the studies of congenital malformations and their aetiology in wild reptiles as a powerful tool when considered in a holistic evaluation of the "one health status". It is known that Reptiles, according to their habitat and biological features, can be promising predictors of environmental contamination. However, research in this area remains scarce and dispersed. The purpose of this review is to compile some documentation that describes and discusses the deleterious effects of pesticides. Further research in this area is urgently needed in order to implement concrete measures with a view to improving ecosystem health and halting the present decline of reptile populations.

Chapter 7 stresses that irrigation ponds (or farm ponds), as a secondary nature, harbour component species of floodplain ecosystems and hence make a significant contribution to biodiversity. However, biological communities in irrigation ponds are exposed to multiple environmental stressors. Survey data on Japanese irrigation ponds show that reptiles, in this case study, turtles, are significantly different from other taxa comprising fishes, molluscs, crustaceans, insects and annelids, in the estimated combined impact (on taxonomic richness) of multiple stressors, including the insecticide 2-*sec*-butylphenyl methylcarbamate (BPMC, also known as fenobucarb), the fungicide probenazole, concrete bank protection, pond depth reduction, macrophyte decline and invasive alien fish. All these stressors are negatively correlated with the taxonomic richness of all sampled but are positively correlated with that of reptiles. This positive correlation of reptiles with concrete bank protection may stem from the suitability of concrete banks for turtles' basking. The positive correlation of reptiles with insecticide concentration implies that some extent of insecticide pollution might favour turtles relative to other taxa, because of their potentially high tolerance against the pollutant(s). The taxonomic richness of turtles may be useful as a bioindicator for the combined impact of environmental stressors on animal diversities in irrigation ponds.

Continuing with practical examples of non-conventional reptile species as experimental models, Chapter 8 highlights that reptiles remain the least studied vertebrate taxon in ecotoxicology and a lack of standardized methods and government regulations has hindered progress. The authors provided an overview of the history of reptile ecotoxicology with particular emphasis on toxicity testing methods. Oral dosing has received the most attention, but there have been important reports of significant dermal exposures and toxicity in reptiles as well. We report on recent efforts to adapt oral dosing with gelatine capsules in a small gecko, expanding on previous methods for oral dosing as well as husbandry of *Anolis* lizards for use in ecotoxicological experiments. Previously published methods combined with our current efforts provide a framework for repeatable, environmentally relevant and consistent dosing of reptiles. Future efforts should focus on trying to create a standardized testing protocol that can be put through validation methods to be acceptable for government regulations. This would greatly increase the opportunity for performing reptile toxicity testing in the future. Finally, the toxicity testing literature has focused heavily on lizards and more research is needed to establish standardized methods for snakes and turtles.

Chapter 9 shows that the expansion of agriculture induces constant habitat fragmentation and degradation. These losses and the concomitant use of pesticides and herbicides play an important role in the decline of reptiles, and among them, lizards are particularly endangered. This chapter describes the toxic effects of pure glyphosate on the liver of the common field lizard *Podarcis siculus*. Glyphosate (GLY) is the most widely used active ingredient in agrochemical formulations worldwide to maximize crop yields. The onset of liver damage was evaluated after oral administration of GLY to simulate a realistic situation of herbicide exposure in the lizard habitat. The results demonstrate a severe liver condition, with fibrotic formations and scarring with increased exposure. Hepatocytes enhance their defence mechanism by activating their antioxidant systems and increasing the expression of *p53*. The results also confirm the xeno-oestrogenic nature of GLY. Altogether, the effects of GLY must be considered biologically relevant and could endanger the viability and survival of lizard populations, also unbalancing the trophic cascade. A more controlled and less intensive use of the herbicide GLY in areas devoted to crop production would therefore be advisable as remarked by the authors.

Chapter 10 illustrates the case of endocrine-disrupting chemicals (EDCs), which include a large variety of man-made chemicals frequently used around the world. These are substances that behave as endogenous hormones or as inhibitors of their actions. The negative impact of EDCs on laboratory rodents, humans and aquatic and avian species has been deeply studied for years. The aim of this chapter is to highlight the huge value of the native South American crocodilian species *Caiman latirostris* as a sentinel species of EDC pollution and to summarize how much we have learned from this caiman in the past two decades. The authors expected that the results they presented can contribute not only to the knowledge of the effects of EDCs on wild species, but also as a warning to government control agencies and EDC users and producers, acknowledging their responsible role in the preservation of natural ecosystems.

In the last chapter of this section, *i.e.*, Chapter 11, the authors also employed the *C. latirostris* as a sentinel organism to evaluate the effect induced by different pesticides, since many populations living in the north-central region of Argentina are located in the vicinity of areas with high agricultural activity. Most of the biomarkers that have been used to date to identify the impact of pesticides on this species included markers of oxidative and DNA damage, imbalances in antioxidant enzymes and immunological and lipid alterations. The possibility of applying gene expression markers in *C. latirostris* will allow the understanding of the specific significance of many pesticide-induced alterations observed by means of traditional biomarkers. The usefulness of these molecular tools applied in *C. latirostris* blood will shed some more light on the understanding of many physiological alterations previously detected that were caused by traditional toxicological biomarkers.

In spite of the fact of having dealt with many diverse topics, we have tried to compile this "wealth of information" into two major parts for the sake of clarity and order. The first part, *Bird Species as Experimental Models in Environmental Risk Assessment Studies*, and the second part, *Reptile Species as Experimental Models in Environmental Risk Assessment Studies*, take into consideration whether different vertebrate species of one or more populations of a living organism or organisms have been selected as experimental matrices, respectively.

This book will be part of a series of four volumes along similar lines utilizing different species rather than Birds and Reptiles (Mammals, Fish and Amphibians).

Without running the risk of being repetitive, we would like to recap on some important concepts as previously mentioned in the two volumes previously published by the Royal Society of Chemistry in 2017, *i.e.*, *Ecotoxicology and Genotoxicology: Non-traditional Terrestrial Models* and *Ecotoxicology and Genotoxicology: Non-traditional Aquatic Models*, by the same editor group. We strongly recommend the perusal of both volumes, which are not overlapping subjects, in order to gain the full benefit of this series and have a holistic and stimulating approach to the matter.

New emerging studies in several reputable publications of common worldwide human diseases such as arterial hypertension, among others, seem to suggest that not only classical and lifestyle factors should be addressed in the search for adequate treatment.[1] From a medical perspective, arterial hypertension is not only a disease of affluence. One in eight deaths worldwide is due to this condition and its associated health conditions (heart disease, kidney disease and stroke). These studies seem to suggest that a closer look of other, so far unrelated factors is required. Early-life nutrition and exposure to air pollution, heavy metals and even noise are factors implicated, which may push blood pressure up later in life. Thinking out of silos and crosslinking medical with applied science will lead to a faster pace of understanding all the factors related to the emergence of a disease and thus finding a cure.

Many scientists feel that time has come for a shift in emphasis away from basic towards applied research. While basic research is needed in order to shed more light on the fundamentals, applied research is required in order to find solutions to the many problems the world faces, *e.g.*, overpopulation, excessive use of the Earth's natural resources and pollution.

The chapters included in this book are just a mere enumeration of some practical examples. There are many more species that can be used as experimental models and the list should be expanded by different research groups and academics all over the world. We hope that many more scientists realize that it is more important to tackle subjects related to the status of the environment using autochthonous, non-target species, which are truly exposed to locally used xenobiotic agents. On the other hand, this research plan would also increase the chances of independent scientists and research institutions in getting access to grants and attracting the attention of local authorities, who would be more interested in financing projects that really are within their sphere of political interest and pride.

Globally, 41% of amphibian species and 19% of reptile species are threatened or dying out. This is the result of the adding up of their natural habitat destruction for land use, disease, pollution and climate change. In many species, these factors have added so much pressure that they have been completely wiped out.

It is our hope that with the emergence of the new legal figure of ecocide, governments, politicians, business corporations and producers will have to weigh and analyse more carefully the consequence of their goals. It is certainly a first important step towards the accountability of their actions.

The UN Biodiversity Conference Intactness Index (BII) shows a world average of 75%; this is a value certainly lower than the 90% required to be considered as the safe limit for not falling in "ecological recession".

The Stone Age, Copper Age, Iron Age, Medieval Ages and Modern Age constitute a few of the developmental stages of mankind. The Anthropocene will be probably followed by the Extinction Age, if we remain quiet bystanders. While the clock is ticking, we still have some valuable time left to be used diligently.

Many researchers have contributed to the publication of this book. We hope that it serves as a herald in order to bolster enthusiasm for the use of native or easily available local species in order to widen our knowledge on the subject. Last but not least, we would like to especially thank the authors who have responded to our call, for their time, contribution and feedback, making possible the compilation of this book.

Let us hope that the Anthropocene will not be the last Age of Humanity.

Reference

1. NCD Risk Factor Collaboration, *Lancet*, 2016, **389**(10064), 37–55.

Subject Index

agricultural expansion, 152–153
  environmental problems associated with, 198–200
agricultural intensification, 73–74, 75
  and bird declines, 74–76
  through seed treatment, 79–80
albumen, 16–17, 20–22
alligators, 11, 177, 180–182
Alligator sinensis, 204
American alligators (*Alligator mississippiensis*), 11, 102, 204
AMO. *See* Atlantic Multidecadal Oscillation (AMO)
Anolis species in reptile ecotoxicology, 138–139
  background, 137–138
  developing a reproductive assay with *Anolis*, 140–142
  dosing studies to assess impacts on survival and growth, 143
  egg studies, 143
  oral dosing experiments, 139–140
  per- and polyfluoroalkyl substances, 138
  reproductive endpoints, 142
    breeding and egg production, 142–143
    egg production, 142
    eggs directly exposed to contaminants, 142
  reproductive studies, 143
aquatic bird indicators of sub-lethal mercury exposure, 6
  Everglades, 7–11
    mercury contamination in, 11–12
  evidence and estimation of Hg effects in natural populations, 18–19
  evidence of mercury effects in field conditions, 16–18
  experimental studies, results from, 15
  field sampling protocol, development of, 12–13
  indicator tissues, 19–24
  local emissions and worldwide contamination of mercury, 6–7
  mercury exposure indicators, development of, 13–14
artificial fibres, 2
Atlantic Multidecadal Oscillation (AMO), 46
atrazine (ATZ), 174

BChE activity. *See* butyrylcholinesterase (BChE) activity
Bermuda petrel (*Pterodroma cahow*), 39, 42
Biomphalaria glabrata, 121
birds. *See also* aquatic bird indicators of sub-lethal mercury exposure; pesticide-treated seeds, bird exposure to; seabird vulnerability
  native seabirds, 44

birds (*continued*)
pesticides' impact on, 77
direct effects, 77–78
indirect effects, 78–79
planktivorous seabirds, 46
wading birds, 7–8, 10, 16, 220
wild birds, pesticides in eggs of, 57
bisphenol A (BPA), 174
black bass (*Micropterus salmoides*), 115
black-browed albatrosses (*Thalassarche melanophris*), 35–37
black-capped petrel (*Pterodroma hasitata*), 39
blackheaded ibis (*Threskiornis melanocephalus*), 65
bluegill (*L. macrochirus*), 115, 120
BPA. *See* bisphenol A (BPA)
BPMC. *See* 2-*sec*-butylphenyl methylcarbamate (BPMC)
breeding and egg production, 142–143
breeding cycle, 12, 18, 19, 21
broad-snouted caiman (*Caiman latirostris*), 132, 169, 170, 196, 197
ecophysiological characteristics of, 169
feeding habits, social behavior and longevity, 171
hatchling growth and development, 171
reproduction, 170
sex determination, 170
sexual dimorphism, 171–172
endocrine-disrupting chemicals (EDCs), 172
effects on reproductive features, 175–182
effects on thyroid histofunctional characteristics, 182–183
mechanism of action of, 173
types of, 173–174
environmental problems, 198–200
evidence of pesticide effects on, 200–202
future perspectives, 209
gene expression markers
importance of the technique and advantages over other tools, 202
optimization of protocols, 203–209
geographical distribution and sustainable use programs, 197–198
as a sentinel of environmental pollution, 174–175
Brook's house gecko (*Hemidactylus brookii*), 134, 135, 145
brown tree snakes (*Boiga irregularis*), 132
bullfrog (*Lithobates catesbeianus*), 115
butyrylcholinesterase (BChE) activity, 104
2-*sec*-butylphenyl methylcarbamate (BPMC), 115–116, 120, 223

caiman eggshell, 177
capillarity, 135
carbamate, 76
carnivorous reptiles, 223
catalase (CAT) enzymes, 200
CAT enzymes. *See* catalase (CAT) enzymes
cerulean bunting (*Emberiza citrinella*), 78
Chelonia mydas, 204
chicken (*Gallus gallus domesticus*), 12
climate change and variability, 44–47
common gulls (*Larus canus*), 62
common loons (*Gavia immer*), 40
common murre (*Uria aalge*), 45
common snapping turtle (*Chelydra serpentina*), 99, 100

concrete bank protection, 114–119, 223
conspicuous flights inland, 45
correlation *vs.* causation, 43
Cory's shearwaters (*Calonectris borealis*), 46
Crocodilia (order), 102–103
CYP superfamily. *See* cytochrome p450 (CYP) superfamily
cytochrome p450 (CYP) superfamily, 203

dark-spotted anaconda (*Eunectes deschauenseei*), 104
DDT. *See* dichlorodiphenyltrichloroethane (DDT)
DEE (1,1,-dichloro-2,2-bis-(p-chlorophenyl)ethylene), 18
deltamethrin, 132
dichlorodiphenyltrichloroethane (DDT), 56, 131, 176

EDCs. *See* endocrine-disrupting chemicals (EDCs)
EDEN. *See* Everglades Depth Estimation Network (EDEN)
egg albumen, 16, 20
egg production
    assessing impacts on, 142
    breeding and, 142–143
eggs directly exposed to contaminants, 142
eggshell thinning and reproductive impairment, 66
Egretta, 12
endocrine-disrupting chemicals (EDCs), 172, 224–225
    mechanism of action of, 173
    reproductive features, effects on, 175
        early postnatal exposure, 182
        experimental exposure, 177–182
        in ovo exposure, 180–182
        natural exposure, 175–177
    thyroid histofunctional characteristics, effects on, 182–183
    types of, 173–174
5-enolpyruvylshikimate-3-phosphate synthase (EPSPS), 151
environmental health, concept of, 4
EPSPS. *See* 5-enolpyruvylshikimate-3-phosphate synthase (EPSPS)
ERs. *See* estrogen receptors (ERs)
estrogen receptors (ERs), 155
Eudocimus albus, 7
European starling (*Sturnus vulgaris*), 12
eutrophication and pesticide pollution, 119–120
Everglades, 7–11, 14, 22
    mercury contamination in, 11–12
Everglades Depth Estimation Network (EDEN), 16

female-producing temperatures (FPTs), 170
fence lizards, 133, 134, 144
field lizard (*Podarcis siculus*) liver, 151
    effects of pure GLY on specimens of, 154–155
    GLY-induced changes in histology of, 155–158
    GLY-induced changes in the expression and synthesis of proteins in, 158–161
    lizards in contaminated environments, 152–154
FMN. *See* frequency of micronuclei (FMN)
food production, 1, 73
FPTs. *See* female-producing temperatures (FPTs)
frequency of micronuclei (FMN), 201

giant petrels (*Macronectes* spp.), 37
global warming, 2
glutathione peroxidases (GPxs), 159

glutathione *S*-transferases (GSTs), 203
GLY. *See* glyphosate (GLY)
glyphosate (GLY), 198, 224
  GLY-based herbicides (GBHs), 152
  -induced changes
    in the expression and synthesis of proteins in *Podarcis siculus* liver, 158–161
    in histology of *Podarcis siculus* liver, 155–158
  on specimens of *Podarcis siculus* liver, 154–155
GPxs. *See* glutathione peroxidases (GPxs)
grayish baywing (*Agelaioides badius*), 82, 84
great egrets, 8, 9, 11, 12
great shearwater (*Ardenna gravis*), 39, 40
great tits (*Parus major*), 18
green anoles (*Anolis carolinensis*), 132, 139
Green Revolution, 55, 73
green sea turtle (*Chelonia mydas*), 99
grey junglefowl (*Gallus sonneratii*), 62
GSTs. *See* glutathione *S*-transferases (GSTs)

hawksbill sea turtle (*Eretmochelys imbricata*), 99
HCH, 56, 62
herbicides, 78, 119, 120, 224
herring gull (*Larus argentatus smithsonianus*), 40
heterogeneous xenobiotics, 3
hormone-induced sex determination (HSD), 170
house sparrows (*P. domesticus*), 62, 63, 64, 65, 221
HSD. *See* hormone-induced sex determination (HSD)

Ibiza wall lizard (*Podarcis pityusensis*), 104
illegal, unreported and unregulated (IUU) fisheries, 33
insecticides, 56, 76–80
intraperitoneal (IP) injection, 132
invasive alien species, 115, 120–121
inverse inference, 43
IP injection. *See* intraperitoneal (IP) injection
Italian field lizard (*Podarcis siculus*), 153
IUU fisheries. *See* illegal, unreported and unregulated (IUU) fisheries

Japanese pond turtle (*Mauremys japonica*), 113
Japanese quail (*Coturnix japonica*), 12, 63, 81

kelp gull (*Larus dominicanus*), 40

lacertids (*Gallotia galloti palmae*), 104
lizards. *See also* field lizard (*Podarcis siculus*) liver
  fence lizards, 133, 134, 144
  Ibiza wall lizard (*Podarcis pityusensis*), 104
  Italian field lizard (*Podarcis siculus*), 153
  Southern sagebrush lizard (*Sceloporus graciosus vandenburgianus*), 104
  western fence lizard (*Sceloporus occidentalis*), 133
loggerhead sea turtle (*Caretta caretta*), 99

macrophyte decline, 115–119
Maderian/Fea's petrel (*Pterodroma feae*), 40
male-producing temperatures (MPTs), 170
mallard duck (*Anas platyrhynchos*), 12, 81
maternal blood, 20

melanin-pigmented cells, 158
Melanoides tuberculatus, 121
mercury
  contamination, in Everglades, 11–12
  experimental studies, results from, 15
  in field conditions, 16–18
  field sampling protocol, development of, 12–13
  indicator tissues, 19–24
  local emissions and worldwide contamination of, 6–7
  in natural populations, 18–19
mercury exposure indicators, development of, 13–14
methylation, 11
methylmercury, 7
MPTs. *See* male-producing temperatures (MPTs)

native seabirds, 44
neoniconicotinoid insecticides, 76
nestling feather, 14–17, 19, 21–24, 220, 221
NOAEL. *See* no-observed adverse effect level (NOAEL)
non-treated seeds, availability of, 83
no-observed adverse effect level (NOAEL), 83, 173
northern bobwhite (*Colinus virginianus*), 12, 81
northern gannet (*Morus bassanus*), 39, 40

OCCs. *See* organochlorine compounds (OCCs)
OCPs. *See* organochlorine pesticides (OCPs)
olive ridley sea turtle (*Lepidochelys olivacea*), 99
OLS estimator. *See* ordinary least squares (OLS) estimator
Ophidia (suborder), 103–104
ordinary least squares (OLS) estimator, 43
organochlorine compounds (OCCs), 173, 175–176, 178
organochlorine pesticides (OCPs), 54
  in eggs of wild birds in India, 57
    eggshell thinning and reproductive impairment, 66
    OCP residues in birds' eggs, 57–63
    variation in residue levels based on species and food habits, 64–66
  Indian scenario, pesticides in, 56–57
  marketing and consumption, 56
organophosphate insecticides, 76, 132

painted stork (*Mycteria leucocephala*), 65
painted turtle (*Chrysemys picta*), 99, 100
PCA. *See* principal component analysis (PCA)
Pearson's correlation test, 43
pesticides. *See also* organochlorine pesticides (OCPs)
  environmental problems associated with use of, 198–200
  in Indian scenario, 56–57
pesticides and environmental stressors on reptile diversity, 110
  biological communities in irrigation, 112–114
  concrete bank protection, 116–119
  eutrophication and pesticide pollution, 119–120
  invasive alien species, 120–121
  macrophyte decline, 116–119
  multiple stressors, combined impact of, 114–116
  turtles as bioindicators, 121–122
  water depth reduction, 116–119

pesticides' impact on birds, 77. *See also* wild birds, pesticides in eggs of
direct effects, 77–78
indirect effects, 78–79
pesticides' teratological effects in reptiles, 97
Crocodilia (order), 102–103
Rhynchocephalia (order), 102
sentinel species, reptiles as, 99
Squamata (order), 103
Ophidia (suborder), 103–104
Sauria (suborder), 104
Testudinata (order), 99–102
pesticide-treated seeds, bird exposure to, 80–81
future directions, 85
regulatory environmental toxicity testing, 81–82
Tier I risk assessment, 83
weight-of-evidence risk assessment, 83
dehusking behavior, 84
focal species, 83
food item preference and avoidance, 84
foraging area, 84
historical data on poisoning incidents and field studies, 85
meal size approach, 84
metabolism and body burden modeling, 84
non-treated seeds, availability of, 83
planktivorous seabirds, 46
Pogona vitticeps, 204
pollutant-induced deleterious environmental effects, 4
principal component analysis (PCA), 38
propylene glycol, 132–133
Proyecto Yacare, 197
pseudo-gavage method, 132, 137, 139, 142, 143, 145
pyrethroids, 111, 132
quantitative real-time polymerase chain reaction (qPCR), 202, 204

Rana clamitans, 155
reactive oxygen species (ROS), 201, 203
red-eared slider turtle (*Trachemys scripta elegans*), 113, 120–121, 204
red swamp crayfish (*Procambarus clarkii*), 115
red-whiskered bulbul (*Pycnonotus jocosus*), 62
Reeves' pond turtle (*Mauremys reevesii*), 113
regulatory environmental toxicity testing, 81–82
reptile ecotoxicology/ecological risk assessment
dermal exposure and toxicity, potential importance of, 144
developing breeding assay using *Anolis* species in, 138–139
background, 137–138
dosing studies, 143
egg studies, 143
oral dosing experiments, 139–140
per- and polyfluoroalkyl substances, 138
reproductive assay development, 140–142
reproductive endpoints, 142–143
reproductive studies, 143
early efforts in reptile toxicology, 131–133
oral dosing methods, 133–137
reptilian toxicity testing, 130
lack of reptile toxicity data in the literature, 130–131
lack of standardized methods, 131
Rhynchocephalia (order), 102
ringed turtle doves (*Streptopelia risoria*), 18

ring-necked pheasant (*Phasianus colchicus*), 63
Rissa brevirostris, 46
Rissa tridactyla, 46
ROS. *See* reactive oxygen species (ROS)

Sauria (suborder), 104
scaly-breasted munia (*Lonchura punctulata*), 62
seabird vulnerability, 33
    anthropogenic and natural risks, vulnerability to, 34–35
    challenges, 43
        climate change and variability, 44–47
        correlation *vs.* causation, 43
        default of no risk, 43–44
        ecological regulation, 44
    observational study, challenges for, 35–36
    standard approach, failings of, 34
    trait-based approach, 37
        ecological dimension reduction technique, 37–39
        ecological risk evaluation, useful traits for, 39–41
        trait-based prediction, 41–42
seed treatment, agricultural intensification through, 79–80
seed treatment with pesticides
    agricultural intensification, 79–80
    bird exposure to pesticide-treated seeds, 80–81
sentinel species, reptiles as, 99
Siamese crocodile (*Crocodylus siamensis*), 204
skink (*Marisora brachypoda*), 104
snakes, 120, 132
snapping turtle (*Chelydra serpentina*), 99, 100
SODs. *See* superoxide dismutases (SODs)
sooty shearwaters (*Ardenna grisea*), 36, 39
South American pit vipers (*Bothrops jararaca*), 103
South American rattlesnakes (*Crotalus durissus*), 103
Southern sagebrush lizard (*Sceloporus graciosus vandenburgianus*), 104
Spanish pond turtles (*M. leprosa*), 119, 120, 204
SPEAR, 122
spot-billed pelican (*Pelecanus philippensis*), 65
spurious correlation, 43
Squamata (order), 103
    Ophidia (suborder), 103–104
    Sauria (suborder), 104
superoxide dismutases (SODs), 158, 200
Synapsida, 218

temperature-dependent sex determination (TSD), 170
Tengmalm's owl (*Aegolius funereus*), 18
TER. *See* toxicity exposure ratio (TER)
Testudinata (order), 99–102
thyroid disruption, 182–183
Tier I risk assessment, 83
toxicity exposure ratio (TER), 83
Trindade petrel (*Pterodroma arminjoniana*), 40
TSD. *See* temperature-dependent sex determination (TSD)
turtles as bioindicators, 121–122

vitellogenin (VTG), 154

wading birds, 7–8, 10, 16, 220
water depth reduction, 116–119
water snakes (*Nerodia fasciata*), 120
web-based data-sharing, 41

weight-of-evidence risk assessment, 83
dehusking behavior, 84
focal species, 83
food item preference and avoidance, 84
foraging area, 84
meal size approach, 84
metabolism and body burden modeling, 84
non-treated seeds, availability of, 83
poisoning incidents and field studies, historical data on, 85
western fence lizard (*Sceloporus occidentalis*), 133
white ibises, 7, 9, 12, 13, 17
white-rumped vultures (*Gyps bengalensis*), 62
wild birds, pesticides in eggs of, 57
eggshell thinning and reproductive impairment, 66
OCP residues in birds' eggs, 57
residues of cyclodiene compounds, 63
residues of DDT and its metabolites, 63
residues of HCH and its isomers, 62–63
residue levels variation based on species and food habits, 64
raptorial and scavenging birds, 65
terrestrial birds, 64–65
water birds, 65–66
wood storks (*Mycteria americana*), 8, 10

xenobiotics, 2, 4, 200, 202, 203, 207, 220
Xenopus laevis, 174

zebra finch (*Taeniopygia guttata*), 12